Cellular and Molecular Alterations in the Failing Human Heart

G. Hasenfuss, Ch. Holubarsch, H. Just,
N. R. Alpert (Eds.)

Cellular and Molecular Alterations in the Failing Human Heart

Steinkopff Verlag Darmstadt
Springer-Verlag New York

The Editors:
PD Dr. G. Hasenfuss
PD Dr. Ch. Holubarsch
Prof. Dr. H. Just
Albert-Ludwigs-Universität
Medizinische Universitätsklinik
Innere Medizin III, Kardiologie
Hugstetter Straße 55
7800 Freiburg, FRG

N. R. Alpert, Ph. D.
Department of Physiology and Biophysics
University of Vermont
Burlington, VT 05405, USA

Die Deutsche Bibliothek – CIP-Einheitsaufnahme

Cellular and molecular alterations in the failing human heart /
G. Hasenfuss ... (ed.). – Darmstadt : Steinkopff ; New York :
Springer, 1992
  (Supplement to Basic research in cardiology ; Vol. 87,1)
  ISBN-13: 978-3-642-72476-3   e-ISBN-13: 978-3-642-72474-9
  DOI: 10.1007/978-3-642-72474-9
NE: Hasenfuss, Gerd [Hrsg.]; Basic research in cardiology / Supplement

Basic Res. Cardiol, ISSN 0300-8428
Indexed in Current Contents.

Typesetting: Macmillan India Ltd., Bangalore, India

Contents

Extracellular matrix

Mitochondrial function

Achievements of the Symposium

Introduction

The myocardium in heart failure:
Cellular and subcellular alterations in the
failing human myocardium.

H. Just

Medizinische Universitätsklinik Freiburg i. Br., Innere Medizin III – Kardiologie, FRG

The syndrome of heart failure continues to be a major challenge to clinicians and scientists. Incidence and mortality of the disease are high, the patient is disabled, and is permanently threatened by the high morbidity and mortality.

The clinician faces a syndrome of complex pathophysiology. Multiple causes or underlying disorders of the heart have to be differentiated from heart failure itself, which often results in exceedingly difficult diagnoses. Likewise, prognostication meets with difficulties due to problems in separating influences of the underlying disease and the heart failure syndrome itself. In chronic refractory failure annual mortality may exceed 50%. If aortic stenosis or ischemic cardiomyopathy with mainstem lesions are present, this percentage may be even higher. The situation becomes particularly threatening to the patient when the reduction in cardiac performance goes along with complex ventricular arrhythmias.

Therapy has remained difficult and of limited effectiveness. Major progress was achieved with the introduction of diuretic substances. Of similar importance was the introduction of vasodilating drugs into the treatment of heart failure. The principle of vasodilation has greatly improved our understanding of the disease, and has brought about a major improvement of symptoms, increase of exercise capacity, and reduction of mortality. This is especially true for the introduction of the angiotensin converting enzyme inhibitors.

However, as effective as diuretics and vasodilators may be, inotropic stimulation of the failing myocardium has met with major difficulties. Today, only the digitalis glycosides are available as safe and effective drugs for acute and chronic treatment. In acute cardiac failure sympathomimetics, occasionally phosphodiesterase inhibitors may be employed. In this field major advances are urgently needed.

As much as the introduction of vasodilator drugs into the treatment of chronic and acute heart failure has improved our understanding of changes in the vascular periphery and of the interactions between the heart and vascular system, successful therapeutic principles have been developed. As far as the heart itself is concerned, a great deal of uncertainty still exists.

In the past, our knowledge of the pathophysiology of the heart was mainly derived from animal experiments. Only recently have the methods for measuring cardiac performance and for analysis of structure and function of the heart in vivo been developed to such an extent that meaningful pathophysiologic insight can be gained beyond the rather indirect evidence obtained with measurement of central hemodynamics. Of probably even greater importance was the possibility to study human myocardium, either obtained at the time of cardiac surgery or with transplantation of the heart. Furthermore, it has become evident that human myocardium needs to be studied under physiological conditions, i.e., temperature, if results are to be obtained that can be directly transferred to the in vivo situation. Finally, modern methods of molecular biology have enabled us to study even minute amounts of cardiac tissue.

In this situation, it was deemed necessary to review our knowledge of structure and function of the failing human myocardium, to reassess our state of knowledge, and to compare the newer results with those obtained previously from animal experimentation.

To this end, researchers around the globe concerned with studies of the failing human myocardium were assembled at the 4th Gargellen-Conference in 1991. The proceedings of this symposium are presented in this Supplement to Basic Research in Cardiology. We, the editors are grateful to Prof. Elzinga and Prof. Jacob from BRC, and to Sabine Müller from the publishers Steinkopff Verlag, Darmstadt, for support and excellent work.

Therefore, the mostly original data presented in this supplement will greatly enhance our understanding of the failing human heart, and will enable us to achieve better treatment results and, hopefully, lead to the development of better drugs (e.g.) inotropes for the treatment of this frequent and life-threatening desease.

At the outset, a definition and description of the syndrome would seem in order:

Heart failure is present if the normal or the diseased myocardium is unable to pump enough blood to fulfill the needs of peripheral organs under stress conditions and/or resting conditions, inspite of a sufficient venous return.

This definition includes that primary abnormalities of the myocardium are not necessarily involved when heart failure occurs. On the other hand, abnormal loading and working conditions of the heart may lead to changes at the cellular and/or the subcellular level, which may be defined as characteristic of heart failure, but are, nevertheless, secondary and, possibly, reversible. We will, therefore, have to be aware of several kinds of alterations:

1) primary ones that lead to heart failure under normal loading conditions of the heart;
2) secondary changes that occur under abnormal working conditions in the state of heart failure imposed upon a primarily normal myocardium;
3) combinations of both;
4) factors basically unrelated to heart failure, but contributing to or aggravating it (e.g., ischemia).

A consideration of the clinical circumstances under which heart failure occurs may further clarify the issue:

Normal myocardium

1) Overload due to pressure (systemic or pulmonary hypertension) or volume overload (valve regurgitation, shunts, hypercirculatory states) may lead to failure in the presence of normally functioning myocardium, if the load conditions change so rapidly that adaptive hypertrophy cannot occur and wall tension exceeds the capabilities of the contractile element 'Especially volume overload (a condition marked by increased ventricular radius and, therefore, intrinsically unfavorable

contractile geometry) may easily lead to failure. Normal muscle has a remarkable ability to rearrange fibers in order to allow for greater ventricular dimensions. However, with increasing chamber radius the shortening distance for each contractile element for the ejection of a given amount of volume shortens. Therefore, the force developed has to increase. The overloaded myocyte may develop specific changes by overstretching if the capacity of the ventricular wall through rearrangement of the fibers is exceeded.

2) External impediments to cardiac filling may induce failure inspite of the presence of normal functioning myocardium. This is the well-known case of pericardial tamponade or pericardial constriction. These conditions, primarily involving diastolic failure, need not be considered any further here, because neither primary nor secondary myocardial alterations are to be anticipated, except for a direct involvement of the myocardium in calcifying constrictive pericarditis which regularly invades the myocardial tissue.

3) Loss of viable myocardium, e.g., myocardial infarction, may lead to failure if more than 20% of the left-ventricular contractile mass is lost. Beyond 40%, cardiogenic shock may occur. Under these conditions the remaining myocardium has to both compensate for the loss as well as for the bulging occurring in the infarcted region. This means that shortening will not be sufficient for the ejection of blood, because the infarcted area will act as an additional series elastic element until fibrosis and scarring occurs. Except for cases where ischemia is present in the residual myocardium, here, a condition of acute overload to normal myocardium leads to the state of failure.

4) Disorders of cardiac rhythm may lead to failure if certain rate limits are exceeded in the presence of normal myocardium. It is well known that, with slowing of the heart rate, increased diastolic filling will lead to increased stroke volume. Thereby, a stroke volume of up to 250 ml can be produced by the normal heart. This means that, at a heart rate of 20 b.p.m., a sufficient cardiac output of 5 L/min can still be maintained. With diseased myocardium, especially if normal myocardial slippage and rearrangement of the fibers cannot occur, the regulatory dilatation with long diastoles cannot occur and failure will appear earlier, i.e., at higher heart rates.

In the case of tachycardia the positive force-frequency relationship implies that up to heart rates of 180–200 b.p.m. contractile performance increases. Thereby, a sufficient stroke volume is maintained inspite of reduced diastolic filling. The incidence of incoordinate depolarization due to ventricular aberrancy of bundle branch block does not add significantly in normal myocardium to the effects of a high heart rate. It has to be anticipated, however, that the decline of cardiac pumping ability with heart rates beyond 180 b.p.m. will occur earlier if a widened QRS-complex appears. It is not known if persistently slow or high heart rates lead to secondary changes which alter the performance of the myocardium itself. It is known, however, that in diseased myocardial states the normal increase of force of contraction with increases in heart rate does not occur. In many cases the force-frequency relationship will be reversed (vide infra).

Abnormal myocardium

1) Let us consider extracellular changes first. Alterations of shape and size of the ventricle influence contractile performance. This seems to be of greatest importance: under normal conditions the arrangement of the myocardial fibers allows slippage with relaxation and contraction. An increase in fibrous tissue with crossed myocyte

linking will prevent slippage and rearrangement. In these cases dilatation will have to occur by stretching of the myocyte. This is a most unfavorable condition, which will very soon lead to failure. Vascular problems and problems of blood supply occur in cases where vascular growth does not parallel the development of hypertrophy, and will not be considered here. A classical example for the occurrence of heart failure due to the inability of the heart to rearrange fibers with relaxation and contraction is the state of structural dilatation, as described by Linsbach. Under these conditions, the increased wall tension with reduced shortening distance makes progression of heart failure inevitable. Under these circumstances of chronic overloading metabolic changes in the myocardium on a subcellular level can be anticipated and have been described.

2) Primary cellular changes accounting for heart failure are seen in hypertension with chronic pressure overload and left-ventricular hypertrophy, and this is seen in coronary artery disease with ischemia. Not as frequent but, nevertheless, clinically important are cases where inflammation of the myocardium occurs, leading to cell destruction and replacement of fibrous tissue. Under conditions of chronic overload or persistent inflammation or recurrent ischemia or inordinate degrees of hypertrophy, single cell or cellgroup necrosis occur, thereby leading to loss of contractile substance and to an increasing amount of myocardial fibrosis. This will lead to both impairment of systolic, as well as diastolic performance. Cellular changes leading to heart failure can also be seen with different metabolic diseases such as hemochromatosis or amyloidosis.

In these diseased states primary involvement of the contractile apparatus due to the underlying disease may occur and present a disease-specific alteration. This may incorporate secondary changes occurring in the failure state itself.

3) Subsequent changes have, in the past, tended to escape detection and description. Today we know that primary abnormalities of the contractile proteins can occur on a genetic level. They can also occur as secondary phenomena. Of particularly high relevance are alterations of the sarcoplasmic reticulum. Here, in particular, the sarcoplasmic reticulum calcium ATPase may be primarily abnormal, or may show secondary changes in structure and function. Changes of the receptor population in the myocardial membrane, e.g., loss of sympathetic receptors or increase of inhibitory G_i-proteins may lead to failure or may aggravate such state. In these circumstances it remains largely unclear if we are dealing with primary or secondary changes. Adaptive processes on this level have only incompletely been studied.

Finally, it has to be considered that age-related changes of contractile performance and possibly certain, as yet undefined subcellular components may reduce systolic and/or diastolic myocardial function and, thereby, lead to or contribute to the development of failure. This will especially be aggravated if, as usually is the case in advanced age, vascular compliance and vasodilatory capacity are reduced and, thereby, afterload to cardiac ejection increases. Therefore, an age-specific group of changes may be described which leads to failure or a state comparable to heart failure.

It will be necessary to more precisely define primary and secondary changes on cellular and subcellular levels in the human myocardium. It has been shown that data from animal experiments cannot necessarily be transferred to the human situation if we proceed to the subcellular level. Therefore, the newer advances achieved in the study of normal and failing human myocardium will inevitably lead to a new definition of the physiology and pathophysiology of heart failure in humans, and will hopefully allows us to develop better modalities of treatment in the future.

Sarcolemma and phosphodiesterases

Receptor systems in the non-failing human heart

O.-E. Brodde, A. Broede, A. Daul, K. Kunde, M.C. Michel

Biochemisches Forschungslabor, Abteilung Nieren- und Hochdruckkrankheiten, Zentrum für Innere Medizin, Medizinische Klinik und Poliklinik, Universitätsklinikum Essen, FRG

Summary: Catecholamines acting through β_1- and β_2-adrenoceptors cause positive inotropic and chronotropic effects in the human heart. In recent years, however, evidence has accumulated that in the human heart also other receptor systems can affect heart rate and/or contractility. Positive inotropic effects can be mediated by receptor systems acting through accumulation of intracellular cAMP (G_s-protein coupled receptors such as 5-HT_4-like, histamine H_2, and vasoactive intestinal peptide) or by receptor systems acting independent of cAMP possibly through the phospholipase C/diacylglycerol/inositol-1,4,5-trisphosphate pathway (such as α_1-adrenergic, angiotensin II, and endothelin). In the non-failing human heart, however, activation of all these receptor systems induces only submaximal positive inotropic effects when compared with those caused by β-adrenoceptor stimulation, indicating that in humans the cardiac β-adrenoceptor-G_s-protein-adenylate cyclase pathway is the most powerful mechanism to increase heart rate and contractility.

On the other hand, at least three receptor systems acting through inhibition of cAMP formation (G_i-protein coupled receptors) exist in the human heart: muscarinic M_2-, adenosine A_1-, and somatostatin-receptors. Activation of M_2- and A_1-receptors causes negative inotropic effects in the non-failing human heart: in atria activation of both receptors causes decreases in basal as well as in isoprenaline-stimulated force of contraction, but in ventricles only isoprenaline-stimulated force of contraction is depressed.

Key words: Human heart; G_s-coupled receptors in human heart; G_i-coupled receptors in human heart; human cardiac β-adrenoceptor regulation

Introduction

Catecholamines acting through β-adrenoceptors produce both positive inotropic and chronotropic effects in the human heart. In recent years, however, evidence has accumulated that also other receptor systems may be involved in regulation of heart rate and/or contractility in the human heart. Among these are receptors acting via accumulation of intracellular cAMP (G_s-protein coupled), via inhibition of cAMP formation (G_i-protein coupled) or receptors acting independently of cAMP formation, possibly involving the phospholipase C/diacylglycerol/inositol-1,4,5-trisphosphate ($PLC/DAG/IP_3$-)-pathway (Fig. 1). In this contribution an attempt is made: a) to discuss properties and functional importance of these receptor systems in the *non-failing human heart*, and b) to give some new insight into drug-induced regulation of the β-adrenoceptor-G_s-protein-adenylate cyclase complex, the most important physiological mechanism to regulate heart rate and contractility in the human heart (8, 10, 14).

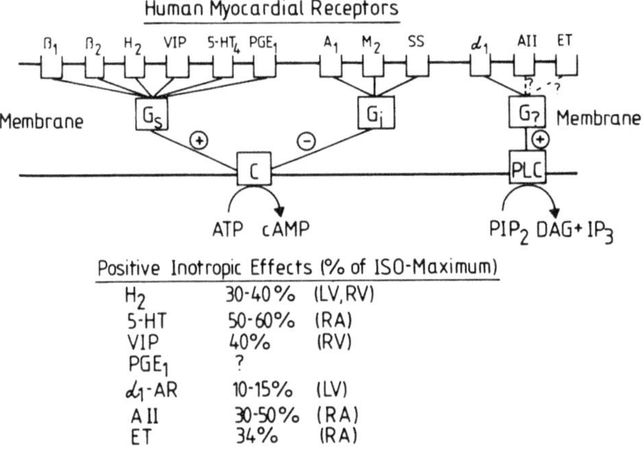

Fig. 1. Receptor systems and their signal-transduction mechanisms in the human heart. For details see text. Abbreviations: β_1, β_2, α_1, $= \beta_1$-, β_2-, and α_1-adrenoceptors; $H_2 =$ histamine H_2-receptors; VIP = vasoactive intestinal peptide receptors; $5\text{-}HT_4 = 5\text{-}HT_4$-receptors; $PGE_1 =$ prostaglandin E_1-receptors; $A_1 =$ adenosine A_1-receptors; $M_2 =$ muscarinic M_2-receptors; SS = somatostatin-receptors; A II = angiotensin II-receptors; ET = endothelin-receptors; $G_s =$ stimulatory guanine nucleotide binding protein; $G_i =$ inhibitory guanine nucleotide binding protein; C = catalytic unit of adenylate cyclase; PLC = phospholipase C; $PIP_2 =$ phosphatidylinositol 4,5-bis-phosphate; DAG = 1,2-diacylglycerol; $IP_3 =$ inositol-1,4,5-trisphosphate; ISO = isoprenaline; + = activation; − = inhibition. Data for positive inotropic effects are from (3, 8, 56) for histamine; from (46, 47) and Brodde, unpublished data for 5-HT; from (37) for VIP; from (6, 58) for α_1-adrenoceptors; from (60) and Brodde, unpublished data for angiotensin II, and from (27) and Brodde, unpublished data for endothelin. RV, LV = positive inotropic effects were determined on isolated electrically driven right- and left-ventricular preparations obtained from potential cardiac transplant donors; RA = positive inotropic effects were determined on isolated electrically driven right atria from patients without apparent heart failure undergoing coronary artery bypass grafting.

G_s-Coupled receptors in the non-failing human heart

β-Adrenoceptors

It is now generally accepted that in the human heart β-adrenoceptors are not a homogeneous population, but both β_1- and β_2-adrenoceptors coexist. This was first demonstrated by radioligand binding studies, and subsequently confirmed in functional experiments (for reviews see (10, 14, 44)). While the number of β-adrenoceptors in the non-failing human heart is quite evenly distributed in right and left atrial and ventricular tissue (amounting to about 80–90 fmol/mg protein in all four tissues; see (15)) the proportion of β_2-adrenoceptors is somewhat higher in the atria (approximately one-third of the total β-adrenoceptor population) than in ventricular myocardium (about 20% of the total β-adrenoceptor population; see (14, 15)).

In the human heart both β_1- and β_2-adrenoceptors couple to adenylate cyclase; this has been clearly demonstrated in broken cell preparations (11, 20, 45), as well as in intact tissues (42). Interestingly, in the human heart adenylate cyclase is preferentially activated by β_2-adrenoceptor stimulation (11, 20, 45), although β_1-adrenoceptors predominate. A possible reason for this may be that β_1-adrenoceptors are partially uncoupled from the adenylate cyclase due to high plasma noradrenaline levels and high sympathetic drive, since human cardiac tissue is usually obtained during open-heart surgery.

In the human heart both β_1- and β_2-adrenoceptors mediate positive inotropic and chronotropic effects of β-adrenoceptor agonists in vivo and in vitro (8, 23, 45, 49, 51, 55). Among the classical catecholamines isoprenaline and adrenaline cause their positive inotropic effects in the human heart via stimulation of β_1- and β_2-adrenoceptors, while noradrenaline, the main transmitter of the sympathetic nervous system, evokes its positive inotropic effect predominantly, if not exclusively via β_1-adrenceptor stimulation (45, 55).

In addition, on right and left atria β_1- *and* β_2-adrenoceptor stimulation can evoke maximum positive inotropic effects, while on right and left ventricles only β_1-adrenoceptor stimulation can evoke maximum positive inotropic effects, β_2-adrenoceptor stimulation only submaximal positive inotropic effects (8, 45, 55). Taking this and the fact that noradrenaline is acting solely at cardiac β_1-adrenoceptors into consideration, it can be concluded that, in man, under normal physiological conditions, contractility and heart rate are under the control of β_1-adrenoceptors (via noradrenaline); in these situations cardiac β_2-adrenoceptors may play only a minor physiological role, if at all. They may come into play only if high amounts of adrenaline (acting at β_1- *and* β_2-adrenoceptors in the human heart) are released from the adrenal medulla (for example in situations of stress) in order to produce additional positive ino- and/or chronotropic effects.

Finally, the human heart does contain only a few spare receptors for β-adrenoceptor-mediated positive inotropic effects and nearly all receptors are needed to evoke maximal positive inotropic effects (24, 57, 61). Thus, if β-adrenoceptor number is reduced (e.g., in heart failure) this automatically leads to a reduced functional response, in contrast to a system with a large receptor reserve.

Other G_s-coupled receptors

As shown in Figures 1 and 2, in addition to β-adrenoceptors several other receptor systems exist in the human heart that couple in an excitatory fashion to adenylate cyclase. Among these are 5-HT, histamine, prostaglandin E_1 (PGE_1), vasoactive intestinal peptide (VIP), and glucagon; stimulation of all these receptors, however, causes only submaximal activation of adenylate cyclase when compared with isoprenaline (Fig. 2). In addition to activation of adenylate cyclase at least 5-HT, histamine, and VIP can evoke positive inotropic effects.

5-HT causes positive inotropic effects on the isolated human right atrium via activation of a 5-HT receptor that resembles the recently described rodent brain 5-HT_4 receptor (29, 30). This was first discovered by Kaumann et al. (46, 47), who demonstrated that the 5-HT-induced increases in contractile force were not antagonized by classical 5-HT_1-, 5-HT_2-, and 5-HT_3-receptor antagonists; only the 5-HT_3-receptor antagonist ICS 205930 antagonized this 5-HT-mediated positive inotropic effect, but in concentrations (pK_B-value 6.7) 1000 times higher than its affinity at 5-HT_3-receptors. Compared to isoprenaline, however, 5-HT induces only submaximal positive inotropic effects (46, 47). Preliminary data from our laboratory confirm these

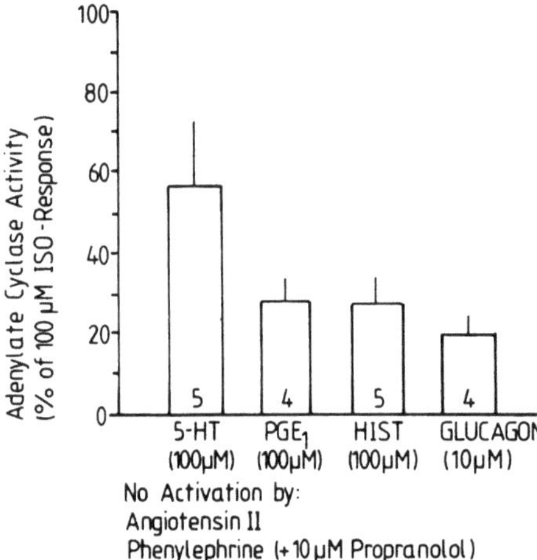

Fig. 2. Activation of adenylate cyclase in membranes from right atria, derived from patients without apparent heart failure undergoing coronary artery bypass grafting; by 100 µM 5-HT, 100 µM prostaglandin E_1 (PGE_1), 100 µM histamine (HIST) and 10 µM glucagon. Ordinate: Adenylate cyclase activity in percent of 100 µM isoprenaline (ISO) response (= 100 %) Means ± SEM; number of experiments at the bottom of the columns. (From Brodde, unpublished data.)

observations; on isolated electrically driven right atria from patients without apparent heart failure undergoing elective coronary artery bypass grafting 5-HT caused maximal positive inotropic effects that were about 50% of those induced by isoprenaline; ICS 205930 (3 µM) antagonized this effect with a pK_B-value of 6.43 (Fig. 3).

Histamine causes in the human heart positive inotropic effects that are mediated predominantly, if not exclusively via H_2-receptors (3, 9). In right- and left-ventricular preparations of the non-failing human heart the maximum positive inotropic effect of histamine was found to be only 30–50% of that of isoprenaline (3, 8, 56). On the other hand, preliminary data from our laboratory on isolated electrically driven human right atria seem to indicate that, in this preparation, histamine induced maximal positive inotropic effects that were about 85–90% of those induced by isoprenaline (Brodde, unpublished data).

Whether glucagon and PGE_1 induce positive inotropic effects in the human heart is not known at present.

Finally, Hershberger et al. (37) recently showed that in ventricular myocardium of the non-failing human heart receptors for VIP exist that activate adenylate cyclase and mediate positive inotropic effects. The maximum positive inotropic effect induced by VIP was, however, only 40% of that brought about by isoprenaline.

Thus, taken together among the G_s-coupled receptors in the non-failing human heart the strongest positive inotropic effect is caused by β-adrenoceptor activation, while obviously all other receptor systems cause only submaximal positive inotropic effects.

G_i-Coupled receptors in the non-failing human heart

In recent years at least two receptor systems in the non-failing human heart have been characterized that couple to adenylate cyclase in an inhibitory fashion and cause negative inotropic effects: the muscarinic cholinoceptor and the adenosine A_1-receptor. We have recently characterized the muscarinic receptor subtype present in

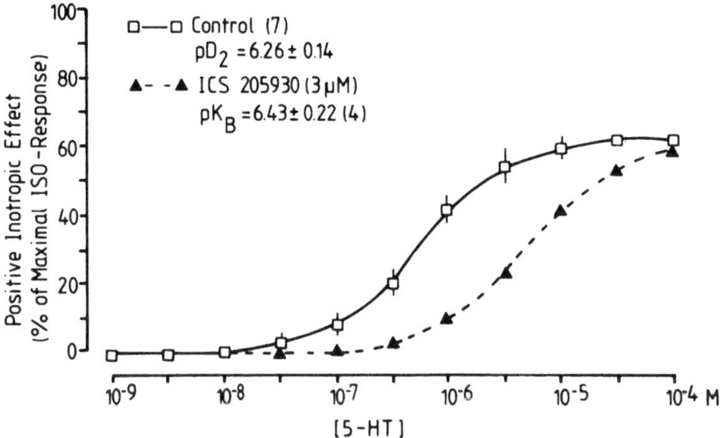

Fig. 3. Effects of ICS 205930 (3 µM) on the positive inotropic effect of 5-HT on isolated electrically driven human right atria, obtained from patients without apparent heart failure undergoing coronary artery bypass grafting. Ordinate: Positive inotropic effect in percent of maximal isoprenaline (ISO) response (= 100 %); Abscissa: molar concentrations of 5-HT. Means ± SEM; number of experiments in parentheses. (From Brodde, unpublished data.)

the human heart by combined radioligand binding and organ bath techniques (28). In the non-failing human heart muscarinic receptors are – in contrast to the β-adrenoceptors (cf. (14, 15)) not evenly distributed (28); the amount of muscarinic receptors in the atria (about 200–250 fmol/mg protein) is approximately 2–2.5-fold higher than in left ventricle (70–100 fmol/mg protein). Using different subtype selective antagonists (non-selective: atropine; M_1-selective: pirenzepine; M_2-selective: AF-DX 116; M_3-selective: hexahydrosiladifenidol [HHSiD]), we could show that in the human heart the muscarinic receptor belongs predominantly, if not exclusively to the M_2-subtype. Activation of this receptor causes inhibition of adenylate cyclase activity and negative inotropic effects: in atria both basal as well as isoprenaline-stimulated force of contraction were reduced, while in the left papillary muscle only isoprenaline-stimulated force of contraction was depressed.

In addition to M_2-receptors, recently, adenosine A_1-receptors have been identified in human atrial and ventricular myocardium that inhibit adenylate cyclase activation and cause negative inotropic effects (7, 38, 58): in the atria both basal as well as forskolin-stimulated force of contraction were reduced, while in the ventricles only forskolin-stimulated force of contraction was reduced.

Finally, somatostatin has been described to inhibit adenylate cyclase activity in human ventricular myocardium (39); whether somatostatin causes also negative inotropic effects in the human heart remains to be elucidated.

Receptors possibly coupled to the phospholipase C/diacylglycerol/inositol-1,4,5-trisphosphate-pathway in the non-failing human heart

It has been known for a long time that in the heart of various species α-adrenoceptor stimulation causes positive inotropic effects that are not accompanied by changes in cAMP or cGMP (19, 26, 31, 64). Recent studies strongly support the view that also in human heart – at least in ventricular tissue – a small proportion of α-adrenoceptors (that belong exclusively to the α_1-subtype) exists that cause positive inotropic effects (58). However, the number of α-adrenoceptors (10–15 fmol/mg protein) and the maximal positive inotropic effect (about 10–15% of that of isoprenaline) is very small (2, 6, 12, 25, 58), thus questioning a physiological role in regulation of contractile force. Evidence is accumulating that human cardiac α_1-adrenoceptors couple via a pertussis toxin insensitive G-protein to the PLC/DAG/IP$_3$-pathway (12, 48, 59).

Very recently, angiotensin II (60, 63) and endothelin (27) have been found to increase contractile force in right atrial preparations of the non-failing human heart. The maximal inotropic effect of both peptides, however, was only 30–40% of that caused by isoprenaline. Although it is still unknown which second messenger system is involved in the positive inotropic action of both peptides in the human heart, it seems to be a cAMP independent pathway: both angiotensin II and endothelin neither affected time-to-peak tension nor rate of relaxation, whereas drugs acting via cAMP shorten both parameters. In addition, angiotensin II did not activate right atrial adenylate cyclase (cf. Fig. 2). Thus, it might be that both peptides couple in the human heart to the PLC/DAG/IP$_3$-pathway; preliminary data from our laboratory indicate that at least endothelin can increase inositol phosphate accumulation in slices of human right atrium derived from patients without apparent heart failure undergoing coronary artery bypass grafting.

Drug-induced regulation of β-adrenoceptors in the non-failing human heart

β-Adrenoceptor agonists

It is now widely accepted that β-adrenoceptors, rather than being static entities are dynamically regulated by a wide variety of drugs, hormones, physiological and pathological conditions. One important example of a clinically relevant consequence of receptor regulation is the phenomenon of desensitization, i.e., the fact that prolonged exposure of the cell to β-adrenoceptor agonists results in a decrease in the functional responsiveness ("desensitization") and, finally, in a decrease in receptor number ("down-regulation").

A down-regulation of β-adrenoceptors in humans following continuous β-adrenergic activation has been directly shown in women undergoing β_2-adrenergic tocolytic therapy, where myometrial β-adrenoceptor number was markedly depressed when compared with that in non-treated women (4, 53). Similarly, the down-regulation of cardiac β-adrenoceptors in patients with chronic heart failure is generally believed to be caused by the chronic exposure of the heart to (cardiac-derived) elevated noradrenaline (for references see Bristow et al., this volume). In addition, chronic treatment of healthy volunteers with a β_1- (xamoterol) or a β_2-adrenoceptor agonist (procaterol) caused desensitization of β-adrenoceptor-mediated physiological in vivo effects, but in a subtype-selective manner: after xamoterol treatment β_1-adrenoceptor mediated effects (isoprenaline-infusion-induced increase in systolic blood pressure and exercise-induced tachycardia) were reduced, after procaterol treatment β_2-adrenoceptor mediated effects (isoprenaline-infusion induced decrease in diastolic blood pressure and increase in plasma noradrenaline levels) (16). Moreover, both agonists desensitized isoprenaline-infusion-induced increase in heart rate (that in humans is mediated by both β_1- and β_2-adrenoceptors; for references see (14, 23, 51)), but to a lesser extent than the pure β_1- (xamoterol) or β_2-adrenoceptor (procaterol)- mediated effects.

Thus, in general, long-term exposure of the human heart to β-adrenoceptor agonists causes β-adrenoceptor desensitization that may occur in a subtype-selective fashion. As mentioned above, the human heart has only a small receptor reserve for β-adrenoceptor agonists. Thus, any decrease in cardiac β-adrenoceptor number induced by chronic β-adrenoceptor agonist treatment will reduce the functional response to β-adrenoceptor agonists. Accordingly, the use of β-adrenoceptor agonists in long-term treatment, e.g., in chronic heart failure, may be of limited value because tolerance to their action will develop.

β-Adrenoceptor antagonists

The opposite of receptor down-regulation is receptor up-regulation, i.e., the fact that reduced exposure of the cell to (endogenous) β-adrenoceptor agonists (by denervation or by treatment with β-adrenoceptor antagonists) leads to an increase in the functional responsiveness of the cell to β-adrenergic stimulation ("supersensitivity") that often is accompanied by an increase in β-adrenoceptor number ("up-regulation"). This has been first observed in rat heart (1, 33), lung and lymphocytes (1), as well as in human lymphocytes (for references see (21)), where chronic administration of propranolol (a non-selective β-adrenoceptor antagonist without partial agonistic activity) produces a substantial increase in β-adrenoceptor density.

A few studies have examined the effects of chronic β-adrenoceptor antagonist treatment on β-adrenoceptors in the non-failing human heart, mainly in right atria from patients without apparent heart failure undergoing coronary artery bypass grafting. Jones et al. (43) found that in these patients chronic treatment with the non-selective β-adrenoceptor antagonists propranolol or sotalol significantly increased right atrial β-adrenoceptor density. Similarly, an increase in right atrial β-adrenoceptors following chronic treatment with propranolol, atenolol, metoprolol or timolol has been also described by Hedberg et al. (36) and Wagner et al. (65). Unfortunately, in these studies no attempt was made to analyze whether the increases in β_1- or β_2-adrenoceptor number were different in patients treated with the selective β_1-adrenoceptor antagonists atenolol or metoprolol compared with those induced by the non-selective β-adrenoceptor antagonists propranolol or timolol. We have recently shown that in patients undergoing coronary artery bypass grafting chronic treatment with different β-adrenoceptor antagonists (non-selective: propranolol, sotalol; β_1-selective: metoprolol, atenolol, bisoprolol) had different effects on right atrial β_1- and β_2-adrenoceptors, as well as on saphenous vein β_2-adrenoceptors (22, 52). Only the non-selective β-adrenoceptor antagonists sotalol and propranolol increased concomitantly right atrial β_1- as well as right atrial and saphenous vein β_2-adrenoceptor density (Fig. 4) while the β_1- selective antagonists atenolol, bisoprolol or metoprolol increased only right atrial β_1-adrenoceptor density but had no effect on right atrial, and saphenous vein β_2-adrenoceptors (Fig. 4).

Surprisingly, the increase in right atrial β_1-adrenoceptor number following chronic treatment with β_1-adrenoceptor-selective antagonists is not accompanied by an increased β_1-adrenoceptor-mediated positive inotropic effect, but is associated with a markedly enhanced β_2-adrenoceptor-mediated effect. This was first shown by

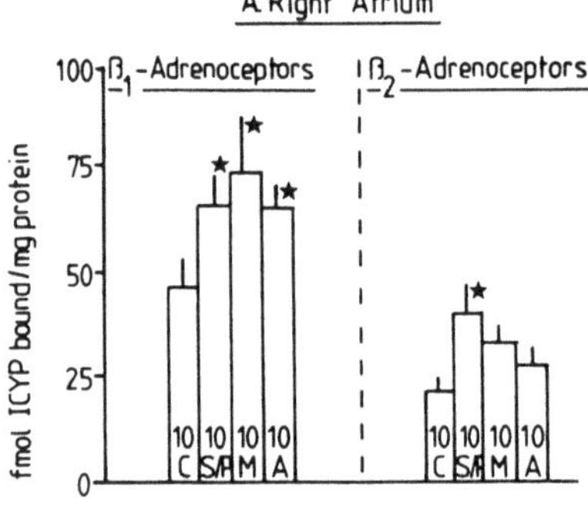

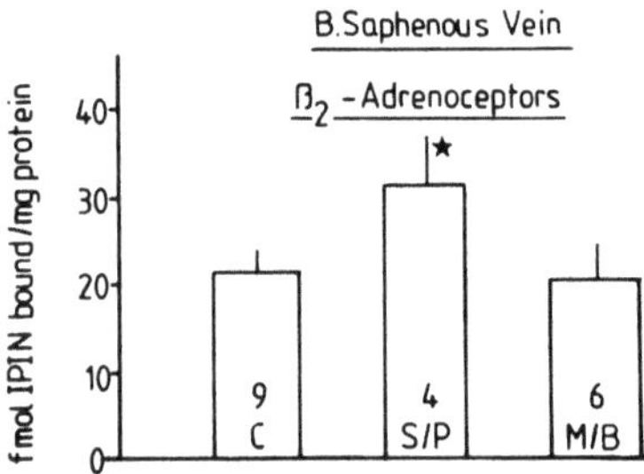

Fig. 4. Effects of chronic β-adrenoceptor antagonist treatment on β_1- and β_2-adrenoceptor density in right atria (A) and on β_2-adrenoceptor density in saphenous veins (B) from patients without apparent heart failure undergoing coronary artery bypass grafting. Ordinates, in A: right atrial β_1- and β_2-adrenoceptor densities in fmol $(-) - [^{125}I]$ iodocyanopindolol (ICYP) specifically bound/mg protein; in B: saphenous vein β_2-adrenoceptor densities in fmol $(-) - [^{125}I]$iodopindolol (IPIN) specifically bound/mg protein. C = control (i.e., patients not treated with β-adrenoceptor antagonists); S = sotalol; P = propranolol; M = metoprolol; A = atenolol; B = bisoprolol. Means ± SEM; number of experiments at the bottom of the columns. *) $P < 0.05$ vs. the corresponding control. (From (22, 52) with modifications.)

Hall et al. (35), who found, in right atria from patients chronically treated with atenolol, that the positive inotropic effect of noradrenaline via β_1-adrenoceptor activation – despite an increased β_1-adrenoceptor number (see Fig. 4) – was not changed, while the inotropic action of adrenaline mediated through activation of both β_1- and β_2-adrenoceptors was markedly enhanced. Moreover, the β_2-adrenoceptor-mediated positive inotropic effect of salbutamol (that amounted in non-β-adrenoceptor-antagonist-treated patients to about 39% of that of isoprenaline) was markedly enhanced in atenolol-treated patients; the maximum inotropic effect increased up to 80% of that of isoprenaline (34). Nearly identical results were obtained with the selective β_2-adrenoceptor agonist procaterol: in non-β-adrenoceptor antagonist treated patients it usually produced 75–80% of the response of isoprenaline, while in patients chronically treated with the selective β_1-adrenoceptor antagonists metoprolol, atenolol or bisoprolol it produced maximal responses identical to those of isoprenaline (54). Thus, it seems that in human right atrium chronic β_1-adrenoceptor antagonist treatment causes a selective up-regulation of β_1-adrenoceptor number, but a sensitization of β_2-adrenoceptor function.

A similar β_2-adrenoceptor sensitizing effect of chronic β_1-adrenoceptor antagonist treatment seems to occur also in vivo: comparison of the effects of acute and chronic (14 days) treatment of healthy volunteers with the selective β_1-adrenoceptor antagonist bisoprolol (13) and the selective β_2-adrenoceptor antagonist ICI 118,551 (5) (both given in doses that produced comparable β_1- and β_2-adrenoceptor occupancies, respectively) on isoprenaline-infusion induced tachycardia (which in humans is mediated equally by β_1- and β_2-adrenoceptors; see above) reveals that acute β_1-adrenoceptor blockade caused an about two to threefold larger rightward shift of the isoprenaline dose-response curve than did chronic β_1-adrenoceptor blockade, where-

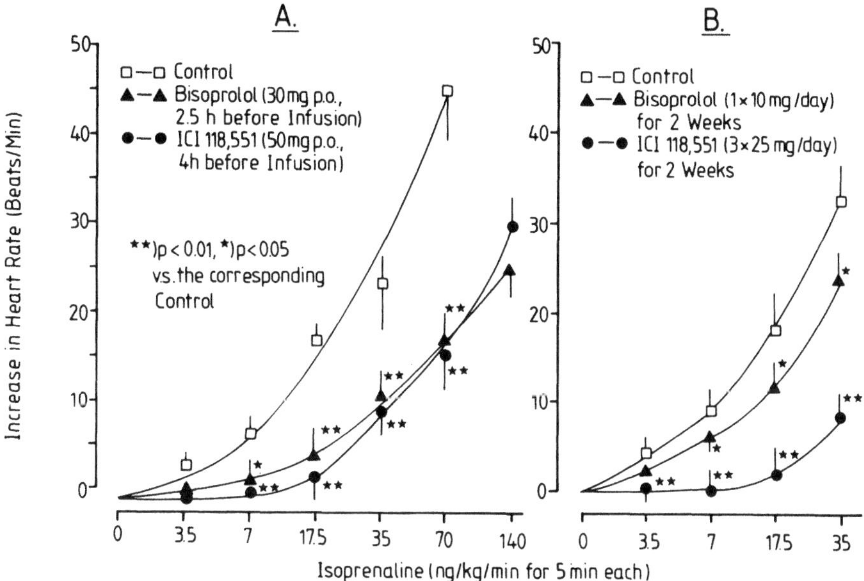

Fig. 5. Effects of acute (A) and chronic (14-day, B) treatment of healthy volunteers with bisoprolol and ICI 118, 551 on isoprenaline-infusion-induced increase in heart rate. Ordinate: increase in heart rate in $\triangle$ beats/min; Abscissae: dose of isoprenaline in ng/kg/min for 5 min each. Each value is the mean $\pm$ SEM of 10 (in A) or 6 (for each antagonist, in B) experiments. (From (17, 55) with modifications.) Note that after acute treatment with bisoprolol or ICI 118, 551 the isoprenaline-dose-response curve is shifted to the right to about the same degree (A), whereas after 2-week treatment the rightward shift of the isoprenaline dose-response curve induced by bisoprolol is significantly less than that evoked by ICI 118, 551 (B).

as the effects of acute and chronic β_2-adrenoceptor blockade were similar (Fig. 5) (17, 55).

These findings could be of clinical importance. If such a β_2-adrenoceptor-sensitizing effect might be true also for ventricular β_2-adrenoceptor-mediated positive inotropic effects it would open the intriguing possibility to treat patients with chronic heart failure who are on low-dose β_1-adrenoceptor-antagonist treatment intermittently with β_2-adrenoceptor agonists to improve cardiac performance, if necessary.

Calcium antagonists

Calcium antagonists (calcium entry blockers, CEB) are often used in many kinds of cardiovascular diseases such as hypertension, angina pectoris, myocardial infarction, and arrhythmia. In recent years evidence has accumulated that, in the heart, β-adrenoceptor number and the number of L-type calcium channels may be coregulated. Ferry and Kaumann (32) observed that in human ventricular myocardium β-adrenoceptor density and the density of calcium channels (determined by [3H]-nimodipine binding) was significantly correlated. In cultured chick embryo ventricular cells a 4-h exposure of the cells to 1 µM isoprenaline caused a decrease in both the number of β-adrenoceptors and calcium channels (assessed by $[^3\text{H}]\text{-}(+)-$ PN 200–110 binding; see (50)). Moreover, in rat heart both β-adrenoceptor and calcium channel number was found to be increased following 6-hydroxy-dopamine treatment (62). In addition, CEB may modulate β-adrenoceptor number. Incubation of cultured cardiac myocytes isolated from neonatal rat ventricle with different CEB (verapamil, diltiazem and nicardipine) produced a time- and concentration-dependent increase in β-adrenoceptor number (66). In vitro experiments have shown that CEB (nifedipine, diltiazem, and verapamil) might prevent isoprenaline-induced desensitization of β_2-adrenoceptors in circulating human lymphocytes (40, 41).

Only a few studies on the effects of chronic CEB treatment on β-adrenoceptor number in the human heart have been published; all these studies were performed on right atria derived from patients without apparent heart failure who underwent coronary artery bypass grafting, but divergent results have been reported: Hedberg et al. (36) found that right atrial β-adrenoceptor number was significantly increased following chronic treatment with nifedipine, whereas Jones et al. (43) and Wagner et al. (65) did not observe any effect of chronic treatment with the CEB verapamil, diltiazem, and nifedipine on right atrial β-adrenoceptor density. We have recently studied the effects of chronic treatment with different CEB (mostly nifedipine, but occasionally diltiazem and verapamil) on right atrial β-adrenoceptor number and subtype distribution in 65 patients without apparent heart failure undergoing coronary artery bypass grafting. We neither observed any significant effect of the CEB-treatment on right atrial β-adrenoceptor density nor on the $\beta_1:\beta_2$-adrenoceptor ratio; moreover, chronic CEB-treatment also did not affect the increase in right atrial β_1-adrenoceptors caused by chronic treatment of these patients with selective β_1-adrenoceptor antagonists (18).

Thus, taking these and the data from the literature together, it is quite likely that chronic CEB treatment has no pronounced effect on β-adrenoceptors in the non-failing human heart.

References

1. Aarons RD, Molinoff PB (1982) Changes in the density of beta adrenergic receptors in rat lymphocytes, heart and lung after chronic treatment with propranolol. J Pharmacol Exp Ther 221:439–443
2. Aass H, Skomedal T, Osnes J-B, Fjeld NB, Klingen G, Langslet A, Svennegig J, Semb G (1986) Noradrenaline evokes an α-adrenoceptor-mediated inotropic effect in human ventricular myocardium. Acta pharmacol toxicol 58:88–90
3. Baumann G, Mercader D, Busch U, Felix SB, Loher U, Ludwig L, Sebening H, Heidecke CD, Hagl S, Sebening F, Blömer H (1983) Effects of the H_2-receptor agonist impromidine in human myocardium from patients with heart failure due to mitral and aortic valve disease. J Cardiovasc Pharmacol 5:618–625
4. Berg G, Andersson RGG, Ryden G (1985) β-Adrenergic receptors in human myometrium during pregnancy: changes in the number of receptors after β-mimetic treatment. Am J Obstet Gynecol 151:392–396
5. Bilski AJ, Halliday SE, Fitzgerald JD, Wale JL (1983) The pharmacology of a β_2-selective adrenoceptor antagonist (ICI 118, 551). J Cardiovasc Pharmacol 5:430–437
6. Böhm M, Diet F, Feiler G, Kemkes, B, Erdmann E (1988) α-adrenoceptors and α-adrenoceptor-mediated positive inotropic effects in failing human myocardium. J Cardiovasc Pharmacol 12:357–364

7. Böhm M, Pieske B, Ungerer M, Erdmann E (1989) Characterization of A_1 adenosine receptors in human atrial and ventricular myocardium from diseased human hearts. Circ Res 65:1201–1211
8. Bristow MR (1989) Myocardial cell surface membrane receptors in heart failure. Heart Failure 5:47–50
9. Bristow MR, Cubicciotti R, Ginsburg R, Stinson EB, Johnson C (1982) Histamine-mediated adenylate cyclase stimulation in human myocardium. Mol Pharmacol 21:671–679
10. Bristow MR, Hershberger RE, Port JD, Gilbert EM, Sandoval A, Rasmussen R, Cates AE, Feldman AM (1990) β-adrenergic pathways in nonfailing and failing human ventricular myocardium. Circulation 82 (Suppl. I):I-12–I-25
11. Bristow MR, Hershberger RE, Port JD, Minobe W, Rasmussen R (1989) β_1- and β_2-adrenergic receptor-mediated adenylate cyclase stimulation in nonfailing and failing human ventricular myocardium. Mol Pharmacol 35:295–303
12. Bristow MR, Minobe W, Rasmussen R, Hershberger RE, Hoffman BB (1988) Alpha-1 adrenergic receptors in the nonfailing and failing human heart. J Pharmacol Exp Ther 247:1039–1045
13. Brodde O-E (1986) Bisoprolol (EMD 33512), a highly selective β_1-adrenoceptor antagonist: in vitro and in vivo studies. J Cardiovasc Pharmacol 8 (Suppl. 11):S29–S35.
14. Brodde O-E (1991) β_1- and β_2-adrenoceptors in the human heart: properties, function, and alterations in chronic heart failure. Pharmacol Rev 43:203–242.
15. Brodde O-E (1991) Pathophysiology of the β-adrenoceptor system in chronic heart failure: consequences for treatment with agonists, partial agonists or antagonists? Eur Heart J 12 (Suppl. F), in press.
16. Brodde O-E, Daul A, Michel-Reher M, Boomsma F, Man in't Veld AJ, Schlieper P, Michel MC (1990) Agonist-induced desensitization of β-adrenoceptor function in humans. Subtype-selective reduction in β_1- or β_2-adrenoceptor-mediated physiological effects by xamoterol or procaterol. Circulation 81:914–921
17. Brodde O-E, Daul A, Wellstein A, Palm D, Michel MC, Beckeringh JJ (1988) Differentiation of β_1- and β_2-adrenoceptor-mediated effects in humans. Am J Physiol 254:H199–H206
18. Brodde O-E, Hundhausen H-J, Zerkowski H-R, Michel MC (1992) Lack of effect of chronic calcium antagonist treatment on β_1- and β_2-adrenoceptors in right atria from patients with or without heart failure. Br J Clin Pharmacol 33: in press
19. Brodde O-E, Motomura S, Endoh M, Schümann HJ (1978) Lack of correlation between the positive inotropic effect evoked by α-adrenoceptor stimulation and the levels of cyclic AMP and/or cyclic GMP in the isolated ventricle strip of the rabbit. J Mol Cell Cardiol 10:207–219.
20. Brodde O-E, O'Hara N, Zerkowski H-R, Rohm N (1984) Human cardiac β-adrenoceptors: both β_1- and β_2-adrenoceptors are functionally coupled to the adenylate cyclase in right atrium. J Cardiovasc Pharmacol 6:1184–1191.
21. Brodde O-E, Wang X-L (1988) Beta-adrenoceptor changes in blood lymphocytes and altered drug responsiveness. Ann Clin Res 20:311–323
22. Brodde O-E, Zerkowski H-R, Doetsch N, Khamssi M (1989) Subtype-selective up-regulation of human saphenous vein β_2-adrenoceptors by chronic β-adrenoceptor antagonist treatment. Naunyn-Schmiedeberg's Arch Pharmacol 339:479–482.
23. Brown JE, McLeod AA, Shand DG (1986) In support of cardiac chronotropic β_2 adrenoceptors. Am J Cardiol 57:11F–16F.
24. Brown L, Deighton NM, Bals S, Söhlmann W, Zerkowski H-R, Michel MC, Brodde O-E (1992) On spare receptors for β-adrenoceptor-mediated positive inotropic effects of catecholamines in the human heart. J Cardiovasc Pharmacol (submitted)
25. Brückner R, Meyer W, Mügge A, Schmitz W, Scholz H (1984) α-adrenoceptor-mediated positive inotropic effect of phenylephrine in isolated human ventricular myocardium. Eur J Pharmacol 99:345–347

26. Brückner R, Mügge A, Scholz H (1985) Existence and functional role of α_1-adrenoceptors in the mammalian heart. J Mol Cell Cardiol 17:639–645

27. Davenport AP, Kaumann AJ, Hall JA, Nunez DJ, Brown MJ (1989) [^{125}I] Endothelin binding in mammalian tissue: relation to human atrial inotropic effect and coronary contraction. Br J Pharmacol 96 (Suppl):102 P. (abstract)

28. Deighton NM, Motomura S, Borquez D, Zerkowski H-R, Doetsch N, Brodde O-E (1990) Muscarinic cholinoceptors in the human heart: demonstration, subclassification, and distribution. Naunyn-Schmiedeberg's Arch Pharmacol 341:14–21

29. Dumuis A, Bouhelal R, Sebben M, Cory R, Bockaert J (1988) A non-classical 5-hydroxytryptamine receptor positively coupled with adenylate cyclase in the central nervous system. Mol Pharmacol 34:880–887

30. Dumuis A, Sebben M, Bockaert J (1989). The gastrointestinal prokinetic benzamide derivatives are agonists at the non-classical 5-HT receptor (5-HT$_4$) positively coupled to adenylate cyclase in neurons. Naunyn-Schmiedebergs Arch Pharmacol 340:403–410

31. Endoh M (1982) Adrenoceptors and the myocardial inotropic response: Do alpha and beta receptor sites functionally coexist? In: Kalsner S (ed) Trends in autonomic pharmacology: vol 2. Urban & Schwarzenberg, Baltimore Munich, pp 304–322.

32. Ferry DR, Kaumann AJ (1987) Relationship between β-adrenoceptors and calcium channels in human ventricular myocardium. Br J Pharmacol 90:447–457

33. Glaubiger G, Lefkowitz RJ (1977) Elevated beta-adrenergic receptor number after chronic propranolol treatment. Biochem Biophys Res Commun 78:720–725

34. Hall JA, Kaumann AJ, Brown MJ (1990) Selective β_1-adrenoceptor blockade enhances positive inotropic responses to endogenous catecholamines mediated through β_2-adrenoceptors in human atrial myocardium. Circ Res 66:1610–1623

35. Hall JA, Kaumann AJ, Wells FC, Brown MJ (1988) β_2-Adrenoceptor mediated inotropic responses of human atria: receptor subtype regulation by atenolol. Br J Pharmacol 93 (Suppl.): 116 P. (abstract)

36. Hedberg A. Kempf F, Josephson ME, Molinoff PB (1985) Coexistence of β_1 and β_2 adrenergic receptors in the human heart: effects of treatment with receptor antagonists or calcium entry blockers. J Pharmacol Exp Ther 234:561–568

37. Hershberger RE, Anderson FL, Bristow MR (1989) Vasoactive intestinal peptide receptor in failing human ventricular myocardium exhibits increased affinity and decreased density. Circ Res 65:283–294

38. Hershberger RE, Bristow MR (1987) The A$_1$ adenosine receptor pathway in failing and nonfailing human heart Circulation 76 (suppl IV): IV-1726 (abstract).

39. Hershberger RE, Port JD, Anderson FL (1988) Somatostatin inhibits adenylate cyclase activity in human ventricular myocardium. Circulation 78 (Suppl II):II–560. (abstract)

40. Hui KK, Yu JL (1988) The effects of calcium channel blockers on isoproterenol induced cyclic adenosine 3',5'-monophosphate generation in intact human lymphocytes. Life Sci 42:2037–2045.

41. Hui KK, Yu JL (1989) Effects of calcium channel blockers and ketotifen on β_2 adrenergic receptor regulation in intact human lymphocytes. Res Commun Chem Pathol Pharmacol 65:3–19

42. Ikezono K, Michel MC, Zerkowski H-R, Beckeringh JJ, Brodde O-E (1987) The role of cyclic AMP in the positive inotropic effect mediated by β_1- and β_2-adrenoceptors in isolated human right atrium. Naunyn-Schmiedeberg's Arch Pharmacol 335:561–566

43. Jones CR, Fandeleur P, Harris B, Bühler FR (1990) Effect of calcium and β-adrenoceptor antagonists on β-adrenoceptor density and G$_{s\alpha}$ expression in human atria. Br J Clin Pharmacol 30 (Suppl. 1):171S–173S

44. Jones CR, Molenaar P, Summers RJ (1989) New views of human cardiac β-adrenoceptors. J Mol Cell Cardiol 21:519–535.

45. Kaumann AJ, Hall JA, Murray KJ, Wells FC, Brown MJ (1989) A comparison of the effects of adrenaline and noradrenaline on human heart: the role of β_1- and β_2-

adrenoceptors in the stimulation of adenylate cyclase and contractile force. Eur Heart J 10 (Suppl. B):29–37
46. Kaumann AJ, Sanders L, Brown AM, Murray KJ, Brown MJ (1990) Human atrial 5-HT receptors: similarity to rodent neuronal 5-HT_4 receptors. Br J Pharmacol 100 (Suppl):319 P. (abstract)
47. Kaumann AJ, Sanders L, Brown AM, Murray KJ, Brown MJ (1990) A 5-hydroxytryptamine receptor in human atrium. Br J Pharmacol 100:879–885
48. Kohl C, Schmitz W, Scholz H, Scholz J, Toth M, Döring V, Kalimar P (1989) Evidence for α_1-adrenoceptor-mediated increase of inositol trisphosphate in the human heart. J Cardiovasc Pharmacol: 13:324–327
49. Levine MAH, Leenen FHH (1989) Role of β_1-receptors and vagal tone in cardiac inotropic and chronotropic responses to a β_2-agonist in humans. Circulation 79:107–115
50. Marsh JD (1989) Coregulation of calcium channels and β-adrenergic receptors in cultured chick embryo ventricular cells. J Clin Invest 84:817–823
51. McDevitt DG (1989). In vivo studies on the function of cardiac β-adrenoceptors in man. Eur Heart J 10 (Suppl. B):22–28
52. Michel MC, Pingsmann A, Beckeringh JJ, Zerkowski H-R, Doetsch N, Brodde O-E (1988) Selective regulation of β_1- and β_2-adrenoceptors in the human heart by chronic β-adrenoceptor antagonist treatment. Br J Pharmacol 94:685–692
53. Michel MC, Pingsmann A, Nohlen M, Siekmann U, Brodde O-E (1989) Decreased myometrial β-adrenoceptors in women receiving β_2-adrenergic tocolytic therapy: correlation with lymphocyte β-adrenoceptors. Clin Pharmacol Ther 45:1–8
54. Motomura S, Deighton NM, Zerkowski H-R, Doetsch N, Michel MC, Brodde O-E (1990) Chronic β_1-adrenoceptor antagonist treatment sensitizes β_2-adrenoceptors, but desensitizes M_2-muscarinic receptors in the human right atrium. Br J Pharmacol 101:363–369
55. Motomura S, Zerkowski H-R, Daul A, Brodde O-E (1990) On the physiologic role of β_2 adrenoceptors in the human heart: in vitro and in vivo studies. Am Heart J 119:608–619
56. Näbauer M, Böhm M, Brown L, Diet F, Eichhorn M, Kemkes B, Pieske B, Erdmann E (1988) Positive inotropic effects in isolated ventricular myocardium from non-failing and terminally failing human hearts. Eur J Clin Invest 18:600–606
57. Port JD, Bristow MR (1988) Lack of spare β-adrenergic receptors in the human heart. FASEB J 2:A602. (abstract)
58. Schmitz W, Scholz H, Erdmann E (1987) Effects of α- and β-adrenergic agonists, phosphodiesterase inhibitors and adenosine on isolated human heart muscle preparations. Trends Pharmacol Sci 8:447–450
59. Schmitz W, Scholz H, Scholz J, Steinfath M, Lohse M, Puurunen J, Schwabe U (1987) Pertussis toxin does not inhibit the α_1-adrenoceptor-mediated effect on inositol phosphate production in the heart. Eur J Pharmacol 134:377–378.
60. Schomisch Moravec Ch, Schluchter MD, Paranandi L, Czerska B, Stewart RW, Rosenkranz E, Bond M (1990) Inotropic effects of angiotensin II on human cardiac muscle in vitro. Circulation 82:1973–1984.
61. Schwinger RHG, Böhm M, Erdmann E (1990) Evidence against spare or uncoupled β-adrenoceptors in the human heart. Am Heart J 119:899–904
62. Skattebol A, Triggle DJ (1986) 6-Hydroxydopamine treatment increases β-adrenoceptors and Ca^{2+} channels in rat heart. Eur J Pharmacol 127:287–289
63. Urata H, Healy B, Stewart RW, Bumpus FM, Husain A (1989) Angiotensin II receptors in normal and failing human hearts. J Clin Endocrinol Metab 69:54–66
64. Wagner J, Brodde O-E (1979) On the presence and distribution of α-adrenoceptors in the heart of various mammalian species. Naunyn-Schmiedeberg's Arch Pharmacol 302:239–254.
65. Wagner JA, Sax FL, Weisman HF, Porterfield J, McIntosh C, Weisfeldt ML, Snyder SH, Epstein SE (1989) Calcium-antagonist receptors in the atrial tissue of patients with hypertrophic cardiomyopathy. N Engl J Med 320:755–761

66. Yonemochi H, Saikawa T, Takakura T, Ito S, Takaki R (1990) Effect of calcium
    antagonists on β-receptors of cultured cardiac myocytes isolated from neonatal rat
    ventricle. Circulation 81: 1401–1408

Author's address:
Prof. Dr. Otto-Erich Brodde
Biochem. Forschungslabor
Abtlg. Nieren- und Hochdruckkrankheiten
Zentrum für Innere Medizin
Medizinische Klinik und Poliklinik
Universitätsklinikum Essen
Hufelandstr 55
D-4300 Essen 1, FRG

Changes in the receptor-G protein-adenylyl cyclase system in heart failure from various types of heart muscle disease

M. R. Bristow[1], A. M. Feldman[2]

[1] Division of Cardiology, University of Colorado School of Medicine, Denver, Colorado, USA
[2] Division of Cardiology, The Johns Hopkins School of Medicine, Baltimore, Maryland, USA

Summary: The abnormalities of the receptor-G protein-adenylyl cyclase (RCG) system in failing human myocardium as the result of 1) idiopathic dilated cardiomyopathy (IDC), 2) ischemic dilated cardiomyopathy (ISCDC), and 3) primary pulmonary hypertension (PPH) were investigated. Depending on the etiology of heart failure, abnormalities of the RCG system result from a reduced number of β_1 receptors, uncoupling of β_1 or β_2 receptors, alteration of G protein function, or decreased catalytic subunit activity of adenylyl cyclase. Compared to IDC, β_1 receptor down-regulation is less pronounced in ISCDC, and slightly more pronounced in PPH. Preliminary data suggest that β_1 receptor down-regulation results from alteration in steady-state receptor mRNA levels. Increased functional activity of G_i protein, which seems to result from posttranslational modification, is observed in IDC and ISCDC. Altered G_i protein function may be the basis for β-receptor uncoupling in IDC and ISCDC, whereas in PPH, this phenomenon may result from altered adenylyl cyclase function. Catalytic subunit activity of adenylyl cyclase is decreased in order of increasing pulmonary hypertension in right-ventricular preparations from PPH > IDC > ISCDC. However, catalytic subunit activity is similar in LV preparations from all three groups. The decrease in adenylyl cyclase catalytic subunit activity may be the result of the marked cellular injury produced by pressure overload.

In summary, numerous desensitization phenomena occur in the failing human heart that are etiology- or model-dependent. To a certain extent, these changes are teleologically beneficial, as they are able to partially protect the failing heart from potentially toxic adrenergic stimuli.

Key words: β receptor; G protein; adenylyl cyclase; receptor down-regulation

Introduction

When the human heart begins to fail powerful compensatory mechanisms are activated in an attempt to preserve central blood pressure and cardiac output to the brain and heart. The two most primarily activated compensatory mechanisms are the renin-angiotensin and the adrenergic nervous systems. Activation of the latter results in an increase in cardiac neurotransmitter activity, as coronary sinus, cardiac interstitial and synaptic cleft levels of norepinephrine rise in order to stimulate contractility and increase heart rate.

The adrenergic effector systems responsible for transducing and converting the adrenergic signal to a biologic response are complex, highly regulated pathways which provide for rapid (within several seconds) modulation of cardiac function. As

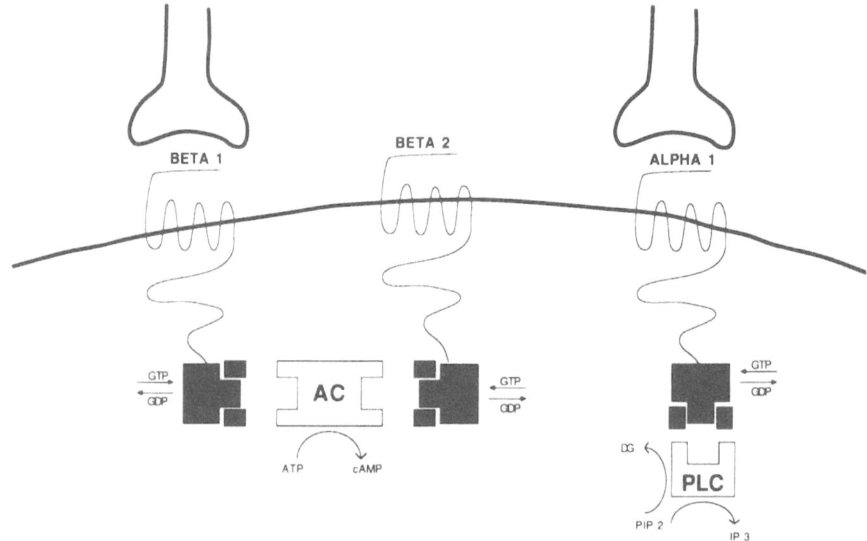

Fig. 1. Schematic representation of the adrenergic receptor-G protein-effector enzyme pathways in the human heart.

shown in Fig. 1, released norepinephrine encounters three signal transduction pathways which begin with specific cell surface receptors. These three pathways begin with the β_1, α_1, and β_2 adrenergic receptors. The α_2 receptor, which is found in platelets and vascular smooth muscle, is not apparently present in human ventricular myocardium.

As shown in Fig. 2, the ability of myocardial adrenergic receptors to bind norepinephrine is not the same among the three subtypes. The rank order of potency is $\beta_1 > \alpha_1 > \beta_2$, with the β_1 receptor having an affinity for norepinephrine that is 10-fold > the α_1 receptor and 60-fold > the β_2 receptor (6). Additionally, it is likely that β_2 receptors are extrasynaptic in their location (31), in order to interact with the hormone epinephrine; β_1 receptors are likely located both intra- and extrasynaptically in order to interact with both norepinephrine and epinephrine. These differences in positioning, affinity, and pathway coupling provide the potential for differential behavior in response to the increased adrenergic drive that accompanies heart failure. In other words, adrenergic receptors exhibit a heterogeneous biological profile despite their structural similarities (35).

Heart failure is a clinical disease or syndrome which can have multiple underlying causes. The most common cause of heart failure in industrialized societies is *systolic dysfunction* from dilated cardiomyopathy (10), which accounts for > 90% of heart failures referred to our own institutions. However, even within the classification of heart failure from dilated cardiomyopathy there are multiple causes, as shown in Table 1. The three most common in the Western United States are ischemic heart disease, idiopathic causes, and valvular heart disease.

Given the diverse nature of these etiologies, it might be expected that the respective pathophysiologic differences would produce fundamentally different effects on the myocardial RGC system. Therefore, it would be important to examine the

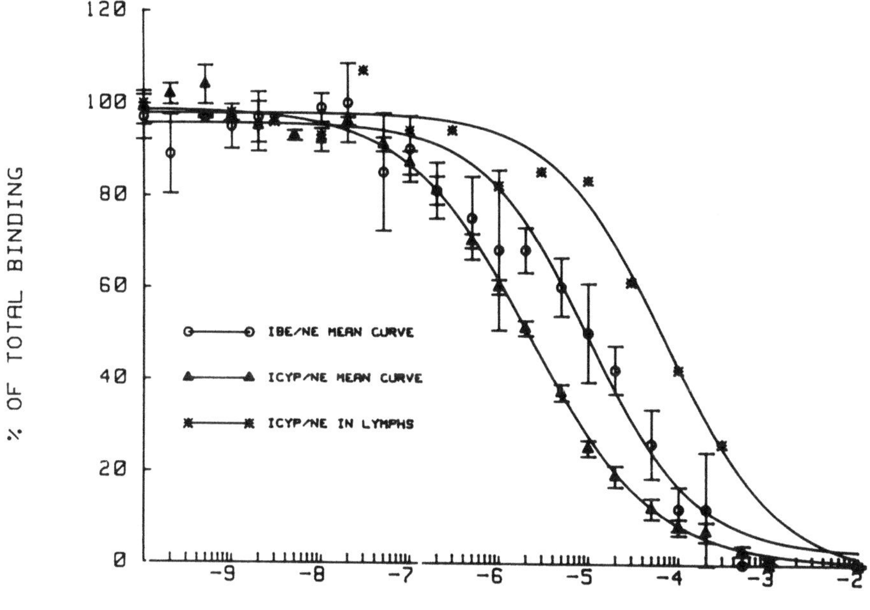

Fig. 2. Affinity of norepinephrine (NE) for β_1 (▲–▲), β_2 (*–*), and α_1 (O–O) adrenergic receptors in membranes prepared from human tissues. (Figure is reproduced with permission of the Journal of Pharmacology and Experimental Therapeutics).

Table 1. Causes of dilated cardiomyopathy in industrialized societies (most common → least common)

1.	Ischemic heart disease
2.	Idiopathic dilated cardiomyopathy
3.	Alcoholic cardiomyopathy
4.	Valvular heart disease
5.	Hypertensive heart disease
6.	Myocarditis
7.	Anthracycline cardiomyopathy
8.	Post partum cardiomyopathy

RGC system in different types of heart muscle disease. Such an examination requires a consistent source of gram quantities of failing (and nonfailing) human ventricular myocardium. This type of material can only be provided by the explanted hearts available in large cardiac transplant programs, where cardiac transplant recipients provide a source of failing hearts and organ donors whose hearts are not able to be placed by computer-based organ matching systems provide a source of nonfailing myocardium.

Another major issue concerning any abnormality detected in the RGC system in the failing human heart is the contribution of systemic or hormonal factors vs local factors such as neurotransmitter activity. This question may be addressed by examining situations in which one chamber is failing and the other is not, such as in isolated right-ventricular failure from primary pulmonary (PPH). In PPH abnormalities which occur only in the right ventricle are the result of local factors, whereas abnormalities occurring in the left or both ventricles are the result of systemic processes. In addition, failing PPH right ventricle provides an opportunity to examine the effect of uncomplicated pressure overload on the RGC system, as situations in which chronic afterloading per se is the cause of left ventricular failure are not commonly encountered in a heart transplant population.

This review will describe the abnormalities of the RGC system that are present in the ventricles of human hearts failing as the result of 1) ischemic dilated cardiomyopathy, 2) idiopathic dilated cardiomyopathy, and 3) primary pulmonary hypertension. In these conditions, six different RGC pathway abnormalities have been identified, and none of them are the result of systemic factors. These abnormalities, listed in Table 2, will now be individually discussed.

Disease-related abnormalities in β receptor-G protein-adenylyl cyclase function in the failing human heart

1. *Down-regulation of β_1 adrenergic receptors*

   a. Idiopathic dilated cardiomyopathy (IDC)
   Down-regulation of the total β receptor population was the first detected abnormality of the ventricular myocardial RGC system in the failing human heart (2).

Table 2. Abnormalities of the β receptor-G protein-adenylyl cyclase pathway in various kinds of heart muscle disease

	Abnormality	Degree of change (0–3 +)	Type of heart muscle disease
1.	β_1 receptor down-regulation	+ +	IDC
		+	ISCDC
		+ + +	PPH RV
2.	β_2 receptor uncoupling	+	IDC
		+ +	ISCDC
3.	β_1 receptor uncoupling	+	ISCDC
4.	Increased activity of $G_{i\alpha}$	+	IDC
		+	ISCDC
5.	Decreased activity of C	+ +	PPH RV
		+	IDC RV
6.	Unspecified G protein abnormality (decreased stimulation of AC by guanine nucleotide)	+	ISCDC

With the advent of highly selective β_1 and β_2 antagonists and computer-assisted techniques for determining binding subtypes, it became possible to distinguish between β_1 and β_2 receptors within the total population identified by nonselective radioligands. When this was done in IDC an extremely interesting observation was made (5, 14), namely that the down-regulation of the total population was entirely due to the loss of the β_1 receptor subtype. The β_2 receptor subtype remained stable in density throughout all kinds of heart failure examined with the exception of papillary muscles harvested from mitral valve disease patients (15) or subjects exposed to therapeutic doses of β agonists (34). However, the β_2 receptor typically exhibits a different kind of desensitization response (see below).

This selective loss in β_1 adrenergic receptors results in a relative increase in the β_2 fraction in failing myocardium, from 70–80% β_1/20–30% β_2 to 60–70% β_1/30–40% β_2 in heart failure. This can be observed in Fig. 3, which gives an analysis of 90 IDC left and right ventricles from patients with endstage heart failure who have not been exposed to β agonist, antagonists, or phosphodiesterase inhibitors, compared to an age-matched control population of nonfailing organ donors. As can be observed, the percent β_1 decreases from 79% to 63.8% in failing chambers, while the percent β_2 increases from 20.7% to 35.8%. These changes in subtype percentage are due to a selective down-regulation of the β_1 subtype by 55%, as shown in Fig. 3.

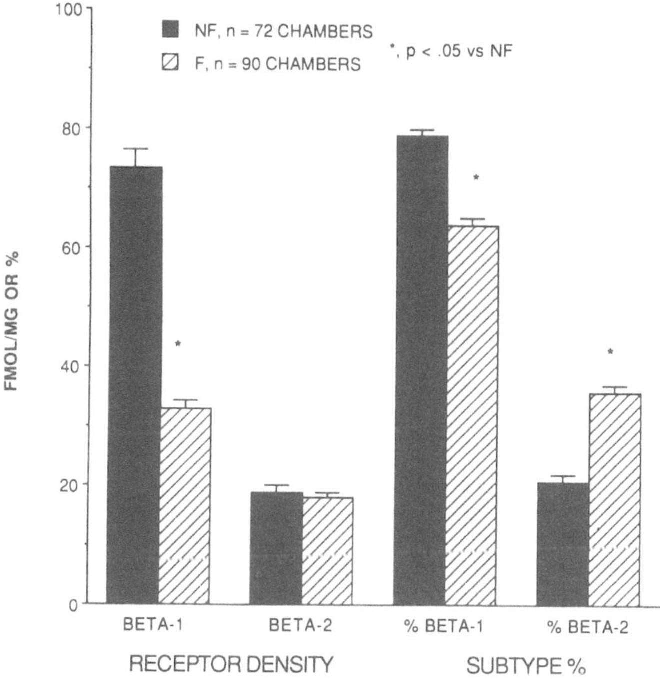

Fig. 3. β-adrenergic receptor subtype density and percentage in membranes prepared from left- and right-ventricular myocardium taken from subjects with endstage heart failure from idiopathic dilated cardiomyopathy (IDC) and organ donors with echocardiography proven normal left-ventricular function (nonfailing).

This selective down-regulation in the β_1 population leads to a decrease in the functional response to selective β_1 agonists, as shown in Fig. 4. Because of the tighter coupling of β_2 receptors to adenylyl cyclase selective activation of adenylyl cyclase by β_1 compounds is, in fact, difficult to detect (8). Nevertheless, under "β_1 conditions" that include the presence of a β_2 blocking agent the stimulation of adenylyl cyclase is markedly attenuated in failing hearts (8).

Figure 5 gives endomyocardial biopsy data that indicate that down-regulation of total β, and by inference β_1 adrenergic receptors, is related to the degree of severity of heart failure. This means that β receptor down-regulation is not something that is present only in endstage failing hearts; it begins in mild-moderate heart failure and becomes more pronounced as heart failure progresses. This progressive decrease is accompanied by, or perhaps more likely caused by, a progressive increase in cardiac adrenergic activation as assessed by coronary sinus norepinephrine levels (Fig. 5).

b. Ischemic dilated cardiomyopathy
Although the noninfarcted areas of hearts with global cardiac dysfunction from ischemic dilated cardiomyopathy (ISCDC) also exhibit down-regulation of β_1 adrenergic receptors, in age and cardiac function matched material the degree of receptor loss is less than in IDC (Fig. 6). As can be observed in Fig. 6, β_1 receptor density is significantly higher in ISCDC left or right ventricles than in IDC, despite levels of adrenergic neurotransmitter depletion that are comparable (12).

c. Pressure overloaded, failing ventricular myocardium
Right-ventricular failure in the setting of primary pulmonary hypertension (PPH) is an example of heart failure due to a marked increase in afterload, or an uncomplicated pressure overload cardiomyopathy. In this disorder the degree of β_1 receptor down-regulation and the degree of neurotransmitter depletion (11) is slightly greater than in IDC (Fig. 7).

d. Summary of β_1 adrenergic receptor down-regulation in various types of dilated cardiomyopathy
In subjects of comparable age who have a similar degree of right-ventricular failure as indicated by elevation in right atrial pressure and decrease in cardiac index

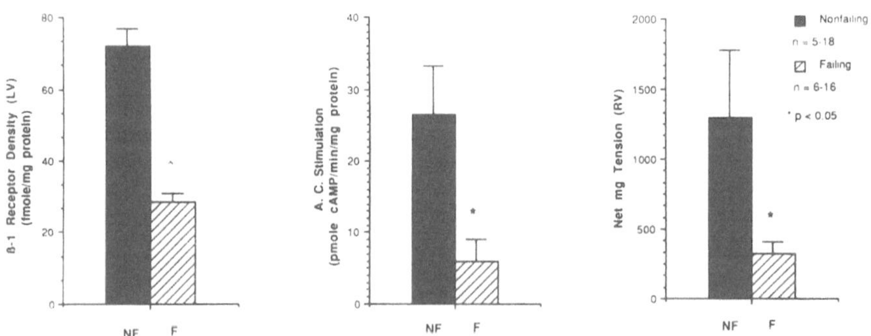

Fig. 4. β_1-adrenergic receptor density and selective β_1-agonist stimulation of adenylate cyclase and muscle contraction in preparations of nonfailing and failing (IDC) in human ventricular myocardium.

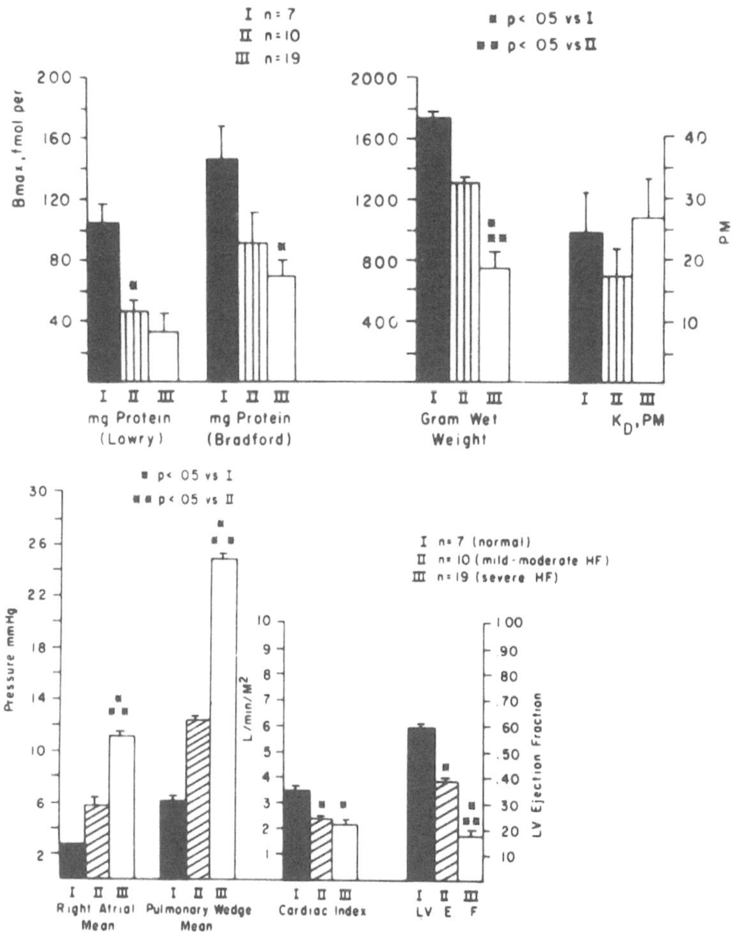

Fig. 5. β receptor density (top panel) and hemodynamics (bottom panel) in endomyocardial biopsy tissue obtained from subjects with normal ventricular function (I), moderate myocardial dysfunction (II), and severe myocardial dysfunction (III).

(Fig. 8), it can be observed that the degree of β_1 receptor down-regulation varies from severe in PPH, to moderate in IDC, to mild in ISCDC. In Fig. 8 it can also be seen that the degree of neurotransmitter depletion in the three types of heart muscle disease does not vary directly with the degree of down-regulation, being equal in IDC and ISCDC, and more pronounced in PPH.

2. β_2 Receptor uncoupling

a. IDC

In contrast to the loss of receptor protein that characterizes the behavior of β_1 adrenergic receptors, in full thickness aliquots of left- or right-ventricular free-wall

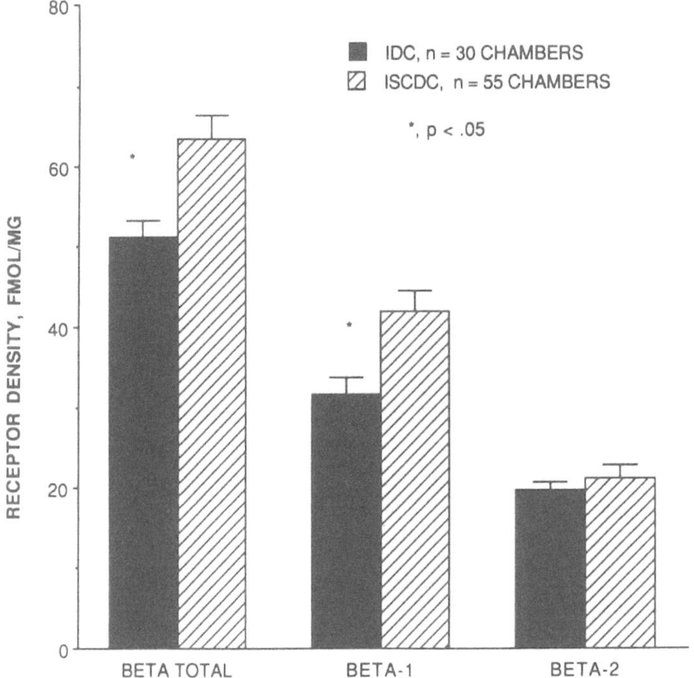

Fig. 6. β receptor densities in pooled left- and right-ventricular preparations taken from subjects with endstage heart failure from idiopathic dilated cardiomyopathy (IDC) or ischemic dilated cardiomyopathy (ISCDC), with IDC subjects taken from a subpopulation with mean pulmonary artery pressures ≤ 35 mm Hg and age ≥ 40 years.

myocardial β_2 receptors do not change in density in heart failure (5, 8, 12, 14, Figs. 3 and 9). This is in contradistinction to lymphocyte β_2 receptors, which exhibit a mild degree of down-regulation in heart failure (16, 20). However, despite the preserved density of β_2 receptors in the failing human heart, they are not capable of transmitting a normal adrenergic signal. As shown in Fig. 9, activity of a β_2 selective agonist (zinterol) was decreased by approximately 30% in heart failure (8). Additionally, β_2 receptor-mediated contractile responses are decreased by a similar degree (6, Fig. 9), although the relatively greater intragroup variation in the muscle bath data preclude the achievement of statistical significance.

b. ISCDC

Left-ventricular ISCDC β_2 receptors exhibit uncoupling from adenylyl cyclase to a greater extent than do IDC β_2 receptors (12). This may be seen in Fig. 10, which illustrates that ISCDC left-ventricular preparations have a greater blunting of the isoproterenol $(\beta_2 + \beta_1)$ or zinterol (β_2) stimulation of adenylyl cyclase, despite having a higher total β receptor density and a β_2 receptor density that is 11% higher than the value in IDC LVs.

As can be seen in Fig. 11, LV ISCDC ventricles also exhibit less stimulation of adenylyl cyclase by fluoride relative to IDC ventricles. On the other hand, ISCDC

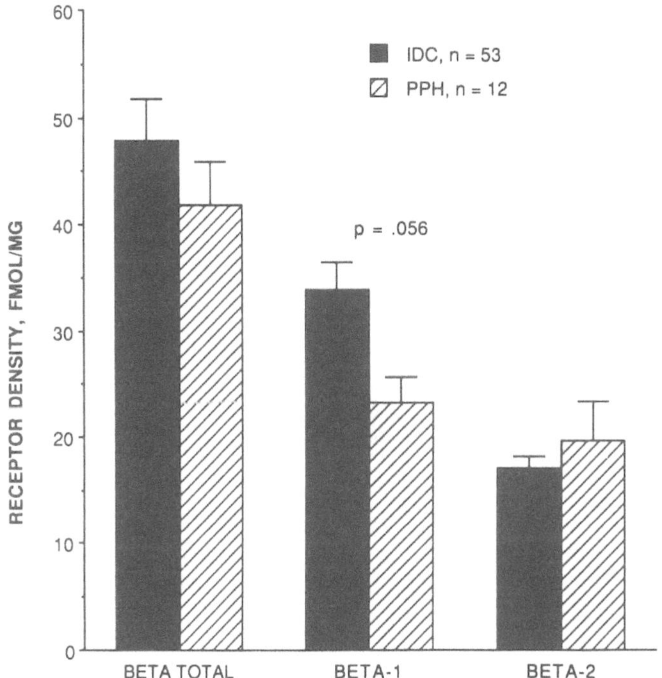

Fig. 7. β receptor densities in right ventricle taken from subjects with endstage heart failure from idiopathic dilated cardiomyopathy (IDC) or severe right-heart failure from primary pulmonary hypertension (PPH).

right ventricles exhibit decreased Gpp (NH)p stimulation of adenylyl cyclase relative to IDC right ventricles.

As shown in Fig. 11, despite these decreases in G protein related stimulation of adenylyl cyclase in ISCDC ventricles, Mn^{2+} stimulation is not significantly decreased in ISCDC ventricles relative to IDC (or to nonfailing controls) (12). These data suggest that ISCDC ventricles have an intact catalytic subunit, but have defects in G protein mediated responses that are not present to a greater degree than in IDC ventricles. However, despite such pharmacologic evidence for G protein abnormalities in ISCDC vs IDC, pertussis toxin catalyzed ADP ribosylation is quantitatively increased to a similar degree in ISCDC and IDC (Fig. 12). Additionally, cholera toxin catalyzed ADP ribosylation is not altered in either group (Fig. 12). These data suggest that the observed functional abnormality in G proteins present in ISCDC that is greater than the abnormality in IDC is due to defects in G protein function that are not detectable by toxin-catalyzed ADP ribosylation.

c. PPH RVs

PPH RVs exhibit an abnormality of the catalytic subunit of adenylyl cyclase (11, see below) which reduces their ability to couple β_2 receptor, guanine nucleotide, forskolin, and Mn^{2+} stimulation. β_2 receptor density is intact in PPH RVs (11, Fig. 8), and it is not possible to pharmacologically detect a functional G protein change in the presence of the catalytic subunit abnormality.

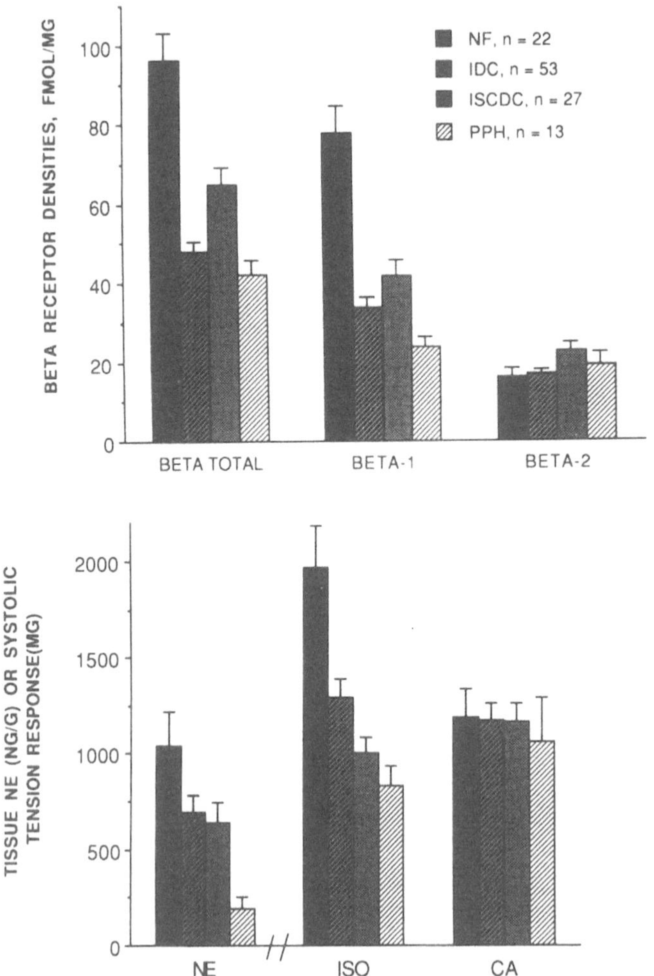

Fig. 8. Top panel: right-ventricular β receptor density in membranes prepared from ventricles taking from nonfailing (NF) and endstage failing (IDC, idiopathic dilated cardiomyopathy; ISCDC, ischemic dilated cardiomyopathy; PPH, primary pulmonary hypertension) hearts. Bottom panel: tissue norepinephrine (NE) and contractile response to isoproterenol (ISO) or calcium (CA) in right-ventricular preparations from the same hearts shown in the top panel.

3. Increased activity of G_i

a. IDC

As can be observed in Fig. 13, under tightly controlled conditions myocardial membranes prepared from IDC LVs exhibit decreased stimulation of adenylyl cyclase by Gpp(NH)p and forskolin. This decrease can be abolished by treatment of membranes with pertussis toxin (23), suggesting that it is due to increased activity of the inhibitory G protein, G_i (Fig. 12). Additionally, activity of G_i, as assessed by ADP

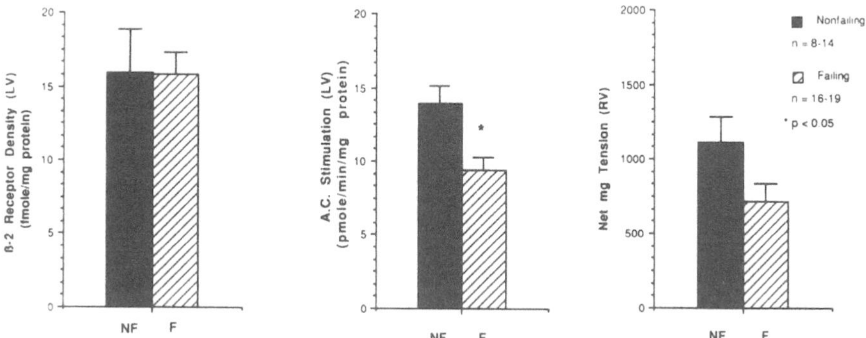

Fig. 9. β_2-receptor density and β_2-selective agonist stimulation of adenylyl cyclase (AC) or contractile response in preparations of nonfailing and failing (IDC) human ventricles.

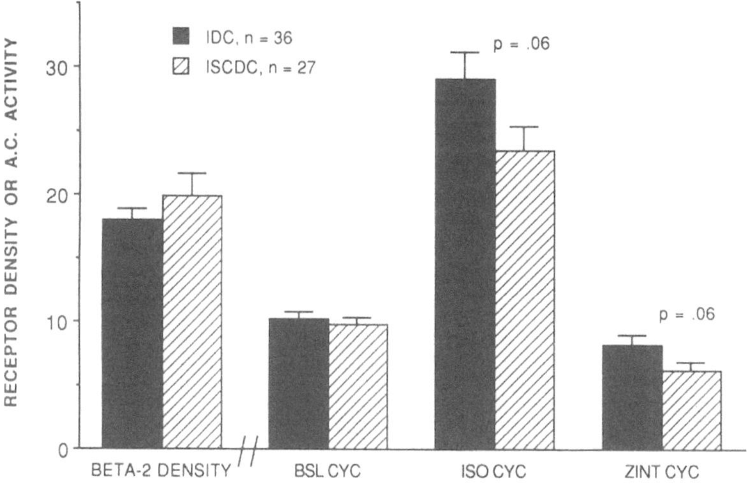

Fig. 10. β_2 receptor density and adenylyl cyclase stimulation in myocardial membranes prepared from left ventricles of subjects with idiopathic dilated cardiomyopathy (IDC) or ischemic dilated cardiomyopathy (ISCDC).

ribosylation, is increased in membranes prepared from IDC left ventricles, as shown in Fig. 12. However, G_i quantity, as assessed by immunoblotting with $G_{i\alpha 3}$ specific antiserum (Fig. 14) revealed no increase in immunodetectable G_i. These data suggest that a posttranslational modification of $G_{i\alpha 3}$ accounts for the increased functional activity of G_i, without altering the amount of immunodetectable protein.

b. ISCDC

ISCDC LVs exhibit a decrease in stimulation by NaF, as shown in Fig. 11. As shown in Fig. 12, these same ventricles also exhibit an increase in pertussis toxin catalyzed ADP ribosylation. As shown in Fig. 11, the right ventricles of ISCDC

                                                    M.R. Bristow, A.M. Feldman

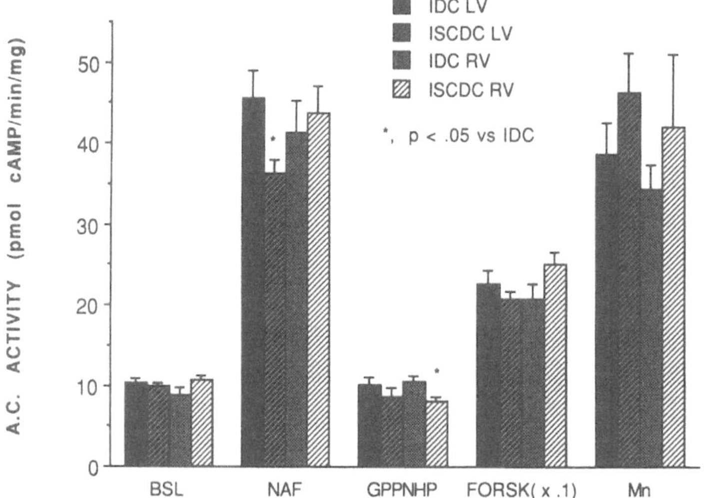

Fig. 11. Adenylyl cyclase activities in left and right ventricles of failing human hearts (BSL = basal activity, other designations are net stimulation in response to various agonists). Forskolin stimulation is given as 0.1 of the actual value.

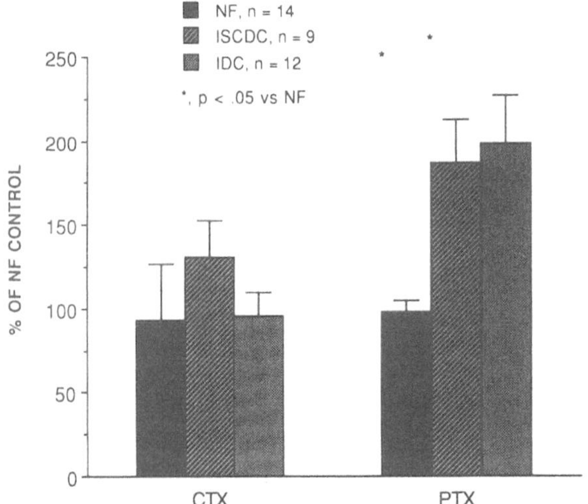

Fig. 12. Toxin catalyzed ADP ribosylation in membranes prepared from nonfailing and failing human left ventricles (CTX = cholera toxin, PTX = pertussis toxin). The data are given as percent of the average of membranes extracted from nonfailing ventricles run on the same gel.

hearts exhibit decreased Gpp(NH)p stimulation. These data suggest that alterations of G protein function are present in ISCDC left and right ventricles, but that different types of abnormalities may be present.

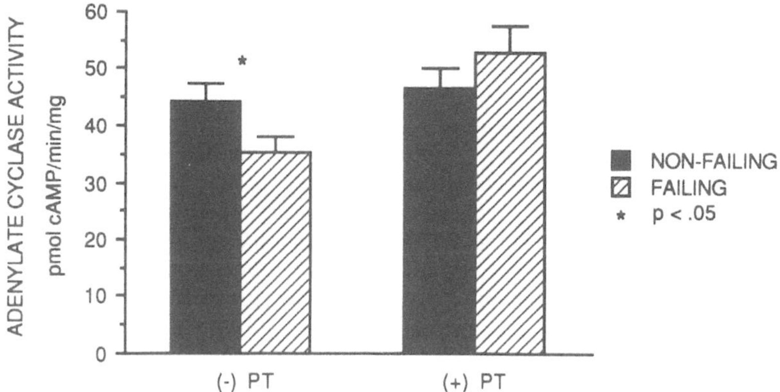

Fig. 13. Effect of pertussis toxin pretreatment on adenylyl cyclase response to forskolin plus GppNHp in myocardial membranes extracted from nonfailing and failing (IDC) human ventricles. PT = pertussis toxin.

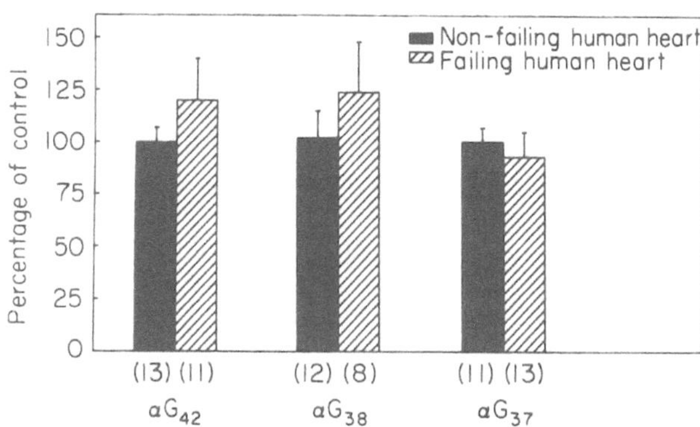

Fig. 14. Data from quantitative immunochemical quantitation of various G proteins in membranes taken from nonfailing and failing human hearts. Data were obtained using antiserum "VFB 5" (αG_{42} and αG_{38}) which recognizes $G_{i\alpha1}$ and $G_{i\alpha3}$. The αG_{37} data were from antiserum G 12, which recognizes $G_{i\alpha2}$. (Figure is reproduced with permission from the Journal of Cellular and Molecular Biology).

c. Pressure overloaded myocardium

PPH right ventricles also exhibit a decrease in Gpp(NH)p and forskolin stimulation of adenylyl cyclase, but this is apparently due to a decrease in adenylyl cyclase catalytic subunit activity (see below). This in turn may contribute to uncoupling of β_1 adrenergic receptors from mechanical response (see below).

4. Uncoupling of β_1 receptor from mechanical response

As shown in Fig. 4, there is a tendency for β_1 stimulated adenylyl cyclase or inotropic responses to be decreased by a greater amount than is β_1 receptor density in IDC ventricles. However, these trends in relative degrees of abnormality are not statistically significant. When β_1 receptor down-regulation is marked, such as in idiopathic dilated cardiomyopathy, it is quite difficult to distinguish an additional uncoupling abnormality.

Figure 15 gives the contractile response to isoproterenol in RV trabeculae taken from IDC and ISCDC. Isoproterenol as a nonselective β agonist will produce a receptor-mediated inotropic response mediated by both β_1 and β_2 receptors, in direct relation to their relative densities (5). Since β_1 receptors predominate in both nonfailing and failing ventricular myocardium, the response of the nonselective agonist isoproterenol is a pharmacologic probe of β_1 adrenergic function. In contrast, the response to isoproterenol for adenylyl cyclase stimulation is a pharmacologic probe for β_2 receptor function, in view of the much tighter coupling of the β_2 receptor to cAMP generation (8, 33). As shown in Fig. 15, in hearts which are age and cardiac function controlled the isoproterenol contractile response is significantly less in ISCDC right ventricles compared to IDC right ventricles, despite the total β and β_1 receptor density being higher. These data indicate that, relative to IDC, ISCDC ventricular β_1 receptors are partially uncoupled from mechanical response (12).

5. Decreased catalytic subunit activity of adenylyl cyclase

Figure 16 compares the hemodynamics of subjects undergoing transplantation for endstage heart failure due to ISCDC, IDC, and PPH. As can be observed, in this data set the degree of pulmonary hypertension varies from mild in ISCDC, to moderate in IDC, to severe in PPH. This creates an opportunity to assess the effect of pressure overload per se on RGC function in the human heart. As can be observed in Fig. 17, there is a progressive decline in Mn^{2+} stimulation of adenylyl cyclase going from mild to moderate to severe pulmonary artery pressure. Also shown in Fig. 17 are data from the left ventricles of these hearts, which are not subjected to differences in

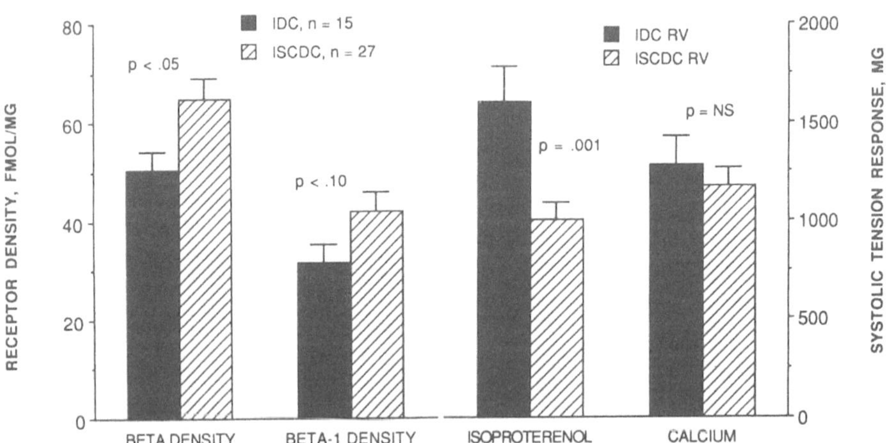

Fig. 15. β receptor densities and contractile responses in age and cardiac function matched right ventricles taken from subjects with endstage heart failure from idiopathic dilated cardiomyopathy (IDC) and ischemic dilated cardiomyopathy (ISCDC).

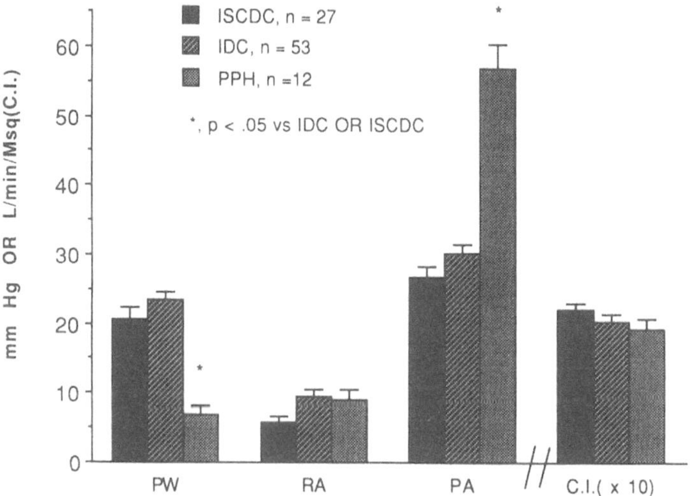

Fig. 16. Hemodynamics of subjects with endstage heart failure from idiopathic dilated cardiomyopathy (IDC), ischemic dilated cardiomyopathy (ISCDC), and primary pulmonary hypertension (PPH). Cardiac index is given as 10 times the actual value.

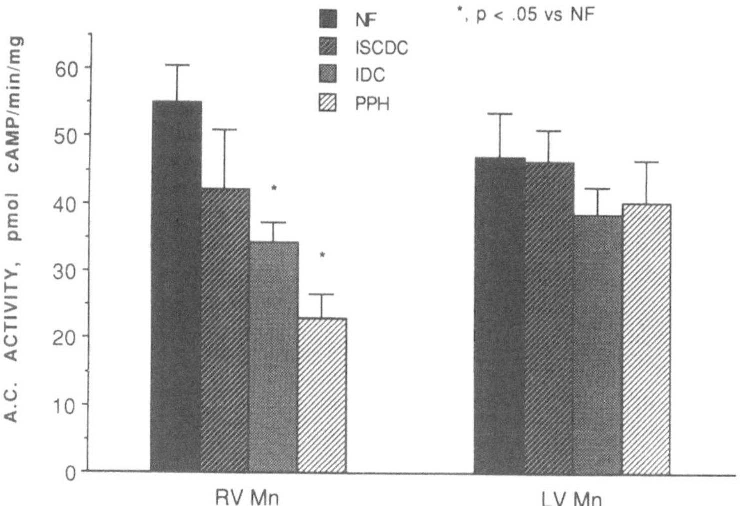

Fig. 17. Manganese (Mn) stimulation of adenylyl cyclase in membranes prepared from nonfailing and failing human left and right ventricles. NF = nonfailing, ISCDC = ischemic dilated cardiomyopathy, IDC = idiopathic dilated cardiomyopathy, and PPH = primary pulmonary hypertension.

pressure overload; in the left ventricles of these hearts there is no difference in Mn^{2+} stimulation of adenylyl cyclase. These data indicate that failing human ventricular myocardium subjected to pressure overload exhibits an abnormality of the catalytic subunit (16), which is similar to what is found in animal heart failure models subjected to pressure overload (19, 40)

Molecular basis for RGC abnormalities in the failing human heart

Some of the molecular abnormalities responsible for the RGC desensitization phenomena noted above are now known and are outlined in Table 3. As discussed, it is likely that at least some of the uncoupling of β_2 receptors from adenylyl cyclase is due to increased activity of G_i. This could result from increased dissociation of β-γ subunits on exchange of GTP for GDP at the $G_{i\alpha 3}$ binding site, which would complex free $G_{s\alpha}$ and uncouple adenylyl cyclase-coupled β adrenergic receptors, alternatively, $G_{i\alpha 3}$ could have direct effects on β_2 receptors. Since β_2 receptors are much more tightly coupled to adenylyl cyclase stimulation than are β_1, it may be that β_2 receptor uncoupling is explained on this basis. However, direct evidence for the association of increased $G_{i\alpha}$ and β_2 receptor uncoupling has, thus far, not been obtained. Since β_1 receptors may be directly coupled to calcium channels by $G_{s\alpha}$ (18), it is also possible that increased G_i is the basis for uncoupling of β_1 receptors in ISCDC.

In model systems increased G_i may be produced by chronic exposure to β agonists (27, 28), including norepinephrine exposure in cultured heart cells (39). This "cross-regulation" on exposure to β agonists typically results in β adrenergic receptor down-regulation and uncoupling, along with an increase in amount of activity of G_i in the failing human heart, in part explained by posttranslational modification of $G_{i\alpha 3}$ (22, 25). The evidence for a posttranslational modification of the major molecular species of G_i in the failing human heart is that, on an absolute basis, the steady-state level of $G_{i\alpha 3}$ mRNA is not different (22) (although the amount normalized to β-actin is) (25), and the amount of $G_{i\alpha 3}$ measured by immunochemical methods is also not different in nonfailing vs failing human heart (25). Since the steady-state message level and the amount of quantifiable protein is unchanged, some kind of posttranslational modification of the gene product must occur to explain the increased functional activity of G_i in the failing human heart (1, 23, 36).

In cardiac cells mRNA for a β_1-like receptor is decreased at the same time or before β receptor down-regulation occurs (13), suggesting that the steady-state message level may be one control point for expression of β_1 adrenergic receptors. Preliminary data in failing human heart suggest that the β_1 message (but not the β_2)

Table 3. Molecular/pharmacologic explanation for heart failure-associated abnormalities in the RGC complex

	Abnormality	Cause(s)
1.	β_1 receptor down-regulation	↓ receptor mRNA ? other
2.	β_2 receptor uncoupling	↑ Gi ? other
3.	β_1 receptor uncoupling	? ↑ Gi ? ↓ Gs
4.	VIP receptor down-regulation	?
5.	Increased VIP receptor affinity	?
6.	↑ αGi	Post-translational modification of αGi-3
7.	↓ C	?

is decreased (Bristow MR and Feldman AM, unpublished data). Therefore, different molecular regulatory processes may control gene expression of the various components of the RGC complex.

The decrease in adenylyl cyclase catalytic subunit activity may be the result of the marked cellular injury produced by pressure overload (16). Although such membrane constituents as β_2 and α_1 receptor density are unchanged in pressure overload IDC and PPH right ventricles, adenylyl cyclase catalytic subunit activity has long been considered an enzyme marker of cell membrane integrity, and declines when membranes are damaged (21). How this occurs at a molecular level is not clear; the general possibilities are increased breakdown of translated protein, decreased steady-state levels of mRNA, decreased translation frequency, or posttranslational modification of enzyme activity.

Pathophysiological significance of β-RGC changes in the failing human heart

The above changes in the β-RGC system have two important general implications. The first is that they are a marker of the various insults that affect the RGC system, most of which have the potential to damage the contractile apparatus. Shown in Fig. 18 is a schema of how this damage might occur, with β-RGC signaling leading to several cellular events which can culminate in both decrease in stimulated contractile function, as well as abnormalities in the intrinsic contractile apparatus. At least in IDC the desensitization phenomena reflect the degree of exposure to adrenergic drive, as interruption of this drive by a β_1 receptive blocking agent reverses both the β_1 down-regulation (32, 41) and the β_2 receptor uncoupling (29, 30). In ISCDC the additional coupling of β_1 and β_2 receptors probably reflects an ischemic mechanism (12), while in IDC and PPH RV pressure overload produces degrees of catalytic subunit abnormality that are in direct relation to the afterloading (12, 16). Thus, the extent to which these various abnormalities are expressed is an indicator of the extent of the exposure to pharmacologic/toxic stimulus or tissue injury. From the standpoint of receptor density, the regulatory changes present probably confer a survival

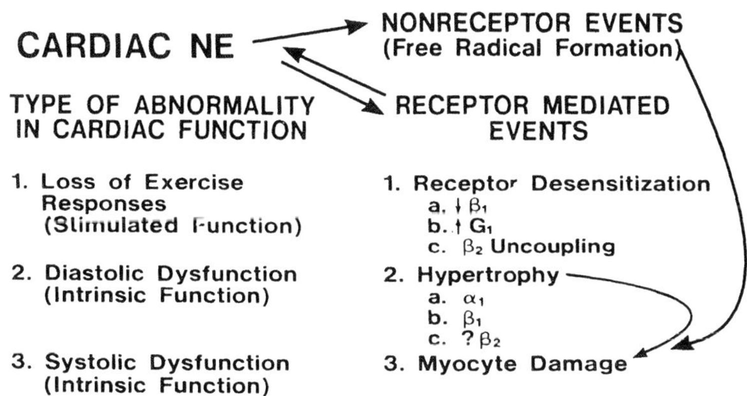

Fig. 18. Schematic of consequences of long-term exposure to increased levels of cardiac norepinephrine in the failing human heart.

advantage in that desensitization, in effect, partially withdraws the cardiac cell from exposure to toxic adrenergic stimulation, producing a type of "endogenous" form of β blockade. However, these regulatory adjustments are only partially successful, as 40–60% of the receptor pathway remains intact and may be unregulatable (9).

The second general implication of these changes is that maximal exercise capacity is lost in direct relation to receptor down-regulation (42). This is because there are no or few spare β receptors in the human heart (17, 38), and maximal inotropic stimulation of the β-RGC pathway will be compromised in direct relation to receptor desensitization (4). Since intrinsic cardiac function is not influenced by the level of adrenergic activity and submaximal exercise is probably influenced very little (37), receptor desensitization only affects the ability to perform higher levels of exercise activity. Since it can be argued that such levels of activity might be potentially harmful to subjects with heart failure, receptor desensitization in heart failure should be viewed as something that is teleologically favorable.

One area in which receptor desensitization may pose a problem for the individual is in compromising the support that the β adrenergic pathway may deliver in times of extreme stress, such as in conditions of shock or weaning from cardiopulmonary bypass. In such situations in patients with β receptor desensitization, strategies to augment signal transduction such as a use of phosphodiesterase inhibitors (7, 26) or forskolin (3) may be necessary.

In summary, numerous desensitization phenomena occur in the failing human heart that are etiology or model dependent. On balance these changes are teleologically beneficial, as they are able to partially withdraw the failing heart from potentially toxic stimuli. However, since part of the β-RGC pathway appears to be unregulatable, complete withdrawal from adrenergic stimuli is not possible by desensitization phenomena. There are clinical situations in which receptor desensitization and attenuation of receptor function is important, but these are confined to instances of extreme requirement for inotropic reserve.

References

1. Bohm M, Gierschik P, Jakobs KH, Schnabel P, Kemkes B, Erdmann E (1989) Localization of a "postreceptor" defect in human dilated cardiomyopathy. Am J Cardiol 64:812–814
2. Bristow MR, Ginsburg R, Minobe WA, Cubicciotti RS, Sageman WS, Lurie K, Billingham ME, Harrison DC, Stinson EB (1982) Decreased catecholamine sensitivity and β-adrenergic receptor density in failing human hearts. New Engl J Med 307:205–211
3. Bristow MR, Ginsburg R, Strosberg A, Montgomery W, Minobe W (1984) Cardiovascular pharmacology and inotropic potential of forskolin in the human heart. J Clin Invest 74:212–223,
4. Bristow MR, Kantrowitz NE, Ginsburg R, and Fowler MB (1985) β-adrenergic function in heart muscle disease and heart failure. J Mol Cell Cardiol 17 (Supp 2):41–52
5. Bristow MR, Ginsburg R, Fowler M, Minobe W, Rasmussen R, Zera P, Menlove R, Shah P, Stinson E (1986) β_1 and β_2-adrenergic receptor subpopulations in normal and failing human ventricular myocardium: Coupling of both receptor subtypes to muscle contraction and selective β_1 receptor down-regulation in heart failure. Circ Res 59:297–309
6. Bristow MR, Minobe W, Rasmussen R, Hershberger RE, Hoffman BB (1988) α_1 adrenergic receptors in the nonfailing and failing human heart. J Pharmacol Exp Ther 247:1039–1045

7. Bristow MR, Lee HR, Gilbert EM, Renlund DG, Hegewald MG, Hershberger RE, O'Connell JB (1989) Use of enoximone in patients awaiting cardiac transplant. Br J Clin Practice 64:69–72
8. Bristow MR, Hershberger RE, Port JD, Rasmussen R (1989) β_1 and β_2 adrenergic receptor mediated adenylate cyclase stimulation in nonfailing and failing human ventricular myocardium. Mol Pharm 35:295–303
9. Bristow MR, Hershberger RE, Port JD, Gilbert EM, Sandoval A, Rasmussen R, Cates AE, Feldman AM (1990) β-adrenergic pathways in nonfailing and failing human ventricular myocardium. Circulation 82:(Suppl I) 12–25
10. Bristow MR, O'Connell JB (1990) Myocardial Diseases. In: W. N. Kelly (ed). Textbook of Internal Medicine. J. B. Lippincott Company, Philadelphia, pp 168–176
11. Bristow MR, Minobe W, Rasmussen R, Larrabee P, Skerl L, Klein JW, Anderson FL, Murray J, Mestroni L, Karwande SV, Fowler M, Ginsburg R (1992) β-adrenergic neuroeffector abnormalities in the failing human heart are produced by local, rather than systemic mechanisms. J Clin Invest (in press)
12. Bristow MR, Anderson FL, Port JD, Skerl L, Hershberger RE, Larrabee P, O'Connell JB, Renlund DG, Volkman K, Murray J, Feldman AM (1991) Differences in β-adrenergic neuroeffector mechanisms in ischemic vs idiopathic dilated cardiomyopathy. Circulation 84:1024–1039
13. Bristow MR, Durham C, Kline J, Rasmussen R, Port JD, Gesteland RR, Barry WH, Feldman AM (1991) Down-regulation of β-adrenergic receptors and receptor mRNA in heart cells chronically exposed to norepinephrine. Clin Res 39:256A
14. Brodde OE, Schuler S, Kretsch R, Brinkmann M, Borst HG, Hetzer R, Reidemeister JC, Warnecke H, Jerkowski HR (1986) Regional distribution of β-adrenoceptors in the human heart: Coexistence of functional β_1- and β_2-adrenoceptors in both atria and ventricles in severe congestive cardiomyopathy. J Cardiovasc Pharmacol 8:1235–1242
15. Brodde OE, Zerkowski HR, Doetsch N, Motomura S, Khamssi M, Michel MC (1989) Myocardial β-adrenoceptor changes in heart failure: concomitant reduction in β_1- and β_2-adrenoceptor function related to the degree of heart failure in patients with mitral valve disease. JACC 14:323–331
16. Broode OE, Michel MC, Gordon EP, Sandoval A, Gilbert EM, Bristow MR (1989) β-adrenoceptor regulation in the human heart: can it be monitored in circulating lymphocytes? Europ Heart J 10:2–10
17. Brodde OE, Brown L, Deighton NM, Zerkowski HR (1989) The human heart exhibits only a small receptor reserve for β-adrenoceptor agonists (abst.). Circulation 82 (suppl. 1):135–144
18. Brown AM, Yatani A, Imoto Y, Codina J, Mattera R, Birnbaumer L (1989) Direct G-protein regulation of Ca^{2+} channels. Ann NY Acad Sci 373
19. Chen L, Vatner DE, Vatner SF, Hittinger L, Homcy JC (1991) Decreased $G_{s\alpha}$ mRNA levels accompany the fall in G_s and adenylyl cyclase activities in compensated left ventricular hypertrophy. In heart failure, only the impairment in adenylyl cyclase activation progresses. J Clin Invest 87:293–298
20. Colucci WS, Alexander RW, Williams GH, Rude RE, Holman BL, Konstam MA, Wynne J, Mudge GH, Braunwald E (1981) Decreased lymphocyte β-adrenergic-receptor density in patients with heart failure and tolerance to the β-adrenergic agonist pirbuterol. N Engl J Med 305:185–190
21. Dhalla NS, Das PK, Sharma GP (1978) Subcellular basis of cardiac contractile failure. J Mol Cell Cardiol 10:363–385
22. Expression of α-subunits of G proteins in failing human heart: A reappraisal utilizing quantitative polymerase chain reaction. J Mol Cell Cardiol (in press)
23. Feldman AM, Cates AE, Veazey WB, Hershberger RE, Bristow MR, Baughman KL, Baumgartner WA, Van Dop C (1988) Increase of the M_r 40,000 pertussis toxin substrate in the failing human heart. J Clin Invest 82:189–197

24. Feldman AM, Cates AE, Bristow MR, Van Dop C (1989) Altered expression of α-subunits of G proteins in failing human hearts. J Mol Cell Cardiol 21:359–365
25. Feldman AM, Jackson DG, Bristow MR, Cates AE, Van Dop C (1991) Immunodetectable levels of the inhibitory guanine nucleotide binding proteins in failing human heart: discordance with measurements of adenylate cyclase activity and levels of pertussis toxin substrate. J Mol Cell Cardiol 23:439–452
26. Gage J, Rutman H, Lucido D, LeJemtel TH (1986) Additive effects of dobutamine and amrinone on myocardial contractility and ventricular performance in patients with severe heart failure. Circulation 74:367–373
27. Hadcock JR, Ros M, Watkins DC, Malbon CC (1990) Cross regulation between G-protein-mediated pathways: stimulation of adenylyl cyclase increases expression of the inhibitory G-protein $G_{i\alpha 2}$. J Biol Chem 265:14784–14790
28. Hadcock JR, Port JD, Malbon CC (1991) Cross-regulation between G-protein-mediated pathways: activation of the inhibitory pathway of adenylyl cyclase increases the expression of β_2-adrenergic receptors. J Biol Chem 266:11915–11922
29. Hall JA, Kaumann AJ, Brown MJ (1990) Selective β_1-adrenoceptor blockade enhances positive inotropic responses to endogenous catecholamines mediated through β_2-adrenoceptors in human atrial myocardium. Circ Res 66:1610–1623
30. Hall JA, Petch MC, Brown MJ (1991) In vivo demonstration of cardiac β_2-adrenoreceptor sensitization by β_1-antagonist treatment. Circ Res 69:959–964
31. Hawthorn MH, Broadley KJ (1982) Evidence from use of neuronal uptake inhibition that β_1-adrenoceptors, but not β_2-adrenoceptors, are innervated. J Pharm Pharmacol 34:664–666
32. Heilbrunn SM, Shah P, Bristow MR, Valantine HA, Ginsburg R, Fowler MB (1989) Increased β-receptor density and improved hemodynamic response to catecholamine stimulation during long-term metoprolol therapy in heart failure from dilated cardiomyopathy. Circulation 79:483–490
33. Kaumann AJ, Lemoine H (1987) β_2-adrenoceptor-mediated positive inotropic effect of adrenaline in human ventricular myocardium: quantitative discrepancies with binding and adenylate cyclase stimulation. Naunyn-Schmeideberg's Arch Pharmacol 335:403–411
34. Lee HR, Hershberger RE, Port JD, Rasmussen R, Renlund DG, O'Connell JB, Gilbert EM, Mealey PC, Volkman K, Menlove R, Bristow MR (1991) Low dose enoximone in subjects awaiting cardiac transplantation: clinical results and effects on β-adrenergic receptors. J Thoracic Cardiovasc Surg 102:246–258
35. Lefkowitz RJ, Caron MG (1988) Adrenergic receptors: models for the study of receptors coupled to guanine nucleotide regulatory proteins. J Biol Chem 263:4993–4996
36. Neumann J, Schmitz W, Scholz H, Meyerinck LV, Doring V, Kalma P (1988) Increase in myocardial G_i-proteins in heart failure. Lancet 2:936–937
37. Olsen SL, Yanowitz FG, Gilbert EM, Mealey PC, Volkman AK, Renlund DG, Bristow MR (1992) β-blocker related improvement in submaximal exercise tolerance in heart failure from idiopathic dilated cardiomyopathy (IDC). JACC (in press)
38. Port JD, Bristow MR (1988) Lack of spare β-adrenergic receptors in the human heart. FASEB J 2(4):A602
39. Reithman C, Gierschik P, Sidiropoulos D, Werdan K, Jakobs KH (1989) Mechanism of noradrenaline-induced heterologous desensitization of adenylate cyclase stimulation in rat heart muscle cells: increase in the level of inhibitory G-protein α-subunits. Eur J Pharmacol 172:211–221
40. Sobel BE, Henry PD, Robison A, Bloor C, Ross J (1969) Depressed adenyl cyclase activity in the failing guinea pig heart. Circ Res 24:507–512
41. Waagstein F, Caidahl K, Wallentin I, Bergh C-H, Hjalmarson A (1989) Long-term β-blockade in dilated cardiomyopathy: effects of short- and long-term metoprolol treatment followed by withdrawal and readministration of metoprolol. Circulation 80:551–563

42. White M, Gilbert EM, Yanowitz F, Mealey P, Olsen S, Larrabee P, Bristow MR (1991) Relation of β-adrenergic receptor status to peak exercise in idiopathic dilated cardiomyopathy. Circulation 84:II–53

Author's address:
Michael R. Bristow, MD, PhD
Division of Cardiology, Campus box B130
University of Colorado School of Medicine
4200 East Ninth Avenue
Denver, CO 80262
USA

Quantification of $G_{i\alpha}$-proteins in the failing and nonfailing human myocardium

M. Böhm, P. Gierschik, E. Erdmann

Medizinische Klinik I der Universität München, FRG

Summary: Heterotrimeric G_i-proteins play an important role in the regulation of cardiac adenylate cyclase. Besides a downregulation of β-adrenoceptors with an accompanying reduction of the positive inotropic effects of cAMP-dependent positive inotropic agents, an increase of pertussis toxin substrates ($G_{i\alpha}$-proteins) has been observed. The increase of $G_{i\alpha}$ has been reported to be associated with a reduced adenylate cyclase activity in dilated cardiomyopathy from hearts with heart failure class NYHA IV. Since the quantification of $G_{i\alpha}$-proteins with the pertussis toxin labeling method is hampered by a number of biological and technical factors, $G_{i\alpha}$-proteins were quantified radioimmunologically using the iodinated C-terminus ^{125}I-KENLKDCGLF as tracer, purified retinal transducin α as standard, and an antiserum (DS 4) raised against the same peptide. With this technique $G_{i\alpha}$-proteins were increased by 118% in dilated cardiomyopathy and 48% in ischemic cardiomyopathy, although pertussis toxin substrates were only increased by 40% in dilated cardiomyopathy and no change was observed in ischemic cardiomyopathy. In cardiomyopathic tissue, an inverse relationship was observed between the increase of $G_{i\alpha}$ and the positive inotropic effects of isoprenaline or milrinone. These data provide evidence for a functional role of $G_{i\alpha}$ in the reduced positive inotropic effects of cAMP-dependent positive inotropic agents. In addition, results obtained with pertussis toxin labeling for quantification of $G_{i\alpha}$-proteins do not necessarily reflect the expression of $G_{i\alpha}$-proteins in the human myocardium.

Key words: Heterotrimeric G_i-proteins; adenylate cyclase; cardiomyopathy, dilated; cardiomyopathy, ischemic

Introduction

In the failing human myocardium, reduced numbers of β-adrenoceptors have been reported (1, 2, 6–8, 10, 17). This reduction is likely to be involved in the reduced positive inotropic effects of β-adrenoceptor agonists (1, 6–8, 10) and cAMP-phospho-diesterase inhibitors (2, 16). The stimulatory guanine-nucleotide binding protein ($G_{s\alpha}$) has been measured to be unchanged by cholera toxin labeling (14, 38) and reconstitution experiments into $G_{s\alpha}$-deficient cyc$^-$ S49 mouse lymphoma cells (14). An increase by 35– 40% of pertussis toxin-sensitive $G_{i\alpha}$ proteins has been observed by three independent groups of investigators in membranes from human hearts of patients with dilated cardiomyopathy (2, 14, 30). The situation in human ischemic cardiomyopathy is less clear. We have previously reported an unchanged amount of $G_{i\alpha}$-proteins determined with the pertussis toxin-induced ADP-ribosylation reaction (2), whereas Neumann et al. (26) reported a 40% decrease and Bristow et al. (9) reported a 100% increase of pertussis toxin-substrates in this condition. Using Western blotting techniques, Feldman et al. (15) reported no significant increase of $G_{i\alpha}$ in dilated cardiomyopathy although ^{32}P-ADP-ribosylation was enhanced. In contrast, our group reported increased levels of $G_{i\alpha}$ in dilated cardiomyopathy, but failed to detect any differences in ischemic cardiomyopathy on Western blots (2). The method of pertussis toxin-catalyzed ^{32}P-ADP-ribosylation to quantify $G_{i\alpha}$-proteins is dependent on a number of influences such as biophysical membrane properties, posttranslational modifications of $G_{i\alpha}$-proteins and numerous cofactors necessary for the ADP-ribosylation reaction. In addition, determination of immunodetectable $G_{i\alpha}$ on western blots is hampered by a lack of precision, when attempts are made to quantify immunostained bands by two-dimensional densitometry on nitrocellulose sheets.

Herein, we should like to report on the difficulties in quantifying $G_{i\alpha}$-proteins in the human myocardium with pertussis toxin-labeling and immunochemical methods. In order to improve the quantitative measurement of $G_{i\alpha}$-proteins, a radioimmunoassay was developed using the ^{125}I-iodinated synthetic C-terminal decapeptide (KENLKDCGLF), an antiserum (DS 4) raised against the same peptide, and retinal transducin α as standard.

ADP-ribosylation of $G_{i\alpha}$

G-proteins are heterotrimeric, membrane-associated proteins which are involved in the signal transduction of receptor generated signals to cellular effectors (18). Pertussis toxin-sensitive $G_{i\alpha}$-proteins occur as at least three different subtypes (α_{i1}, α_{i2}, α_{i3}) in various cells and tissues, each of which is a single gene product (5). The α-subunits of one family of G-proteins (G_{i1}, G_{i2}, G_{i3}, G_0 and retinal transducin) can be covalently modified by the mono ADP-ribosyltransferase activity of pertussis toxin (18, 28). When the radioactively labeled substrate ^{32}P-NAD is used these $G_{i\alpha}$-proteins can be identified and quantified after separation of membrane proteins by SDS-PAGE and autoradiography. With this technique an increase by 35–40% of pertussis toxin substrates was observed in myocardial membranes from patients with dilated cardiomyopathy (2, 14, 15, 22, 30) compared to nonfailing myocardium. In ischemic cardiomyopathy, no change of incorporated radioactivity was observed (Fig. 1), whereas other investigators observed a decrease (30) or even an increase (9) of pertussis toxin labeling in this condition.

The ADP-ribosyltransferase activity of pertussis toxin covalently links an ADP-ribosyl moiety to the cysteine residue at the fourth amino acid position from the C-terminus (28, 46). There are a number of factors which influence the pertussis toxin-catalyzed ADP-ribosylation of $G_{i\alpha}$-proteins and limit their usefulness to quantify $G_{i\alpha}$ in membranes or intact cells. These factors are summarized in Figure 2. Pertussis toxin-induced ADP-ribosylation is facilitated by GTP, GDP, and GDPrS, and least by nonhydrolyzable GTP derivatives (35–37, 43) and $\beta\gamma$-subunits (29, 31). This indicates that the GDP-liganded $\alpha\beta\gamma$-complex is the most susceptable substrate for pertussis toxin-induced ADP-ribosylation. In addition ATP enhances pertussis toxin labeling by activating the toxin itself (24) by binding to the B-oligomer (21) and facilitating dissociation of A-subunit and B-oligomer of the toxin (11). Endogenous NADase activity metabolizing ^{32}P-NAD necessary for ADP-ribosylation can reduce pertussis toxin labeling of membrane proteins (26). Moreover, the nonionic detergent Lubrol PX has been shown to concentration-dependently increase labeling in thyroid slices (36), indicating that the biophysical membrane properties play an important role. In addition, phosphorylation (44) or endogenous ADP-ribosylation might limit the ability of pertussis toxin to incorporate ^{32}P-ADP-ribose into $G_{i\alpha}$ in membranes.

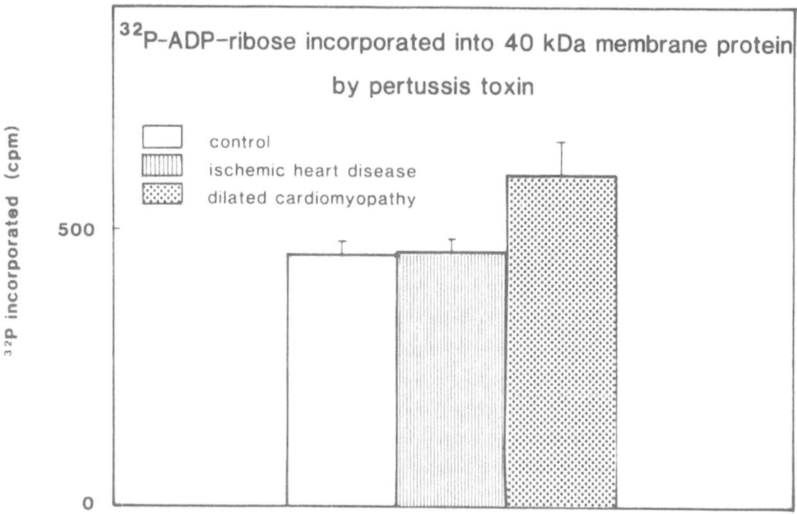

Fig. 1. ^{32}P-ADP-ribosylation by pertussis toxin of a 40 kDa membrane protein in nonfailing myocardium (control), ischemic cardiomyopathy and dilated cardiomyopathy. Samples from nonfailing myocardium were studied on each gel as controls. Values represent mean ± SEM. Modified from Böhm et al. (2).

In fact, an endogenous ADP-ribosyltransferase activity has been identified in human erythrocytes (41), which is capable to ADP-ribosylate G-proteins in vitro. Moreover, an endogenous inhibitor of ADP-ribosylation (20) and an endogenous ADP-ribosyl-hydrolase C activity cleaving mono ADP-ribosyl linkages from G$_{i\alpha}$ (40) have been identified. Finally, lipid modifications, such as myristoylation of G-protein α-subunits can influence the affinity of the α-subunits to βγ-subunits and could indirectly influence pertussis-toxin labeling (25). Taken together, all mentioned factors influence the capability of pertussis toxin to determine G$_{i\alpha}$ in membranes.

The possible differences between pertussis-toxin substrates and immunodetectable G$_{i\alpha}$ content are demonstrated in Figure 3. Pertussis toxin substrates comigrating with the α-subunit of G$_i$/G$_o$ from bovine brain in membranes from human heart membranes are more strongly labeled compared to the 40 kDa membrane protein in human lung. On the contrary, immunodetectable G$_{i\alpha}$ using an antiserum raised against the C-terminal decapeptide of retinal transducin α, clearly showed more immunodetectable G$_{i\alpha}$ in the same samples in human lung than in human heart. These examples emphasize that the intensity of pertussis toxin labeling does not necessarily corresponds to the amount of expressed G$_{i\alpha}$ proteins.

Figure 4 shows that pertussis toxin labeling in the human myocardium is critically dependent on biophysical membrane which can be influenced experimentally. The nonionic detergent Lubrol PX increased labeling concentration-dependently with a maximum at 0.1% (v/v) and decreased incorporation at higher concentrations. These findings show that G$_{i\alpha}$ determination by pertussis toxin labeling is hampered by numerous technical and biological uncertainties and that alternative techniques should be used to reliably quantify G$_{i\alpha}$-proteins.

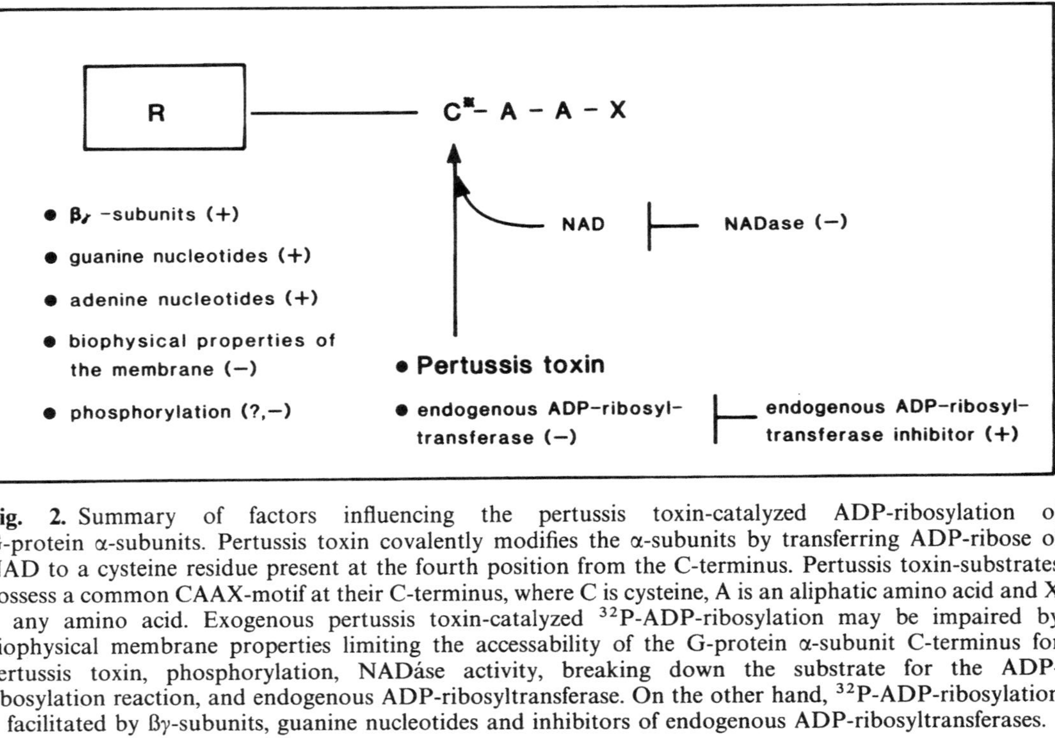

Fig. 2. Summary of factors influencing the pertussis toxin-catalyzed ADP-ribosylation of G-protein α-subunits. Pertussis toxin covalently modifies the α-subunits by transferring ADP-ribose of NAD to a cysteine residue present at the fourth position from the C-terminus. Pertussis toxin-substrates possess a common CAAX-motif at their C-terminus, where C is cysteine, A is an aliphatic amino acid and X is any amino acid. Exogenous pertussis toxin-catalyzed ^{32}P-ADP-ribosylation may be impaired by biophysical membrane properties limiting the accessability of the G-protein α-subunit C-terminus for pertussis toxin, phosphorylation, NADáse activity, breaking down the substrate for the ADP-ribosylation reaction, and endogenous ADP-ribosyltransferase. On the other hand, ^{32}P-ADP-ribosylation is facilitated by ßγ-subunits, guanine nucleotides and inhibitors of endogenous ADP-ribosyltransferases.

Immunochemical quantification of $G_{i\alpha}$

In order to study whether the increase by 35–40% of pertussis toxin-substrates in dilated cardiomyopathy is due to an increased amount of $G_{i\alpha}$, immunoblotting techniques were employed in two previous studies to directly quantify $G_{i\alpha}$-proteins on immunoblots. One study reported an increase of immunoreactive material (2), whereas another report showed only a slight insignificant increase of $G_{i\alpha}$, despite a significant increase of pertussis toxin labeling (15). In ischemic cardiomyopathic tissue, we were unable to detect changes with this technique (2).

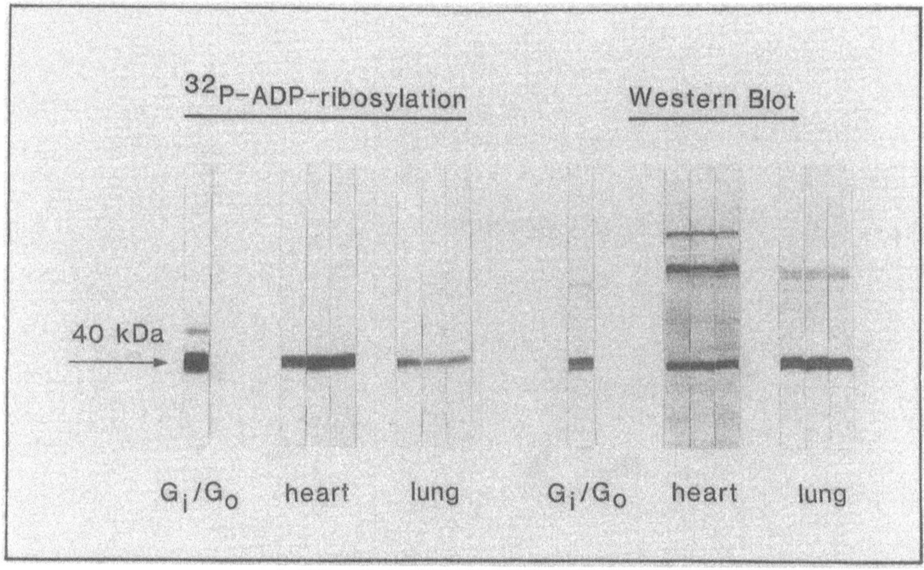

Fig. 3. Pertussis toxin-catalyzed ^{32}P-ADP-ribosylation (left, autoradiogram) and immuno-chemical detection (right, Western blots) of $G_{i\alpha}$ subunits (Mr $\approx$ 40 kDa) in human myocardial membranes and human lung membranes. G_i/G_o (1.5 µg) purified from bovine brain is shown for comparison. Each lane contained 50 µg of membrane protein in ^{32}P-ADP-ribosylation experiments and 150 µg of membrane protein in Western blots. Samples from the same specimens were used. Experimental procedures are described elsewhere (13). Note that ^{32}P-ADP-ribose incorporation by pertussis toxin was more pronounced in myocardial membranes compared to lung membranes, whereas the intensity of immunostaining of $G_{i\alpha}$ was stronger in lung membranes than in myocardial membranes.

In order to improve the immunochemical quantification of $G_{i\alpha}$-protein, a novel radioimmunoassay was developed (3). An antiserum (DS 4) was raised against the C-terminal decapeptide of retinal transducin α (Fig. 5), which strongly recognizes $G_{i\alpha1}$ and $G_{i\alpha2}$ and is less reactive against $G_{i\alpha3}$ and $G_{o\alpha}$ (19). Figure 6 demonstrates that immunostaining of the 40 kDa $G_{i\alpha}$-related protein with DS 4 can be completely antagonized by the KENLKDCGLF peptide, whereas nonspecific staining above and below the 40 kDa band was unaffected. The transducin α C-terminus KENLKDCGLF was iodinated according to Bolton and Hunter (4) and used as tracer. Isolated retinal transducin α was purified (33) and used as standard. Lung and myocardial membranes were completely solubilized and $G_{i\alpha}$ was quantified with a radioimmunoassay as described earlier (3).

Figure 7 shows concentration-response curves for the competition of solubilized membranes from human peripheral lung and human left-ventricular myocardium of patients with dilated or ischemic cardiomyopathy and membranes from one non-failing donor heart. The standard curve with retinal transducin α is shown for comparison. All competitors completely displaced antiserum binding to DS 4 (^{125}I-KENLKDCGLF). The displacement by lung membrane extracts and solubilized membranes from dilated cardiomyopathy occurred at lower protein concentrations

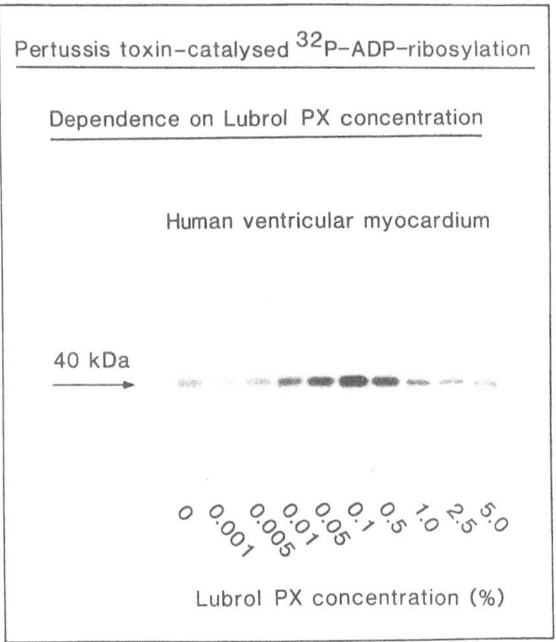

Fig. 4. Representative autoradiography showing the effect of the nonionic detergent Lubrol PX (0.001-5% v/v) on ^{32}P-ADP-ribosylation by pertussis toxin of a 40 kDa membrane protein. Membrane proteins were separated on a 10% SDS-polyacrylamide gel electrophoresis before autoradiography. Each lane contained 50 µg/lane. Exposure time of the film was 4h.

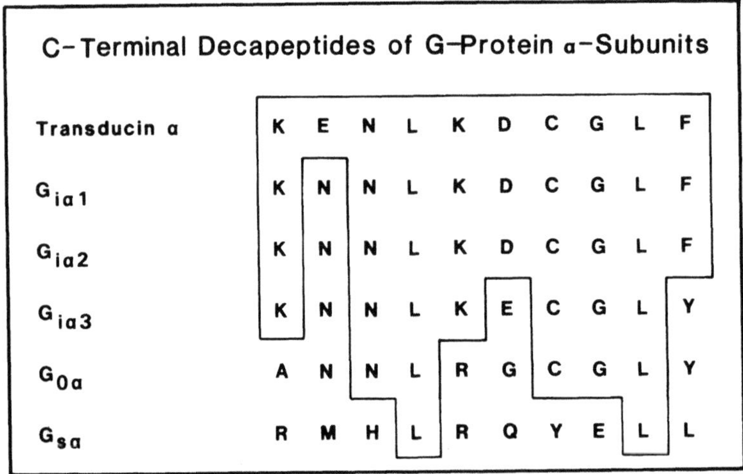

Fig. 5. Amino acid sequences of the C-terminal decapetides of pertussis toxin-sensitive G-protein α-subunits of transducin α, $G_{i\alpha 1}$, $G_{i\alpha 2}$, $G_{i\alpha 3}$ or $G_{0\alpha}$. The C-terminus of $G_{s\alpha}$ is shown for comparison. The residues identical to those of transducin α are enclosed.

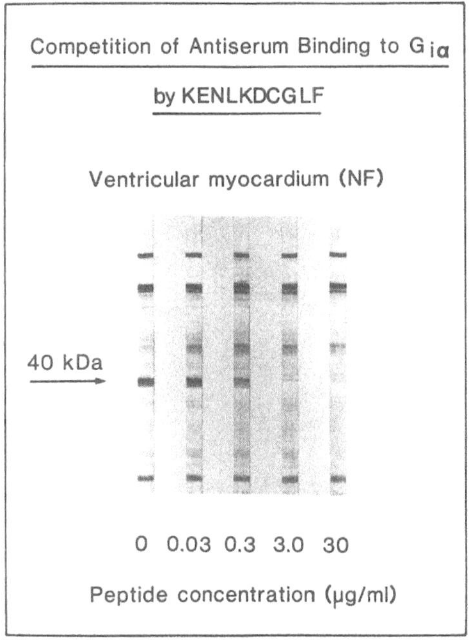

Fig. 6. Inhibition of immunostaining of $G_{i\alpha}$ (Mr ≈ 40 kDa) by the C-terminal decapeptide of retinal transducin α (KENLKDCGLF) in nonfailing (NF) left-ventricular membranes. Membrane proteins were separated by 10% SDS-PAGE before electrophoretic transfer to nitrocellulose membranes. Each lane contained 100 µg of membrane protein. Note that the immunostaining of the Mr ≈ 40 kDa band was concentration-dependently inhibited by KENLKDCGLF, whereas nonspecific staining was unaffected. Modified from Böhm et al. (3a).

than by that obtained with membranes from ischemic cardiomyopathy or nonfailing myocardium. This indicates a higher $G_{i\alpha}$-content in the former tissues. With this technique the amount of $G_{i\alpha}$ was quantified in left-ventricular myocardium of patients with different cardiac diseases. The results are summarized in Figure 8. In dilated cardiomyopathy, an increase of $G_{i\alpha}$ by 116% compared to nonfailing myocardium was observed. In ischemic cardiomyopathy, the increase was less pronounced (47%) than in dilated cardiomyopathy, although this increase was still significant. In the left ventricles from one patient with aortic stenosis and one patient with myocarditis, the increase of $G_{i\alpha}$ was smaller than in dilated cardiomyopathy, but more pronounced than in ischemic cardiomyopathy. These findings indicate that the increase of $G_{i\alpha}$ can occur independent of the underlying cardiac disease. However, it is more pronounced in dilated cardiomyopathy, myocarditis and aortic stenosis than in ischemic heart disease. Figure 9 demonstrates the relationship between pertussis toxin substrates and immuno-detectable $G_{i\alpha}$-content in membranes from nonfailing myocardium in ischemic and dilated cardiomyopathy. In ICM, a small increase of $G_{i\alpha}$ was not accompanied by an increase of pertussis toxin substrates. In DCM, pertussis toxin substrates were increased by about 40% when the amount of ADP-ribosylated $G_{i\alpha}$ was elevated by 116%. These observations might indicate that, in the

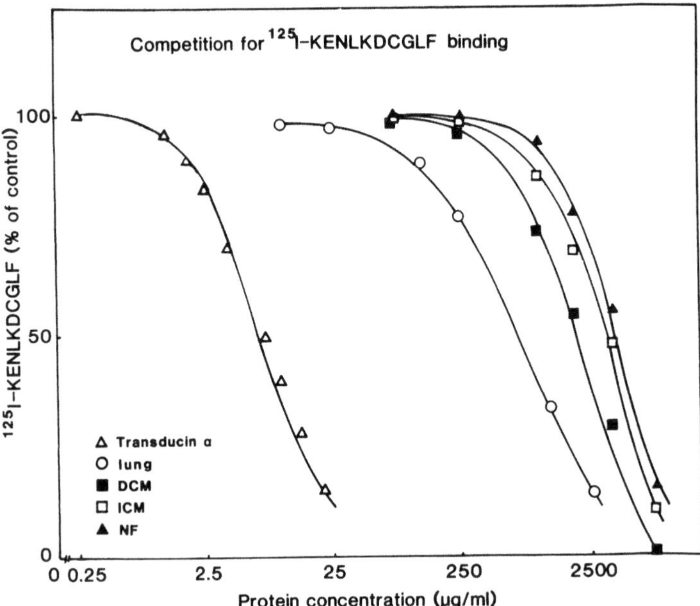

Fig. 7. Competition for ^{125}I-KENLKDCGLF binding to DS 4 of purified retinal tranducin α, and detergent extracts of human lung membranes and myocardial membranes (ICM, ischemic cardiomyopathy, DCM, dilated cardiomyopathy and NF, nonfailing heart).
Ordinate: ^{125}I-KENLKDCGLF bound in % of control; abscissa: protein concentration of competitor.

human heart, the amount of immunodetectable $G_{i\alpha}$ is underestimated by the pertussis toxin-catalysed ADP-ribosylation by about 50%.

Relationship of $G_{i\alpha}$ to positive inotropic responses

Positive inotropic responses to cAMP-dependent positive inotropic agents have been reported to be reduced in isolated cardiac preparations from failing human hearts (1, 2, 6–9, 16, 17). Since both β-adrenoceptors are reduced and $G_{i\alpha}$-proteins are increased, it is not clear to which extent each biochemical alteration contributes to the impaired functional responses in human heart failure. Figure 10 shows concentration response-curves for the β-adrenoceptor agonist isoprenaline (Fig. 10A) and for the cAMP-phosphodiesterase inhibitor milrinone (Fig. 10B). The positive inotropic responses to both agents were reduced in ischemic cardiomyopathy and further blunted in dilated cardiomyopathy compared to nonfailing myocardium. The numbers of β-adrenoceptors were similar in ischemic and dilated cardiomyopathy, but pertussis toxin substrates are increased only in dilated cardiomyopathy (2). Since the immunodetectable $G_{i\alpha}$ is increased in dilated cardiomyopathy, but only slightly elevated in ischemic cardiomyopathy, these findings provide evidence for a functional role of the elevation of $G_{i\alpha}$ in the failing myocardium. Figure 11 shows the relation of

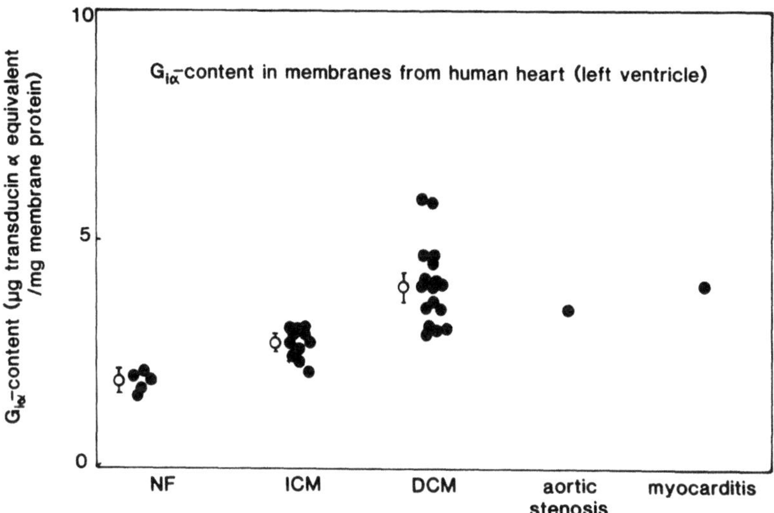

Fig. 8. $G_{i\alpha}$-content in detergent extracts of membranes from the left ventricles of human hearts with various cardiac diseases and nonfailing hearts (NF).
Ordinate: $G_{i\alpha}$-content in µg transducin α equivalents/mg membrane protein.
Abscissa: studied conditions

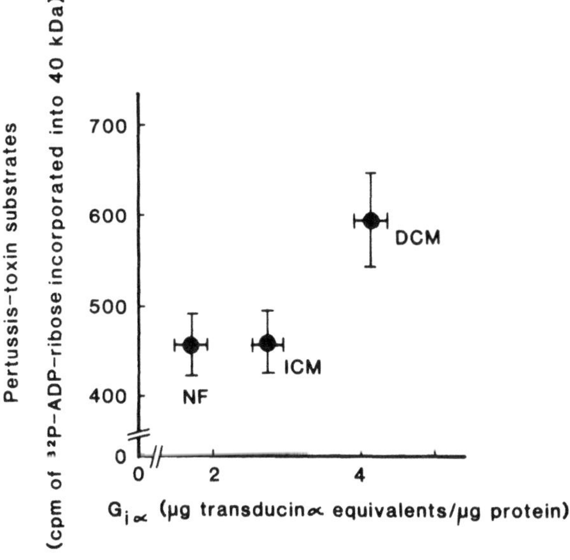

Fig. 9. Relation of pertussis toxin-substrates (ordinate) to the $G_{i\alpha}$-content (abscissa) in left-ventricular membranes from nonfailing myocardium (NF) and failing myocardium from hearts of patients with dilated (DCM) or ischemic (ICM) cardiomyopathy. Note that ^{32}P-ADP-ribosylation by pertussis toxin is only increased by 38% in DCM, in which immunodetectable $G_{i\alpha}$ is increased by 114% compared to NF.

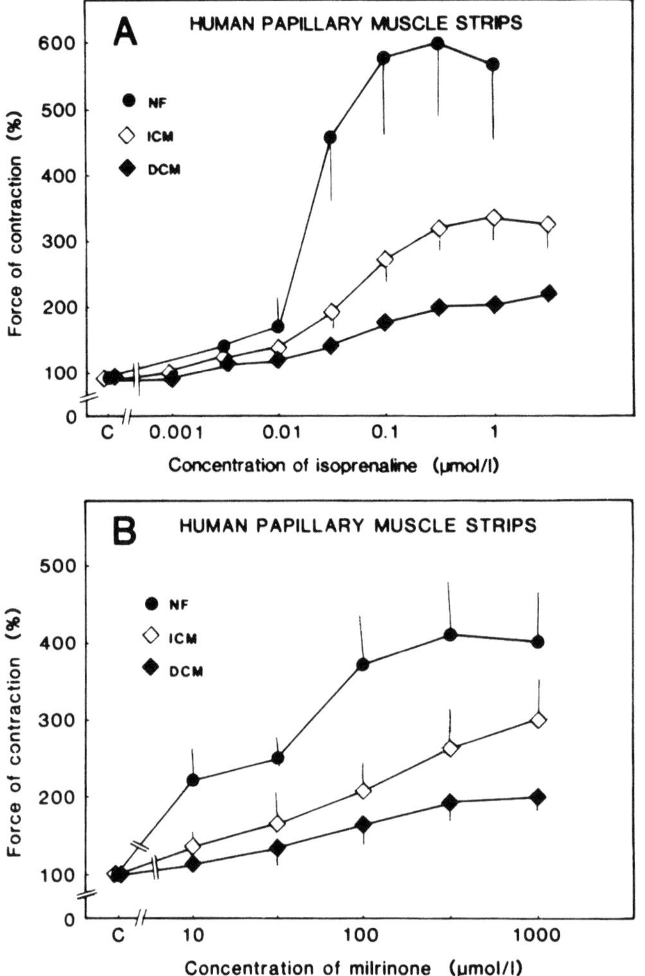

Fig. 10. Positive inotropic effects of isoprenaline (A) and milrinone (B) in isolated electrically driven (1 Hz) paillary muscle strips from hearts of patients with dilated (DCM) or ischemic (ICM) cardiomyopathy, as well as from nonfailing (NF) myocardium. Modified from (2). The density of β-adrenoceptors was similar in ICM and DCM (not shown; see (2)).

the positive inotropic responses to the β-adrenoceptor agonists isoprenaline (Fig. 11A) and the cAMP-phosphodiesterase inhibitor milrinone (Fig. 11B) to the amount of immunodetectable $G_{i\alpha}$. The inotropic responses to milrinone were reduced in parallel with an increase of immunodetectable $G_{i\alpha}$ levels. These observations might indicate that the reduced positive inotropic effects of cAMP-phosphodiesterase inhibitors are-at least in past-related to an increase of $G_{i\alpha}$-proteins.

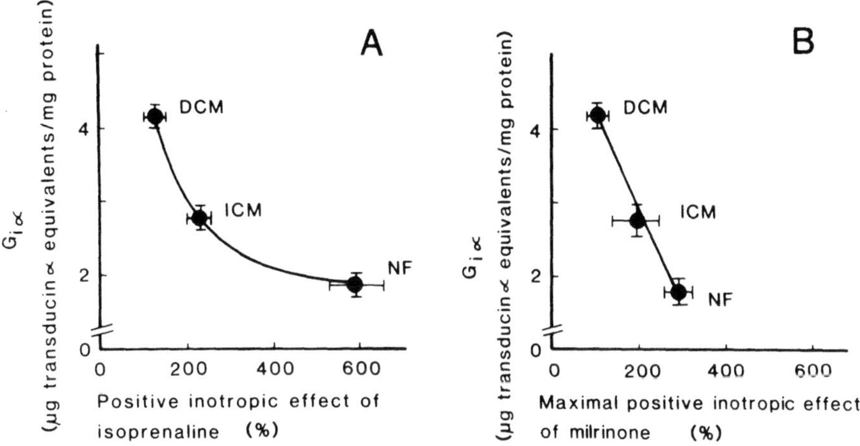

Fig. 11. Relation of immunochemically quantified $G_{i\alpha}$-content in myocardial membranes (ordinates) to the maximal positive inotropic effects (abscissa) of isoprenaline (A) or milrinone (B) in isolated papillary muscle strips from hearts of patients with dilated (DCM) or ischemic (ICM) cardiomyopathy, as well as in nonfailing (NF) hearts. Functional data are derived from previous work (2).

Conclusion

In the failing human heart, two distinct alterations in the signal transduction and regulation of the adenylate cyclase have been identified. The number of β-adrenoceptors and the inotropic responses following their stimulation have been reported to be reduced in numerous previous studies. An increase of $G_{i\alpha}$ proteins has also been observed and is most likely involved in the reduced adenylate cyclase activity (14), and in the depressed effects of β-adrenoceptor independent cAMP-elevating positive inotropic agents such as cAMP-phosphodiesterase inhibitors (2). The increase in $G_{i\alpha}$ involves an increased expression of immunodetectable $G_{i\alpha}$-proteins and not only an increase of pertussis toxin substrates, which is known to be influenced by a variety of processes including posttranslational modifications of G-protein subunits. The decline of the number of β-adrenoceptors has been suggested to represent a sequel of the excessive stimulation of the sympathetic nervous system (32) in heart failure leading to increased serum levels of circulating catecholamines (12, 13, 42), and increased β-adrenoceptor occupation and downregulation (34). The underlying mechanism of increased $G_{i\alpha}$-protein expression is not entirely resolved. Since treatment of rats with isoprenaline (27) or neonatal rat cardiocytes with norepinephrine (34) produces an increase of $G_{i\alpha}$, it is not unreasonable to speculate that excessive β-adrenoceptor stimulation of the failing heart could also be relevant in this receptor independent adenylate cyclase regulation. In this respect, it is noteworthy that the promoter region of the $G_{i\alpha 2}$ gene—which is the predominant $G_{i\alpha}$-subtype in the human-heart (2)—possesses a binding domain for AP-2 (45). This effector is involved in the effects of cAMP on transcription (23). Hence, not only myocardial β-adrenoceptors, but also $G_{i\alpha}$-proteins could be altered by the activity of the sympathetic nervous system. Therefore, it is tempting to speculate that the increase of $G_{i\alpha}$-proteins could represent a general mechanism of long-term regulation of adenylate cyclase activity. Future studies will have to investigate whether this mechanism can be pharmacologically reversed in the condition of heart failure.

Acknowledgement: Experimental work was supported by the Deutsche Forschungsgemeinschaft.

References

1. Böhm M, Beuckelmann D, Brown L, Feiler G, Lorenz B, Näbauer M, Kemkes B, Erdmann E (1988) Reduction of beta-adrenoceptor density and evaluation of positive inotropic responses in isolated, diseased human myocardium. Eur Heart J 9:844–852
2. Böhm, Gierschik P, Jakobs KH, Pieske B, Schnabel P, Ungerer M, Erdmann E (1990) Increase of $G_{i\alpha}$ in human hearts with dilated but not ischemic cardiomyopathy. Circulation 82:1249–1265
3. Böhm M, Larisch K, Erdmann E, Camps M, Jakobs KH, Gierschik P (1991) Failure of (^{32}P)-ADP-ribosylation by pertussis toxin to determine $G_{i\alpha}$ content in membranes from various human tissues. Improved radioimmunological quantification using the ^{125}I-labeled C-terminal decapeptide of retinal transducin. Biochem J 277:223–229
3a. Böhm M, Gierschik P, Larisch K, Erdmann E (1991) Radioimmunochemical quantification of $G_{i\alpha}$ in atria and ventricles from cardiomyopathic human hearts. Circ Res (submitted)
4. Bolton AE, Hunter WN (1973) The labeling of proteins to high specific radioactivities by conjugation to a ^{125}I-containing acylating agent. Biochem J 133:529–539
5. Bray P, Carter A, Guo V, Puckett C, Kamholz J, Spiegel A, Nirenberg M (1987) Human cDNA clones for an α subunit of G_i signal-transduction protein. Proc Natl Acad Sci USA 84:5115– 5119
6. Bristow MR, Ginsburg R, Minobe W, Cubiciotti RS, Sageman WS, Lurie K, Billingham ME, Harrison DC, Stinson EB (1982) Decreased catecholamine sensitivity and beta-adrenergic-receptor density in failing human hearts. N Engl J Med 307 205-211
7. Bristow MR, Ginsburg R, Umans V, Fowler MR, Minobe W, Rasmussen R, Zera P, Menlove R, Shah P, Jamieson S, Stinson SB (1986) β_1- and β_2-Adrenergic receptor subpopulations in nonfailing and failing human ventricular myocardium: coupling of both receptor down-regulation in heart failure. Circ Res 59:297–309
8. Bristow MR, Kantrowitz NE, Ginsburg R, Fowler MR (1985) Beta-adrenergic function in heart muscle disease and heart failure. J Mol Cell Cardiol 17 (Suppl 2):41–52
9. Bristow MR, Skerl L, Jackson DG, Larrabee P, Feldman AM (1990) Uncoupling of β-adrenergic receptors and increased G_i in ischemic dilated cardiomyopathy. Circulation 82 (Suppl 4):III–567
10. Brodde OE, Zerkowski HR, Borst HG, Maier W, Michel MC (1989) Drug- and disease-induced changes of human cardiac β_1- and β_2-adrenoceptors. Eur Heart J 10 (Suppl B):38–44
11. Burns DL, Manclark CR (1985) Adenine nucleotides promote dissociation of pertussis toxin subunits. J Biol Chem 261:4324– 4327
12. Chidsey CA, Harrison DC, Braunwald E (1962) Augmentation of the plasma noradrenaline response to exercise in patients with congestive heart failure. N Engl J Med 267:650–638
13. Cohn JN, Levine TB, Olivari MT, Garberg V, Lura D, Francis GS, Simon AB, Rector T (1984) Plasma noradrenaline as a guide to prognosis in patients with chronic congestive heart failure. N Engl J Med 311:819–823
14. Feldman AM, Cates AE, Veazey WB, Hershberger RE, Bristow MR, Baughman KL,

Baumgartner WA, Van Dop C (1988) Increase of the 40,000-mol wt pertussis toxin substrate (G protein) in the failing human heart. J Clin Invest 82:189–197

15. Feldman AM, Jackson DG, Bristow MR, Van Dop C (1989) Immunologic quantification of G proteins in failing and nonfailing human heart. Circulation 80 (Suppl) :II–293

16. Feldman MD, Copelas L, Gwathmey JK, Phillips P, Warren SE, Schoen FJ, Grossman W, Morgan JP: Deficient production of cyclic AMP (1987) pharmacologic evidence of an important cause of contractile dysfunction in patients with end-stage heart failure. Circulation 75:331–339

17. Fowler MB, Laser JA, Hopkins GL, Minobe W, Bristow MR (1986) Assessment of the β-adrenergic receptor pathway in the intact failing human heart: progressive receptor down-regulation and subsensitivity to agonist response. Circulation 74:1290–1302

18. Gilman AG (1987) G proteins: Transducers of receptor-generated signals. Ann Rev Biochem 56:615–649

19. Goldsmith P, Gierschik P, Milligan G, Unson CG, Vinitsky R, Malech HL, Spiegel AM (1987) Antibodies directed against synthetic peptides distinguish between GTP-binding proteins in neutrophils and brain. J Biol Chem 262:14683–14688

20. Hara-Yokoyama M, Furuyama S (1988) Endogenous inhibitor of the ADP-ribosylation of (a) G-protein(s) as catalyzed by pertussis toxin is present in rat liver. FEBS Lett 234:27–30

21. Hausman SZ, Manclark CR, Burns DL (1990) Binding of ATP by pertussis toxin and isolated toxin subunits. Biochemistry 29:6128–6131

22. Hershberger RE, Feldman AM, Bristow MR (1991) A$_1$-Adenosine receptor inhibition of adenylate cyclase in failing and nonfailing human ventricular myocardium. Circulation 83:1343–1351

23. Imagawa M, Chiu R, Karin M (1987) Transcription factor AP-2 mediates induction of two different signal-transduction pathways: protein kinase C and cAMP. Cell 51:251–260

24. Lim LK, Sekura RD, Kaslow HR (1985) Adenine nucleotides directly stimulate pertussis toxin. J Biol Chem 260:2585–2588

25. Linder ME, Pang IH, Duronio RJ, Gordon JI, Sternweiss PC, Gilman AG (1991) Lipid modifications of G protein subunits. Myristolation of Goα increases its affinity for βγ. J Biol Chem 266:4654–4659

26. Longabaugh JP, Vatner DE, Graham RM, Homcy CJ (1986) NADP improves the efficiency of cholera toxin catalyzed ADP-ribosylation in liver and heart membranes. Biochem Biophys Res Commun 137:328–333

27. Mende U, Geertz B, Scholz H, Schulte am Esch J, Sempell R (1991) Time course of the effect of isoprenaline on the inhibitory G-protein α-subunit in the heart. Naunyn Schmiedeberg's Arch Pharmacol 343 (Suppl): R54

28. Milligan G (1988) Techniques used in the identification and analysis of function of pertussis toxin-sensitive guanine nucleotide binding proteins. Biochem J 255:1–13

29. Neer EJ, Lok JM, Wolf LG (1984) Purification and properties of the inhibitory guanine nucleotide regulatory unit of brain adenylate cyclase. J Biol Chem 259:14222–14229

30. Neumann J, Scholz H, Döring V, Schmitz W, v. Meyerinck L, Kalmar P (1988) Increase in myocardial G$_i$-proteins in heart failure. Lancet II:936–937

31. Ohguro H, Fukada Y, Yoshizawa T, Saito T, Akino T (1990) A specific βγ-subunit of transducin stimulates ADP-ribosylation of the α-subunit by pertussis toxin. Biochem Biophys Res Commun 167:1235–1241

32. Packer M (1988) Neurohormonal interactions and adaptations in congestive heart failure. Circulation 77:721–730

33. Pines M, Gierschik P, Milligan G, Klee W, Spiegel A (1985) Antibodies against the carboxyl-terminal 5-kDa peptide of the α-subunit of transducin crossreact with the 40-kDa but not the 39-kDa guanine nucleotide binding protein from brain. Proc Natl Acad Sci USA 82:4095–4099

34. Reithmann C, Gierschik P, Müller U, Werdan K, Jakobs KH (1990) Pseudomonas exotoxin A prevents β-adrenoceptor-induced up-regulation of G$_i$ protein α-subunits and adenylyl cyclase desensitization in rat heart muscle cells. Mol Pharmocol 37:631–638

35. Ribeiro-Neto F, Mattera R, Grenet D, Sekura RD, Birnbaumer L, Field JB (1987) Adenosine disphosphate ribosylation of G proteins by pertussis and cholera toxin in isolated membranes. Different requirement for and effects of guanine nucleotides and Mg^{2+}. Mol Endocrinol 1:472–481

36. Ribeiro-Neto F, Mattera R, Grenet D, Sekura RD, Birnbaumer L, Field JB (1987) Adenosine disphosphate ribosylation of G proteins by pertussis and cholera toxin in isolated membranes. Different requirements for and effects of guanine nucleotides and Mg^{2+}. Mol Endocrinol 1:472–481

37. Ribeiro-Neto F, Mattera R, Hildebrandt JD, Codina J, Field JB, Birnbaumer L, Sekura RD (1985) ADP-ribosylation of membrane components by pertussis and cholera toxin. Meth Enzymol 109:566–582

38. Schnabel P, Böhm M, Gierschik P, Jakobs KH, Erdmann E (1990) Improvement of cholera toxin-catalyzed ADP-ribosylation by endogenous ADP-ribosylation factor from bovine brain provides evidence for an unchanged amount of $G_{s\alpha}$ in failing human myocardium. J Mol Cell Cardiol 22:73–82

39. Sibley DR, Lefkowitz RJ (1985) Molecular mechanisms of receptor desensitization using the beta-adrenergic receptor-coupled adenylate cyclase system as a model. Nature 317:124–129

40. Tanuma S, Endo H (1990) Identification in human erythrocytes of mono(ADP-ribosyl)protein hydrolase that cleaves a mono (ADP-ribosyl) G_i linkage. FEBS Lett 261:381–384

41. Tanuma S, Kawashima K, Endo H (1988) Eukaryotic mono(ADP-ribosyl)transferase that ADP-ribosylates GTP-binding regulatory $G_{i\alpha}$ protein. J Biol Chem 263:5485–5489

42. Thomas JA, Marks BH (1987) Plasma noradrenaline in congestive heart failure. Am J Cardiol 41:233–243

43. Tsai SC, Adamik R, Kanaho Y, Hewlett EL, Moss J (1984) Effects of guanyl nucleotides and rhodopsin on ADP-ribosylation of the inhibitory GTP-binding component of adenylate cyclase by pertussis toxin. J Biol Chem 259:15320–15323

44. Watanabe Y, Imaizumi T, Misaki N, Iwankura K, Yoshiba H (1988) Effects of phosphorylation of inhibitory GTP-binding protein by cyclic AMP-dependent protein kinase on its ADP-ribosylation by pertussis toxin, islet activating protein. FEBS Lett 236(2):372–374

45. Weinstein LA, Spiegel AM, Carter AD (1988) Cloning and characterization of the human gene for the α-subunit of G_{i2}, a GTP-binding signal transducin protein. FEBS Lett 232:333–340

46. West RE jr, Moss J, Vaughan M, Liu T, Liu TY (1985) Pertussis toxin-catalyzed ADP-ribosylation of transducin. Cysteine 347 is the ADP-ribose acceptor site. J Biol Chem 260:14428–14430

Author's address:
Dr. M. Böhm
Medizinische Klinik I der Universität München
Klinikum Großhadern
Marchioninistrasse 15
D-8000 München 70, FRG

Regulation and possible functional implications of G-protein mRNA expression in nonfailing and failing ventricular myocardium

T. Eschenhagen, U. Mende, M. Nose, W. Schmitz, H. Scholz, J. Schulte am Esch, R. Sempell, A. Warnholtz, J.-M. Wüstel

Abteilung Allgemeine Pharmakologie, Universitäts-Krankenhaus Eppendorf, Hamburg, FRG

Summary: In human end-stage heart failure an increased amount of inhibitory G-protein α-subunits ($G_{i\alpha}$) is assumed to play a role in desensitization of the adenylyl cyclase signaling pathway. In the present study, northern blot experiments with ^{32}P-labeled cDNA probes in ventricular tissue samples from explanted human hearts revealed that $G_{i\alpha-2}$- and $G_{i\alpha-3}$-mRNA are the predominant $G_{i\alpha}$-mRNA subtypes in human ventricles, whereas $G_{i\alpha-1}$-mRNA was not detectable. The mRNA for the stimulatory G-protein α-subunit ($G_{S\alpha}$) consisted of two mRNA sizes. Quantification of mRNA levels revealed a $103 \pm 38\%$ increase in $G_{i\alpha-2}$-mRNA levels in hearts with idiopathic dilative cardiomyopathy (IDC; n = 8), and a $77 \pm 25\%$ increase in hearts with ischemic cardiomyopathy (ICM; n = 6) as compared to nonfailing controls (NF, n = 8). In contrast, $G_{i\alpha-3}$- and $G_{S\alpha}$-mRNA levels were similar in failing and nonfailing hearts. To investigate whether or not the increased expression of $G_{i\alpha-2}$-mRNA might be due to chronically elevated catecholamine levels, we determined the influence of a 4-day infusion of isoprenaline (Iso; 2.4 mg/kg·d), propranolol (Prop; 9.9 mg/kg·d), Iso + Prop or 0.9% NaCl as control (Ctr) on myocardial $G_{i\alpha}$-mRNA and $G_{i\alpha}$-protein levels in rats. In Iso-treated rats, hybridization experiments revealed a $49 \pm 18\%$ (n = 7) and $27 \pm 7\%$ (n = 8) increase in $G_{i\alpha-2}$ and $G_{i\alpha-3}$-mRNA, respectively. Pertussis toxin-catalyzed ADP-ribosylation revealed a $22 \pm 7\%$ (n = 8) increase in G_i-protein as compared to Ctr (n = 8). These alterations were accompanied by an increased potency for the negative inotropic effect (NIE) of carbachol (mean EC_{50}: 0.04 μM vs. 0.28 μM) in the presence of Iso in isolated electrically driven (1 Hz) papillary muscles. Prop itself had no effect, but it antagonized all Iso-induced effects. We conclude that, in human heart failure due to IDC or ICM, increased $G_{i\alpha-2}$-, but not $G_{i\alpha-3}$-mRNA levels accompany the increased amount of $G_{i\alpha}$-protein, suggesting that this increase is at least in part due to increased de novo synthesis. The experiments in rats demonstrated that chronic β-adrenergic stimulation leads to an increased expression of $G_{i\alpha}$-mRNA and -protein, and to an enhanced potency of the negative inotropic effect of muscarinic agonists. These results support the hypothesis that increased expression of $G_{i\alpha}$ might play a pathophysiological role in human end-stage heart failure as a process of adaptation to an increased adrenergic drive.

Key words: G-proteins; mRNA expression; pertussis toxin substrates; force of contraction

Introduction

The contractile force of failing human hearts in response to β-adrenergic stimulation and phosphodiesterase inhibition is markedly diminished as compared to nonfailing controls. In contrast, the response to increasing concentrations of extracellular calcium or the cardiac glycoside dihydroouabain is preserved (3, 7, 33). This fundamental abnormality led to the suggestion that desensitization of the adenylyl cyclase signaling pathway with deficient cAMP production might play a central role in heart failure (15). Desensitization of the adenylyl cyclase pathway is currently explained by β-adrenoceptor downregulation (9) and an increase in the α-subunit of inhibitory G-proteins ($G_{i\alpha}$), which was originally determined by pertussis toxin-catalyzed ADP-ribosylation (4, 13, 30), and was recently confirmed by a radioimmunoassay with antibodies against the C-terminus common to different $G_{i\alpha}$-proteins (5). In contrast, the amount of the stimulatory G-protein α-subunit ($G_{S\alpha}$) was found to be unchanged, both with cholera toxin-catalyzed ADP-ribosylation, as well as functionally (13, 34). It has been proposed (13) that the altered relation of inhibitory and stimulatory G-protein α-subunits might cause a more pronounced tonic inhibition of the adenylyl cyclase and might, therefore, be responsible for the diminished basal and forskolin-stimulated adenylyl cyclase activity. This was supported by the ability of pertussis toxin-treatment to abolish differences in cyclase activity between failing and nonfailing hearts (13).

Although highly attractive, this hypothesis is still controversial, mainly due to four reasons: 1) Studies on isolated ventricular trabeculae from failing and nonfailing hearts consistently found no difference in the positive inotropic effect of forskolin (3, 8, 15). 2) The number of G-protein α-subunit molecules is believed to far exceed the number of membrane receptors (16) and it is, hence, questionable if an increase in the amount of $G_{i\alpha}$ by about 40–100% indeed has functional impacts. 3) The mechanism of G_i-mediated signal transduction is still not clear, i.e., which proportion is mediated directly by the α-subunit and which is indirectly mediated by relieve of free β-subunits (17, 19). 4) It is not known which of the three known $G_{i\alpha}$-subtypes are expressed in the heart, and which might be responsible for the increase in the pertussis toxin-sensitive or immunoreactive amount of G-proteins in the failing heart. This could be important since there is recent evidence from in vitro reconstitution experiments that the affinity of the different $G_{i\alpha}$-subtypes for receptors (e.g., for the adenosine receptor) might differ (16).

In the present study, we determined the G-protein expression pattern in human ventricular myocardium on the level of messenger RNA with highly specific radio-labeled cDNAs, and investigated whether or not the increase in $G_{i\alpha}$-protein in the failing heart might be due to increased mRNA levels. Furthermore, experiments on rats were performed to address the possible underlying mechanism of $G_{i\alpha}$-upregulation and its influence on contractile response in the heart.

Methods

Human cardiac tissue

Failing human hearts were obtained from patients undergoing heart transplantation because of end-stage heart failure due to idiopathic dilative cardiomyopathy (IDC, n = 8) or ischemic cardiomyopathy (ICM, n = 6). The diagnosis of ischemic heart disease was confirmed by left-heart catheterization and coronary angiography. Patients with other forms of cardiomyopathy like muscular dystrophy or acute myocarditis were excluded from the study. All of the patients were classified as NYHA IV with markedly abnormal pre-transplant hemodynamics (Table 1). Between the two groups (IDC vs. ICM) there were no significant differences in mean age, mean cardiac index, mean pulmonary capillary wedge pressure, or mean LV-

Table 1. Hemodynamic data from patients

Patient	Sex	Age	NYHA	PCW	EF	CI
Dilative cardiomyopathy						
1	m	49	IV	28	10	1.4
2	m	46	IV	19	16	2.1
3	m	61	IV	18	17	2.4
4	m	42	IV	30	14	2.0
5	f	38	IV	11	19	1.7
6	m	34	IV	31	19	2.0
7	m	41	IV	24	22	1.9
8	m	54	IV	21	13	1.6
mean		45.6		22.8	16.3	1.9
$\pm$ SEM		3.1		2.4	1.4	0.11
Ischemic cardiomyopathy						
9	m	41	IV	19	11	1.5
10	m	51	IV	32	14	1.6
11	m	50	IV	25	23	2.3
12	m	56	IV	22	18	2.0
13	m	54	IV	27	23	2.7
14	m	53	IV	35	17	1.7
mean		50.8		26.7	17.7	2.0
$\pm$ SEM		2.1		2.5	2.0	0.2

Abbreviations used: f = female; m = male; age = age in years, NYHA = functional class New York Heart Association; PCW = pulmonary capillary wedge pressure in mm Hg; EF = radionuclide-determined left-ventricular ejection fraction in %; CI = cardiac index in $1/\text{min m}^2$.

ejection fraction. Prior to cardiectomy, all patients received conventional medical treatment, including cardiac glycosides, organic nitrates, diuretics, ACE inhibitors and other vasodilators in varying combinations. No patient received catecholamines, α- or β-adrenoceptor antagonists or phosphodiesterase inhibitors. Nonfailing control hearts (NF, n = 8) were obtained from prospective multiorgan donors, whose hearts could not be transplanted due to surgical reasons, bloodgroup incompatibility, suspicion of malignant neoplasm or a history of drug abuse. Some of these patients had a history of hypertension (n = 1) or were shown to suffer from arteriosclerosis of great vessels on surgical inspection (n = 3). However, none of them had a history of heart failure and they were therefore classified as nonfailing donors. The mean age of organ donors (43.2 $\pm$ 3.2 years) did not differ significantly from heart recipients. Prior to explantation, seven of eight patients received dopamine to maintain physiological blood pressure (blood pressure at time of transplantation ranged from 100/60 to 120/80 mmHg) and renal function, one of them received a 20-mg single dose of dexamethasone additionally. Another patient received a single dose of 1.5 mg/kg phenoxybenzamine. Written informed consent was given by all patients or the families of prospective heart donors before cardiectomy. Tissue samples from the free wall of the left ventricles were obtained at explantation, rapidly frozen in liquid N_2 and, stored at $-80°C$ until use.

Animal model

Male Wistar rats were treated with a 4-day subcutaneous infusion by means of osmotic minipumps (Alzet 2ML2). Treatment groups were: 0.9% NaCl (Ctr), ($\pm$)-isoprenaline (Iso; 2.4 mg/kg·d). ($\pm$)-propranolol (Prop; 9.9 mg/kg·d) and a combination thereof (Iso + Prop). Heart rate (ECG recordings from conscious rats) and body weight were measured daily, 2 days before and during treatment. Minipumps were removed 30 min before sacrifice during short term ether anesthesia. Rats were killed by cervical dislocation and bleeding from carotid arteries. Ventricular wet weight, total RNA, and protein content of the whole tissue were determined for characterization of the model.

Total RNA preparation

Preparation of total RNA was performed according to the method of Auffray and Rougeon (procedure C in (1)) with modifications described in detail elsewhere (11). Total RNA was extracted from frozen tissue samples of the free wall of human left-ventricular myocardium and from fresh tissue samples from both left and right ventricles of the rat, respectively. RNA-concentration was determined photometrically at 260 nm. The mean OD_{260}/OD_{280} was 2.08 $\pm$ 0.04 (n = 22) in human and 2.14 $\pm$ 0.02 (n = 45) in rat RNA samples. Total RNA yield was not significantly different between NF, IDC, and ICM with mean values of 238 $\pm$ 54 (n = 8), 336 $\pm$ 45 (n = 8) and 377 $\pm$ 43 (n = 6) µg/g tissue wet weight, respectively. Total RNA yield was 940 $\pm$ 60 µg/g tissue (n = 9) in control rats and was significantly increased by isoprenaline infusion to 1310 $\pm$ 50 µg/g tissue wet weight (n = 9).

RNA blotting

Twenty µg total RNA was size-fractionated by electrophoresis in 1% agarose gels under denaturing conditions, using formamide/formaldehyde. RNA-baselength was determined with a RNA-ladder from Boehringer Mannheim (Mannheim, FRG). RNA was transferred overnight to Hybond N nylon membranes (Amersham, Braunschweig, FRG) by northern blot capillary transfer using 20 × SSC (3 M NaCl, 0.3 Nacitrate, pH 7.0) as transfer medium. For quantification of specific G_α-mRNA levels, 5–10 µg total RNA was denatured with 50% formamide and 5.9% formaldehyde, and applied to Hybond N nylon membranes by vacuum filtration with a Bio Dot SF-slot blot apparatus (Bio-Rad Laboratories, Richmond, California). Determinations of specific mRNA levels were standardized by constructing six-point standard curves of a total RNA standard in human myocardium, or by constructing six-point standard curves of pure in vitro transcribed sense orientated cRNAs in rats. The latter allowed to express specific mRNA levels in absolute values (pg/µg total RNA).

Hybridization

Hybridizations of northern and slot blots were performed with ^{32}P-nick translated rat cDNA probes (a gift from Dr. R.R. Reed, (23)) for $G_{i\alpha-1}$ (600 bp 3' noncoding

XbaI fragment), $G_{i\alpha-2}$ (full length), $G_{i\alpha-3}$ (625 bp 3' noncoding Eco RV-fragment), and $G_{S\alpha}$ (1100 bp 3' coding and noncoding Eco RI fragment) under highly stringent conditions of hybridization and washing, which have been described in detail before (11). Analysis of the expression of $G_{i\alpha-3}$ in human heart was performed with an Eco RI/HindIII fragment of the human $G_{i\alpha-3}$-cDNA, which was kindly supplied by Dr. E.J. Neer (25). Autoradiographic signals were quantified by two-dimensional densitometry. For normalization of mRNA determinations slot blots were reprobed with a ^{32}P-labeled oligo d(T) probe which was prepared with terminal deoxyribonucleotidyl transferase (Pharmacia Fine Chemicals, Uppsala, Sweden) according to Höppner et al. (20).

Pertussis toxin-catalyzed ADP-ribosylation

These experiments were performed in crude ventricular homogenates prepared from 150 mg tissue wet weight according to Ribeiro-Neto et al. (32) with minor modifications as described previously (29). The proteins (30 µg crude homogenate) were separated by discontinuous SDS-PAGE according to Laemmli (26). Gels were dried and exposed to medical x-ray films. Autoradiographic signals were measured by two-dimensional densitometry.

Contraction experiments

Contraction experiments were performed in electrically (1 Hz) driven left papillary muscles, according to Böhm et al. (2). Cumulative concentration-response curves for carbachol in the presence of 0.1 µM isoprenaline were constructed. The predrug values for the respective treatment groups (in mN/mm^2) did not differ significantly (n = number of papillary muscles): Ctr 4.0 ± 0.9 (n = 11), Iso 5.3 ± 1.0 (n = 14), Prop 4.3 ± 0.5 (n = 10), Iso + Prop 2.4 ± 0.6 (n = 11).

Statistics

Values presented are arithmetic means ± SEM. Statistical significance was determined by Student's *t*-test for paired and unpaired observations; $p < 0.05$ was considered significant.

Results

Northern blot analysis

Northern blots of 20 µg total RNA extracted from human left-ventricular myocardium and rat left- and right-ventricular myocardium are shown in Fig. 1. The radiolabeled rat $G_{i\alpha-2}$-cDNA hybridized to a predominant band at about 2.4 kb in both human and rat myocardium. The EcoRI/HindIII fragment of human $G_{i\alpha-3}$-cDNA hybridized to a band at about 2.8 kb in human myocardium, but revealed only a very faint band in rat myocardium at about 3.5 kb (not readily visible in Fig. 1).

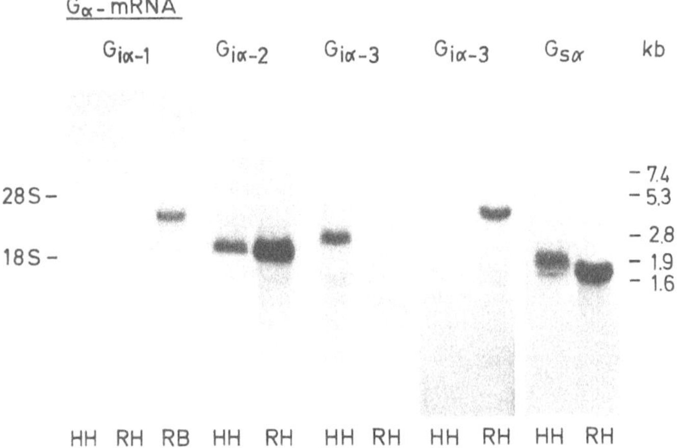

Fig. 1. Northern blot analysis of 20 μg total cellular RNA from failing human left-ventricular myocardium (HH) with idiopathic dilative cardiomyopathy, rat ventricular myocardium (RH) or rat brain (RB). Hybridization was performed with ^{32}P-radiolabeled cDNA encoding rat G$_{i\alpha-1}$ (600 bp XbaI fragment), rat G$_{i\alpha-2}$ (full-length), human G$_{i\alpha-3}$ (550 bp HindIII fragment), rat G$_{i\alpha-3}$ (625 bp Eco RV fragment) and G$_{s\alpha}$ (1100 bp Eco RI fragment) under high stringency conditions. Autoradiographic exposure time was 10 d (G$_{i\alpha-1}$), 2 d (G$_{i\alpha-2}$), 5 d (G$_{i\alpha-3}$ human), 7 d (G$_{i\alpha-3}$ rat) and 1 d (G$_{s\alpha}$). The position of the 28S and 18S ribosomal RNA and of the RNA size ladder in kilo base (kb) are indicated.

Similar results were obtained with an XbaI/EcoRI fragment of the human G$_{i\alpha-3}$-cDNA (not shown). The EcoRV/EcoRI fragment of the rat G$_{i\alpha-3}$-cDNA hybridized to a band at about 3.5 kb in rat myocardium (Fig. 1), but did not detect any band in human myocardium, even after very long exposure time (15 d) and low stringent washing conditions (not shown). G$_{i\alpha-1}$-mRNA was not detectable in human or rat ventricular myocardium, even after long exposure time (10 d; Fig. 1). There was, however, hybridization to a single band at about 3.5 kb in rat brain, indicating that the G$_{i\alpha-1}$-cDNA probe used is able to specifically detect G$_{i\alpha-1}$-mRNA. The rat G$_{s\alpha}$-cDNA probe detected two clearly distinguishable bands in human myocardium at about 1.9 and 1.8 kb, but only one predominant band at about 1.8 kb in rat myocardium. The two G$_{s\alpha}$-mRNAs in human heart probably reflect different splicing products of the single G$_{s\alpha}$-gene (22). The G$_\alpha$-mRNA expression pattern did not differ qualitatively between nonfailing and failing human myocardium (not shown).

Quantification of specific mRNA levels in human heart

Relative amounts of G$_{i\alpha-2}$-, G$_{i\alpha-3}$-, and G$_{s\alpha}$-mRNAs in myocardium of nonfailing and failing human myocardium (Fig. 2A–C) were quantitated on slot blots. mRNA levels were referred to the ^{32}P-oligo d(T) determined amount of poly A$^+$ RNA. Almost identical results were obtained if mRNA values were referred to the applied amount of total RNA (not shown). In failing human hearts the level of G$_{i\alpha-2}$-mRNA

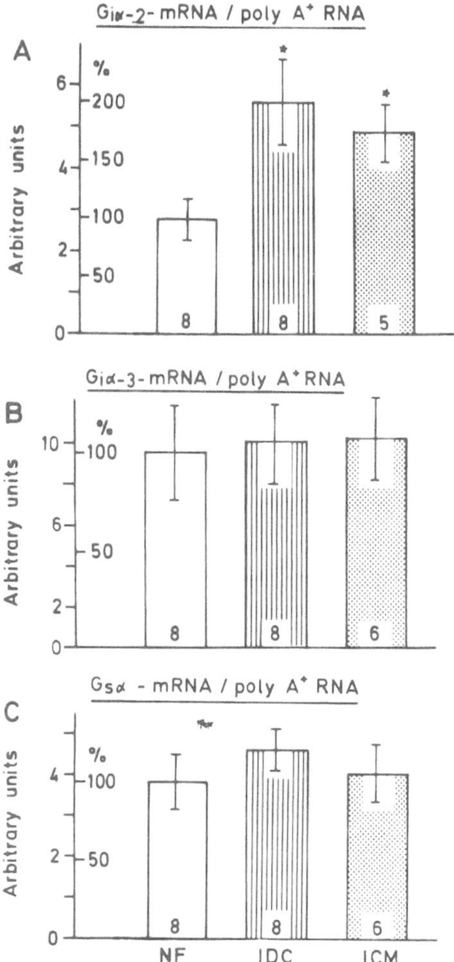

Fig. 2. Levels of $G_{i\alpha-2}$-mRNA (A), $G_{i\alpha-3}$-mRNA (B) and $G_{S\alpha}$-mRNA (C) in failing human hearts with idiopathic dilative cardiomyopathy (IDC) or ischemic cardiomyopathy (ICM) and nonfailing hearts (NF). The ordinates give mRNA levels in arbitrary density units and in percent of the mRNA level in nonfailing hearts. mRNA levels were referred to poly A^+ RNA which was determined by reprobing slot blots with ^{32}P-oligo d(T). Values are means $\pm$ SEM. Number in columns = number of hearts. *p < 0.05 vs. nonfailing hearts.

(Fig. 2A) was increased by $103 \pm 38\%$ in idiopathic dilative cardiomyopathy (IDC), and increased by $77 \pm 25\%$ in ischemic cardiomyopathy (ICM), as compared to nonfailing hearts (NF). In contrast, levels of $G_{i\alpha-3}$- and $G_{S\alpha}$-mRNA were not different within the three groups (Fig. 2B, C).

Animal model

To study possible underlying mechanisms of the increased myocardial expression of $G_{i\alpha-2}$-mRNA in human end-stage heart failure under controlled experimental in vivo conditions, rats (n = 9 in each group) were treated by subcutaneous infusions of isoprenaline, propranolol, and a combination thereof for 4 days. In accordance with other data (35, 36), a 4-day infusion of isoprenaline induced an increase in ventricular weight/body weight ratio by about 40% which was accompanied by a pronounced increase in total cardiac RNA (from 686 ± 75 to 1211 ± 41 µg), and a smaller but significant increase in total cardiac protein (from 152 ± 8 to 181 ± 9 mg), indicating the induction of cardiac hypertrophy. All isoprenaline-effects were antagonized by simultaneously administered propranolol, which itself had no effect on heart weight, but caused a pronounced decrease in heart rate (88 bpm), thus, indicating sufficient dosage of the drug.

The influence of the different treatments on myocardial G_α-mRNA concentrations is shown in Fig. 3A, B. Isoprenaline-infusion increased $G_{i\alpha-2}$-mRNA by $49 \pm 18\%$ (n = 7–8) and $G_{i\alpha-3}$-mRNA by $27 \pm 7\%$ (n = 8). The $G_{S\alpha}$-mRNA concentration was not influenced by any treatment (data not shown). The mRNA determinations revealed a rank order of mRNA levels (in pg/µg total RNA) of $G_{S\alpha}$ (36.9) $> G_{i\alpha-2}$ (10.8) $> G_{i\alpha-3}$ (3.7). In parallel with the increase in $G_{i\alpha-2}$- and $G_{i\alpha-3}$-mRNA concentrations, the pertussis toxin-labeled amount of $G_{i\alpha}$-protein in crude ventricular homogenates was increased in the isoprenaline-treated rats by $22 \pm 7\%$ (Fig. 3C). Propranolol antagonized the effects of isoprenaline when given simultaneously, and had no effect on mRNA or G-protein concentrations when given alone (Fig. 3A–C).

To study whether or not the increase in $G_{i\alpha}$-protein might modulate force of contraction, we investigated the influence of the different treatments on the negative inotropic effect of the M-cholinoceptor agonist carbachol on electrically driven left papillary muscles in the presence of 0.1 µM isoprenaline (Fig. 4). Carbachol exerted a biphasic inotropic effect in papillary muscles from all treatment groups. In each group carbachol completely abolished the positive inotropic effect of isoprenaline. However, the concentration-response curve was shifted to the left in isoprenaline-treated rats by almost one order of magnitude as compared to control (Fig. 4A, mean EC_{50}: 0.04 µM vs. 0.28 µM). Thus, isoprenaline treatment increased the potency of carbachol about eight-fold. Propranolol given in combination antagonized the effect of isoprenaline (Fig. 4C) and had no effect itself (Fig. 4B).

Discussion

The present study revealed four main results: In human ventricular myocardium $G_{i\alpha-2}$- and $G_{i\alpha-3}$-mRNA were the predominant inhibitory G-protein α-subunit mRNAs, whereas $G_{i\alpha-1}$-mRNA was not detectable. In failing human hearts with both dilative or ischemic cardiomyopathy an increased $G_{i\alpha-2}$-mRNA, but not $G_{i\alpha-3}$-mRNA level accompanied the increase in $G_{i\alpha}$-protein. $G_{S\alpha}$-mRNA levels were unchanged. In rats chronic β-adrenergic stimulation induced an increased myocardial expression of $G_{i\alpha-2}$- and $G_{i\alpha-3}$-mRNA and an increase in $G_{i\alpha}$-protein. These alterations were accompanied by an increased potency of carbachol to produce a negative inotropic effect on isolated papillary muscles.

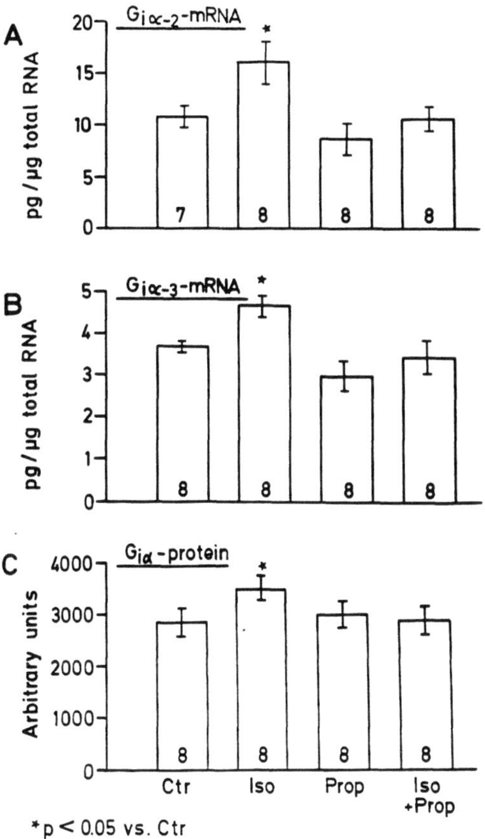

Fig. 3. Influence of a 4-day subcutaneous infusion of 0.9% NaCl (Ctr), isoprenaline (Iso; 2,4 mg/kg·d), propranolol (Prop; 9.9 mg/kg·d) and Iso + Prop on mRNA levels of $G_{i\alpha-2}$ (A) and $G_{i\alpha-3}$ (B) in pg/µg total RNA from ventricular myocardium and on the pertussis toxin-sensitive amount of $G_{i\alpha}$-protein (C) in crude ventricular homogenates in arbitrary density units. Number in columns = number of hearts. *p < 0.05 vs. Ctr.

The expression pattern and the baselength of the different G_{α}-mRNAs in human ventricles reported in the present study are in agreement with data from rat, hamster, and dog myocardium (21, 23, 24), and one study on fetal human tissues (25). They support the general view that $G_{i\alpha-2}$ is the ubiquitous and major $G_{i\alpha}$-subtype in the inhibitory control of the adenylyl cyclase (6, 18), whereas $G_{i\alpha-3}$ and $G_{i\alpha-1}$ exhibit a tissue-specific expression pattern, $G_{i\alpha-1}$ being predominantly expressed in brain and kidney (23). The question if $G_{i\alpha-2}$-mRNA levels are higher than $G_{i\alpha-3}$-mRNA levels in human ventricles (as they are in the rat (10.7 vs. 3.7 pg/µg total RNA)) cannot be answered definitely from the present experiments due to the use of nonhomologous cDNA probes for detection of $G_{i\alpha-2}$-mRNA. The only other study on adult human heart (14) that reports higher $G_{i\alpha-3}$- than $G_{i\alpha-2}$-mRNA levels in human left ventricles used only rat cDNAs for hybridization experiments and the results are

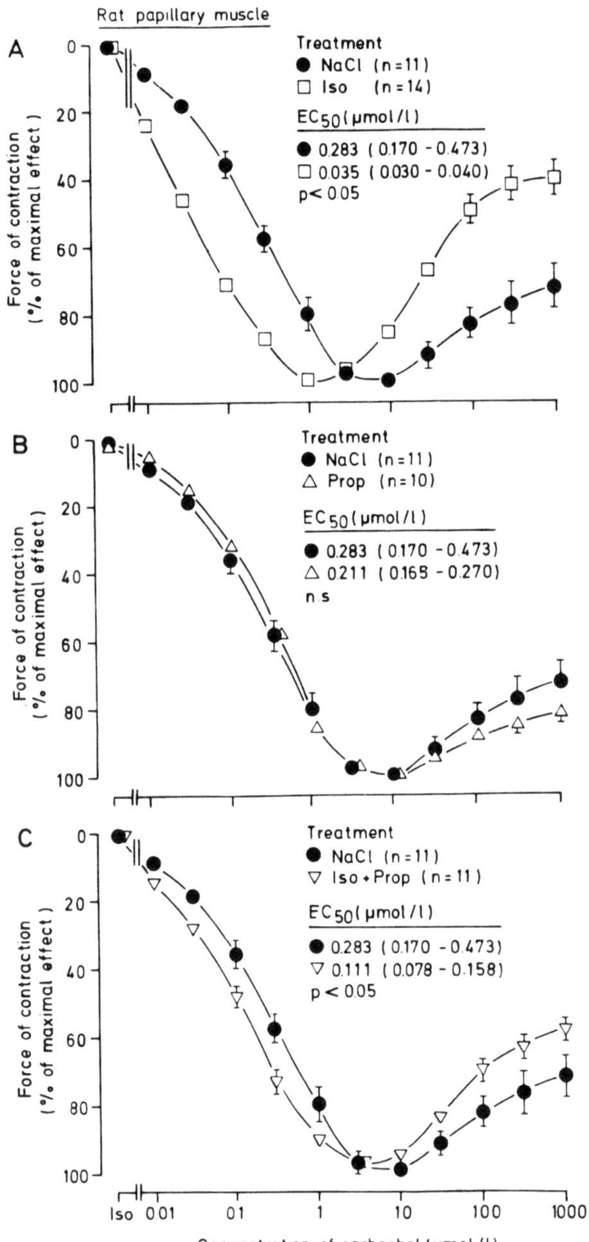

Fig. 4. Influence of a 4-day subcutaneous infusion of 0.9% NaCl (A–C), isoprenaline (Iso; A; 2.4 mg/kg·d), propranolol (Prop; B; 9.9 mg/kg·d) and Iso + Prop (C) on the concentration-response curves for the effect of carbachol in the presence of 0.1 μM isoprenaline on contractile response of papillary muscles. Contractile response to isoprenaline amounted to (in % of predrug value): Ctr 245 $\pm$ 17 (n = 11), Iso 136 $\pm$ 6 (p < 0.05 vs. Ctr; n = 14), Prop 237 $\pm$ 20 (n = 10), Iso + Prop 166 $\pm$ 10 (p < 0.05 vs. Ctr; n = 11). Predrug values in mN/mm^2

impeded by an unusual $G_{i\alpha-3}$-mRNA size of 1.85 kb. Further studies will be necessary to finally answer this question.

Quantitative data of our study also differ from the above-mentioned paper in that we found an increase only in $G_{i\alpha-2}$-, but not in $G_{i\alpha-3}$- or $G_{S\alpha}$-mRNA levels, whereas Feldman et al. (14) described significant increases in $G_{i\alpha-3}$- and $G_{S\alpha}$-mRNAs, and slight increases also in $G_{o\alpha}$-mRNA and rRNA levels; they did not quantitate $G_{i\alpha-2}$-mRNA levels. Differences might be explained by the use of β-actin mRNA as an internal standard, which bears the risk of being downregulated itself. At present, the choice of an ideal internal standard for quantifying mRNA levels is controversial and referring mRNA values on total RNA or poly A^+ RNA might be considered as the safest way. The result of increased $G_{i\alpha-2}$-mRNA but not $G_{i\alpha-3}$-mRNA levels suggests that the increase in $G_{i\alpha}$-protein in human heart failure might be restricted to $G_{i\alpha-2}$, and that the increase in $G_{i\alpha}$-protein is at least in part due to increased de novo synthesis and not only due to posttranslational mechanisms. It should be emphasized, however, that one cannot directly extrapolate from mRNA levels to protein levels and, hence, studies with $G_{i\alpha}$-subtype selective antibodies are needed to investigate if the increase in $G_{i\alpha}$ in human heart failure is indeed restricted to $G_{i\alpha-2}$. Comparable results in dilative and ischemic cardiomyopathy suggest that the increased expression of inhibitory G-proteins is a secondary phenomenon in heart failure, independent of etiology.

The hypothesis that $G_{i\alpha}$-upregulation is secondary to heart failure is supported by results of the animal model in which chronic β-adrenergic stimulation led to increased expression of $G_{i\alpha-2}$-, $G_{i\alpha-3}$-mRNA, and $G_{i\alpha}$-protein. Similar results on the protein level have been recently published from in vitro experiments on cultured neonatal cardiomyocytes (31), S49 lymphoma cells (18) or endothelial cells (27, 28). Taken together, these data support an emerging new concept of "cross-regulation" or "transregulation" of the adenylyl cyclase pathway: chronic activation of the stimulatory pathway leads to facilitated signal transduction of the inhibitory pathway and vice versa. Beside receptor up- and downregulation, altered expression of G-protein α-subunits seems to be involved in both processes.

The contraction experiments of the present study suggest that upregulation of myocardial $G_{i\alpha}$, although quite moderate (22%), might at least partially account for the increased potency of carbachol. The latter cannot be explained by upregulation of muscarinic receptors since the number of ^{3}H-NMS binding sites remained unchanged under isoprenaline infusion (own unpublished results). However, the functional importance of $G_{i\alpha}$-upregulation is strongly supported by recent experiments demonstrating that, in isoprenaline-treated rats, beside the increased potency of carbachol, the potency of the adenylyl cyclase-activating diterpene forskolin was decreased six-fold (unpublished result). Moreover, in accordance with this concept, chronic treatment of rats with the M-cholinoceptor agonist carbachol led to downregulation of $G_{i\alpha}$-protein accompanied by a decrease in the potency of carbachol (12). We are aware of the problem that the results obtained in experiments on young healthy animals cannot easily be extrapolated to the failing human heart, and that results from animal models strongly depend on the species and/or the experimental conditions used. It should be emphasized at this point that studies on human ventricular preparations consistently reported unchanged inotropic effects of forskolin (3, 8, 15) and carbachol or adenosine derivatives ((4) and own unpublished results). These discrepancies are unexplained at present. However, in spite of these limitations, the experiments on rats revealed interesting aspects of the possible mechanisms of $G_{i\alpha}$ upregulation under in vivo conditions and its impact on contractile performance of the heart. In light of these data upregulation of $G_{i\alpha-2}$-mRNA and $G_{i\alpha}$-protein demonstrated in the human heart might serve as a mechanism of adaptation against chronically elevated catecholamine levels which generally accompany human heart failure (10).

Acknowledgements. We are grateful to Dr. R.R. Reed and Dr. E.J. Neer for providing the cDNA probes for the different rat and human $G_{i\alpha}$. We would also like to thank Dr. A. Haverich, Medizinische Hochschule Hannover and Dr. V. Döring, Universitäts-Krankenhaus Eppendorf in Hamburg for providing the human heart tissue. This work was supported by the Deutsche Forschungsgemeinschaft.

References

1. Auffray C, Rougeon F (1980) Purification of mouse immunoglobulin heavy-chain messenger RNAs from total myeloma tumor RNA. Eur J Biochem 107:303–314
2. Böhm M, Brückner R, Hackbarth I, Haubitz B, Linhart R, Meyer W, Schmidt B, Schmitz W, Scholz H (1984) Adenosine inhibition of catecholamine-induced increase in force of contraction in guinea-pig atrial and ventricular heart preparations. Evidence against a cyclic AMP- and cyclic GMP-dependent effect. J Pharmacol Exper Ther 230:483–492
3. Böhm M, Beuckelmann D, Brown L, Feiler G, Lorenz B, Näbauer M, Kemkes B, Erdmann E (1988) Reduction of beta-adrenoceptor density and evaluation of positive inotropic responses in isolated, diseased human myocardium. Eur Heart J 9:844–852
4. Böhm M, Gierschik P, Jakobs KH, Pieske B, Schnabel P, Ungerer M, Erdmann E (1990) Increase of $G_{i\alpha}$ in human hearts with dilative but not ischemic cardiomyopathy. Circulation 82:1249–1265
5. Böhm M, Larisch K, Erdmann E (1991) Quantification of $G_{i\alpha}$ by a novel radioimmunoassay in cardiomyopathic human hearts. Naunyn Schmiedeberg's Arch Pharmacol 343 (Suppl): R54
6. Brann MR, Collins RM, Spiegel A (1987) Localization of mRNAs encoding the α-subunits of signal-transducing G-proteins within rat brain and among peripheral tissues. FEBS Lett 222:191–198
7. Bristow MR, Ginsburg R, Minobe W, Cubiccioti RS, Sageman WS, Lurie K, Billingham ME, Harrison DC, Stinson EB (1982) Decreased catecholamine sensitivity and beta-adrenergic receptor density in failing human hearts. N Engl J Med 307:205–211
8. Bristow MR, Ginsburg R, Strosberg A, Montgomery W, Minobe W (1984) Pharmacology and inotropic potential of forskolin in the human heart. J Clin Invest 74:212–223
9. Bristow MR, Hershberger RE, Port JD, Gilbert EM, Sandoval A, Rasmussen R, Cates AE, Feldman AM (1990) β-adrenergic pathways in nonfailing and failing human ventricular myocardium. Circulation 82 (Suppl I):12–25
10. Daly PA, Sole MJ (1990) Myocardial catecholamines and the pathophysiology of heart failure. Circulation 82 (Suppl I):35–43
11. Eschenhagen T, Mende U, Nose M, Schmitz W, Scholz H, Wüstel JM (1991a) Isoprenaline-induced increase in mRNA levels of inhibitory G-protein α-subunits in rat heart. Naunyn Schmiedeberg's Arch Pharmacol 343:609–615
12. Eschenhagen T, Geertz B, Hertle B, Mende U, Memmesheimer C, Pohl A, Schmitz W, Scholz H, Steinfath M (1991b) In vivo treatment with positive and negative inotropic agents differentially affects myocardial β-adrenoceptors, G-protein expression and force of contraction in rats. Naunyn Schmiedeberg's Arch Pharmacol 344 (Suppl): R 56
13. Feldman AM, Cates AE, Veazey WB, Hershberger RE, Bristow MR, Baughman KL, Baumgartner WA, Van Dop C (1988) Increase in the 40,000-mol wt pertussis toxin substrate (G-protein) in the failing human heart. J Clin Invest 82:189–197

14. Feldman AM, Cates AE, Bristow MR, Van Dop C (1989) Altered expression of alpha-subunits of G-proteins in failing human hearts. J Mol Cell Cardiol 21:359–365
15. Feldman MD, Copelas L, Gwathmey JK, Phillips P, Warren SE, Schoen FJ, Grossman W, Morgan JP (1987) Deficient production of cyclic AMP: pharmacologic evidence of an important cause of contractile dysfunction in patients with endstage heart failure. Circulation 75:331–339
16. Freissmuth M, Selzer E, Schütz W (1991) Interactions of purified bovine brain A_1-adenosine receptors with G-proteins. Biochem J 275:651–656
17. Gilman AG (1990) Regulation of adenylyl cyclase by G proteins. The Biology and Medicine of Signal Transduction, ed. by Y. Nishizuka et al., Raven Press, New York: 51–57
18. Hadcock JR, Ros M, Watkins DC, Malbon CC (1990) Cross-regulation between G-protein-mediated pathways. J Biol Chem 265:14784–14790
19. Hildebrandt JD, Kohnken RE (1990) Hormone inhibition of adenylyl cyclase. J Biol Chem 265:9825–9830
20. Höppner W, Rasmussen UB, Abuerreish G, Wohlrab H, Seitz HJ (1988) Thyroid hormone effect on gene expression of the adenine nucleotide transporter in different rat tissues. Mol Endocrinol 2:1127–1131
21. Holmer SR, Stevens S, Homcy CJ (1989) Tissue- and species-specific expression of inhibitory guanine nucleotide-binding proteins. Circ Res 65:1136–1140
22. Ishikawa Y, Bianchi C, Nadal-Ginard B, Homcy CJ (1990) Alternative promoter and 5' exon generate a novel $G_{S\alpha}$ mRNA. J Biol Chem 265:8458–8462
23. Jones DT, Reed RR (1987) Molecular cloning of five GTP-binding protein cDNA species from rat olfactory neuroepithelium. J Biol Chem 262:14241–14249
24. Katoh Y, Komuro I, Takaku F, Yamaguchi H, Yazaki Y (1990) Messenger RNA levels of guanine nucleotide-binding proteins are reduced in the ventricle of cardiomyopathic hamsters. Circ Res 67:235–239
25. Kim S, Ang SL, Bloch DB, Bloch KD, Kawahara Y, Tolman C, Lee R, Seidman JG, Neer EJ (1988) Identification of cDNA encoding an additional alpha-subunit of a human GTP-binding protein: Expression of three $alpha_i$ subtypes in human tissues and cell lines. Proc Natl Acad Sci USA 85:4153–4157
26. Laemmli UK (1970) Cleavage of structural proteins during assembly of the head of bacteriophage T4. Nature 227:680–685
27. Marquetant R, Bergmaier C, Oehl U, Strasser RH (1991a) Adenosin A_1 Rezeptoren transregulieren β_1-adrenerge Rezeptoren und G-protein. Zeitschrift für Kardiologie 80 (Suppl 3):541
28. Marquetant R, Brehm B, Strasser RH (1992) Transregulation of inhibitory receptors of the adenylyl cyclase system after chronic β-blockade. J Mol Cell Cardiol (in press)
29. Mende U, Eschenhagen T, Geertz B, Schmitz W, Scholz H, Schulte am Esch J, Sempell R, Steinfath M (1992) Isoprenaline-induced increase in the 40/41 kDa pertussis toxin substrates and functional consequences on contractile response in rat heart. Naunyn Schmiedebergs Arch Pharmacol 345: 44–50
30. Neumann J, Schmitz W, Scholz H, von Meyerinck L, Döring V, Kalmar P (1988) Increase of myocardial G_i-proteins in human heart failure. Lancet II:936–937
31. Reithmann C, Gierschik P, Müller U, Werdan K, Jakobs KH (1990) Pseudomonas exotoxin A prevents β-adrenoceptor-induced upregulation of G_i protein α-subunits and adenylate cyclase desensitization in rat heart muscle cells. Mol Pharmacol 37:631–638
32. Ribeiro-Neto F, Mattera R, Grenet D, Sekura RD, Birnbaumer L, Field JB (1987) Adenosine diphosphate ribosylation of G-proteins by pertussis and cholera toxin in isolated membranes. Different requirements for and effects of guanine nucleotides and Mg^{2+}. Mol Endocrinol 1:472–481
33. Schmitz W, Scholz H, Erdmann E (1987) Effects of alpha- and beta-adrenergic agonists, phosphodiesterase inhibitors and adenosine on isolated human heart muscle preparations. Trends Pharmacol Sci 8:447–450

34. Schnabel P, Böhm M, Gierschik P, Jakobs KH, Erdmann E (1990) Improvement of cholera toxin-catalysed ADP-ribosylation by endogenous ADP-ribosylation factor from bovine brain provides evidence for an unchanged amount of $G_{s\alpha}$ in failing human myocardium. J Mol Cell Cardiol 22:73–82
35. Stanton HC, Brenner G, Mayfield ED (1969) Studies on isoproterenol-induced cardiomegaly in rats. Am Heart J 77:72–80
36. Zierhut W, Zimmer HG (1989) Significance of myocardial α- and β-adrenoceptors in catecholamine-induced cardiac hypertrophy. Circ Res 65:1417–1425

Author's address:
Dr. med. T. Eschenhagen
Universitäts-Krankenhaus Eppendorf
Abteilung Allgemeine Pharmakologie
Martinistr, 52
2000 Hamburg 20, FRG

Phosphodiesterase inhibition and positive inotropy in failing human myocardium

W. Schmitz, T. Eschenhagen, U. Mende, F. U. Müller, J. Neumann, H. Scholz

Abteilung Allgemeine Pharmakologie, Universitäts-Krankenhaus Eppendorf, Universität Hamburg, Hamburg, FRG

Summary: Positive inotropic effects of phosphodiesterase inhibitors like 3-isobutyl-1-methylxanthine (IBMX), pimobendan, adibendan, milrinone, saterinone, and enoximone are greatly diminished in isolated heart muscle preparations from human failing myocardium as compared to nonfailing myocardium. This is accompanied by a reduced increase in cAMP content in intact isometrically contracting human trabeculae. With anion exchange chromatography four peaks of phosphodiesterase activities (PDE I–IV) could be separated from both nonfailing and failing human myocardium. Substrate specificity, K_m, and V_{max} were similar in nonfailing and failing myocardium. Furthermore, the PDE inhibitors investigated exhibited similar IC_{50}-values in both tissues, indicating that the sensitivity of the enzymes from nonfailing and failing tissue was unchanged. Thus, changes in PDE are probably not responsible for the reduced positive inotropic and cAMP-increasing effects of PDE inhibitors in human failing heart muscle preparations. Instead, an increase in signal transducing inhibitory G-proteins may keep the adenylyl cyclase at reduced activity, resulting in an attenuated formation of cAMP, even in the presence of PDE inhibitors.

Key words: Phosphodiesterase inhibitors; heart failure; human myocardium; cAMP

Introduction

The positive inotropic response to β-adrenergic agents is markedly reduced in heart muscle preparations from human severely failing, as compared to nonfailing heart muscle preparations (8, 13, 18). This has been attributed to down-regulation of β-adrenoceptors (3, 4, 5). In cardiac tissue from patients with endstage heart failure due to idiopathic dilated cardiomyopathy (IDC) the β_1-adrenoceptor subpopulation is selectively down-regulated, whereas the number of β_2-adrenoceptor is unchanged (5, 14). In contrast, the contractile response to β_2-selective positive inotropic agents is also reduced (5, 14), suggesting that receptor down-regulation may only partially explain the diminished increase in force of contraction. Moreover, positive inotropic cAMP-PDE inhibitors which increase cellular cAMP content, bypassing cell surface receptors, also exhibit markedly reduced inotropic responses (4, 8, 10, 13).

In the present study, we investigated whether or not the reduced positive inotropic response to PDE inhibitors is due to alterations of the PDE isoenzymes. Therefore, we compared the elution profile and kinetic parameters of cardiac PDE isoenzymes from failing and nonfailing heart. Furthermore, the effect of different PDE inhibitors on the isolated enzymes, as well as on isometric force of contraction and on cAMP content in intact contracting ventricular trabeculae were measured.

Methods

Isometric force of contraction and cAMP content were measured in electrically driven (0.5 Hz) right ventricular trabeculae from explanted human hearts of patients undergoing urgent heart transplantation. Preparations from nonfailing human hearts serving as control were obtained from prospective multiorgan donors without heart failure who died of noncardiac causes. Their hearts could not be used for transplantation for surgical reasons and/or because of blood group incompatibility. Written informed consent was obtained from the families of all organ donors and from all recipients before excision or transplantation. After explantation the hearts were placed into aerated, ice-cold bathing solution (modified Tyrode solution) and transferred to the laboratory. Contraction experiments and cAMP content measurements were started immediately. PDE experiments were done on ventricular tissue from the same hearts which were first frozen in liquid nitrogen and stored at $-80°C$ until use. Force of contraction and cAMP content as determined by radioimmunoassay were measured as described previously (10). PDE isoenzymes were separated by DEAE-sepharose anion exchange liquid chromatography by the method of Reeves et al. (12), with minor modifications as described in (2). The PDE activity was determined in a two-step procedure according to Thompson and Appleman (16) and Bauer and Schwabe (1), as described by von der Leyen et al. (10). The values presented are arithmetic means $\pm$ SEM. Concentrations of the drugs producing 50% of the maximal effect (EC_{50}) or reducing the basal activity of PDE to 50% (IC_{50}) were determined graphically and the values are given as geometric means with 95% confidence limits. Statistical significance was estimated using Student's t-test for paired or unpaired observations. A p-value of less than 0.05 was considered significant.

Results

In right-ventricular trabeculae from nonfailing human hearts the three PDE inhibitors exerted a concentration-dependent positive inotropic effects (Fig. 1A). Compared to the positive inotropic effects of calcium, the cardiac glycoside dihydroouabain, and the β-adrenoceptor agonist isoprenaline, the effects of the PDE inhibitors were smaller, amounting to 36–56% of the maximal effect of calcium. As expected for the combined action of a PDE inhibitor and a β-adrenergic agent in the presence of pimobendan the concentration-response curve for isoprenaline was shifted to the left in a parallel manner by a factor of 3.8 (based on EC_{50}-values). Similar results were obtained with pimobendan and saterinone (data not shown). In contrast, in trabeculae from failing hearts (Fig. 1B) the PDE inhibitors did not increase force of contraction significantly. While the positive inotropic effects of the cAMP-independent acting agents dihydroouabain and calcium were unchanged the maximal positive inotropic response to isoprenaline was also reduced by about 40%, and the EC_{50} was increased by factor of about 9, indicating a reduced efficacy, as well as a reduced potency of the β-adrenoceptor agonist. In presence of the PDE inhibitor pimobendan the concentration-response curve for isoprenaline was significantly shifted to the left and the reduced efficacy was restored by pimobendan to the values obtained in nonfailing hearts. Similar results were obtained with adibendan and saterinone (data not shown).

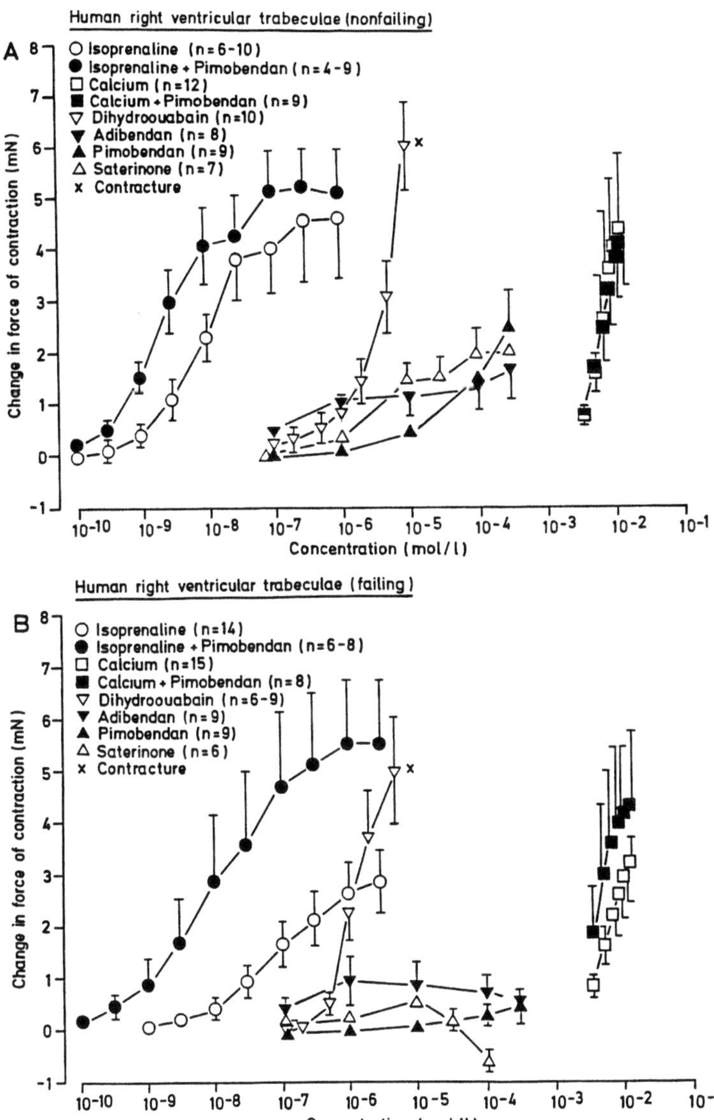

Fig. 1A, B. Cumulative concentration-response curves for the effects of pimobendan, adibendan, saterinone, isoprenaline, dihydroouabain, and calcium on force of contraction in electrically driven (0.5 Hz) right ventricular trabeculae carneae isolated from **A** five nonfailing and **B** six failing (NYHA IV) human hearts (n = number of preparations in each case). *Ordinates,* change in force of contraction (mN). *Abscissae,* drug concentration (mol/l). In all experiments the bathing solution contained 2% (v/v) DMSO. x, Concentration of dihydroouabain at which contractures occurred in some preparations. Pre-drug values were: **A** 0.97 ± 0.11 mN (n = 56; ▲, ▼, △, ○, ▽, □); in the presence of 10^{-4} mol/l pimobendan 1.04 ± 0.22 mN (n = 9, ■) and 0.94 ± 0.27 mN (n = 9, ●); **B** 2.03 ± 0.30 mN (n = 62; ▲, ▼, △, ○, ▽, □); in the presence of 10^{-4} mol/l pimobendan 3.37 ± 1.14 mN (n = 8, ■) and 1.97 ± 0.47 mN (n = 8, ●)

To investigate whether or not the reduced positive inotropic response to the PDE inhibitors may be due to altered PDE activity or altered sensitivity to the inhibitors, we isolated the PDE I-IV isoenzymes from failing and nonfailing human hearts and studied the inhibitory effects of the PDE inhibitors on the different isoenzymes. The K_m and V_{max} values of the four isoenzymes for nonfailing and failing hearts are summarized in Table 1. Except for a lower V_{max} of PDE I with cAMP as substrate in the failing hearts, there were no obvious differences in the kinetics of the enzymes of nonfailing and failing myocardium. Comparing the IC_{50} values of different PDE inhibitors for their inhibitory effects on PDE III, (the isoenzyme assumed to be the most relevant enzyme involved in the positive inotropic action of PDE inhibitors) there were no differences between preparations from nonfailing and failing hearts (Table 2).

Besides the unchanged enzyme kinetics and inhibitory effects of the PDE-inhibitors on the isolated enzymes, the reduced positive inotropic action of pimobendan, which was investigated experimentally, was accompanied by only a marginal increase in cAMP content in intact contracting ventricular trabeculae (Fig. 2).

Table 1. PDE kinetics for cAMP or cGMP as substrate, human ventricular tissue (n = 3–4)

Nonfailing Enzyme	K_m, cAMP (μmol/l)	V_{max}, cAMP (nmol/mg min)	K_m, cGMP (μmol/l)	V_{max}, cGMP (nmol/mg min)
PDE I	0.5 ± 0.1	2.7 ± 0.7	0.6 ± 0.1	3.1 ± 1.0
PDE II	134.0 ± 43.6	55.6 ± 14.9	26.4 ± 8.6	29.8 ± 2.6
PDE III	0.3 ± 0.1	10.0 ± 4.0	0.1 ± 0.01	1.1 ± 0.4
PDE IV	0.3 ± 0.1	7.0 ± 1.0	n.d.	n.d.
Failing				
PDE I	0.3 ± 0.1	$0.6 \pm 0.6*$	0.7 ± 0.01	0.9 ± 0.2
PDE II	90.5 ± 32.0	32.4 ± 11.1	17.7 ± 0.7	22.1 ± 5.6
PDE III	0.2 ± 0.1	3.0 ± 0.9	0.1 ± 0.01	0.4 ± 0.1
PDE IV	0.2 ± 0.1	3.1 ± 1.1	n.d.	n.d.

n.d. = not detectable; $*p < 0.05$ vs. corresponding nonfailing value

Table 2. cAMP PDE III inhibition, human ventricular tissue

Agent	IC_{50} (μmol/l; n = 3–4) nonfailing	failing
Pimobendan	0.70 (0.25–2.00)	0.57 (0.30–1.07)
Adibendan	0.58 (0.40–0.86)	0.61 (0.20–1.90)
Milrinone	1.14 (0.84–1.55)	1.35 (0.59–3.07)
Saterinone	0.05 (0.04–0.07)	0.03 (0.02–0.07)
IBMX	8.23 (5.7–11.8)	6.64 (4.0–11.0)

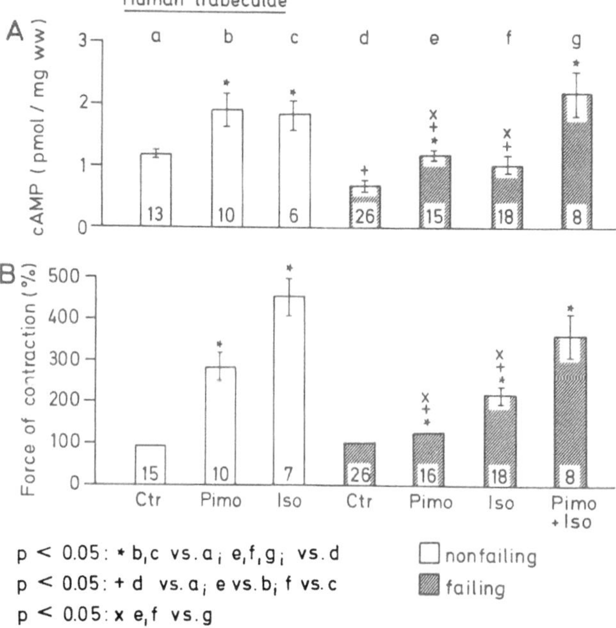

Fig. 2. A Cyclic AMP content in pmol/mg wet weight and **B** force of contraction as percentage of pre-drug value in electrically driven (0.5 Hz) right-ventricular trabeculae carneae isolated from nonfailing and failing human hearts in drug-free bathing solution (*a* and *d*, control) and after addition of pimobendan (*b* and *e*, *Pimo*, 100 µmol/l for 15 min) or isoprenaline (*c* and *f*, *Iso*, 0.2 µmol/l for 5 min). *g*, Isoprenaline was given for a further 5 min after pimobendan had been administered for 10 min. *Numbers* in *columns* denote numbers of experiments. The pre-drug values for force of contraction were 1.26 ± 0.25 mN (32 preparations from five nonfailing hearts) and 1.58 ± 0.12 mN (68 preparations from six failing hearts)

However, the combination of the β-adrenoceptor agonist isoprenaline and pimobendan led to an increase in force of contraction and cAMP content similar to the isoprenaline-induced increase in the nonfailing hearts. Thus, the PDE inhibitor restored the effect of isoprenaline on contractility as well as on cAMP.

Discussion

In the present study the cAMP-increasing β-adrenoceptor agonist isoprenaline and PDE inhibitors pimobendan, adibendan, and saterinone had markedly reduced or no positive inotropic effects in ventricular trabeculae isolated from hearts of patients with endstage heart failure. The impaired inotropic response was accompanied by an attenuated increase in cAMP content in failing as compared to nonfailing preparations. The failure of PDE inhibitors to adequately increase force of contraction and cAMP content cannot be explained by altered properties of the PDE isoenzymes, as kinetic properties or inhibition of the isolated PDE forms were similar in failing and nonfailing hearts.

It is well-established that cAMP-increasing agents like β-adrenergic agents or PDE inhibitors reveal reduced positive inotropic effects in failing as compared to nonfailing heart muscle preparations (3, 6, 8, 13, 18). The impaired response to β-adrenergic agents has at least partially been attributed to a reduced β-adrenoceptor number (3, 4, 5, 14). However, the down regulation of β-adrenoceptors cannot explain the diminished increase in force of contraction and cAMP content observed with PDE inhibitors, which exert their effects by bypassing the cell surface receptors. Although the cAMP content was increased only marginally, the present data suggest that a reduced sensitivity of the PDE is probably also not the cause of the blunted action of PDE inhibitors. Instead, it may be explained by a reduced cAMP formation due to an attenuated adenylyl cyclase activity resulting from the recently described increase in inhibitory signal transducing GTP-binding proteins (G-proteins) and the mRNA encoding for these proteins (7, 11, Eschenhagen et al., this volume).

Interestingly, the reduced increase in force of contraction and cAMP could be restored by the combined application of the PDE inhibitor pimobendan and isoprenaline. Although the molecular mechanism of this effect has not been studied as yet, the finding may be of clinical relevance. It may at least in part explain the acute effectiveness of PDE inhibitors in patients with severe heart failure, as enhanced endogenous catecholamine levels in patients with heart failure (9, 15) are well known. In this situation, therapy with a PDE inhibitor may increase myocardial force of contraction because of the combined action of the PDE inhibitors and the endogenous catecholamines. However, one should keep in mind that such a combination may increase the arrhythmogenic risk.

References

1. Bauer AC, Schwabe U (1980) An improved assay of cyclic 3′,5′-nucleotide phosphodiesterases with QAE-sephadex columns. Naunyn-Schmiedeberg's Arch Pharmacol 311:193–198
2. Bethke T, Klimkiewicz A, Kohl C, von der Leyen H, Mehl H, Mende U, Meyer W, Neumann J, Schmitz W, Scholz H, Starbatty J, Stein B, Wenzlaff H, Döring V, Kalmar P, and Haverich A (1991) Effects of Isomazole on Force of Contraction and Phosphodiesterase Isoenzymes I–IV in Nonfailing and Failing Human Hearts *Journal of Cardiovascular Pharmacology* 18:386–397 1991 Raven Press, Ltd., New York
3. Böhm M, Beuckelmann D, Brown L, Feiler G, Lorenz B, Näbauer M, Kemkes B, Erdmann E (1988) Reduction of β-adrenoceptor density and evaluation of positive inotropic responses in isolated, diseased human myocardium. Eur Heart J 9:844–852
4. Bristow MR, Ginsburg R, Minobe W, Cubicciotti RS, Sageman WS, Lurie K, Billingham ME, Harrison DC, Stinson EB (1982) Decreased catecholamine sensitivity and β-adrenergic-receptor density in failing human hearts. N Engl J Med 307:205–211
5. Bristow MR, Port JD, Sandoval AB, Rasmussen R, Ginsburg R, Feldman AM (1989) β-Adrenergic receptor pathways in the failing human heart. Heart Failure 5:77–90
6. Erdmann E (1988) The effectiveness of inotropic agents in isolated cardiac preparations from the human heart. Klin Wochenschr 66:1–6
7. Feldman AM, Cates AE, Veazey WB, Hershberger RE, Bristow MR, Baughman KL, Baumgartner WA, Dop C van (1988) Increase of the 40,000-mol wt pertussis toxin substrate (G protein) in the failing human heart. J Clin Invest 82:189–197
8. Feldman MD, Copelas L, Gwathmey JK, Phillips P, Warren SE, Schoen FJ, Grossmann W, Morgan JP (1987) Deficient production of cyclic AMP: pharmacologic evidence of an important cause of contractile dysfunction in patients with end-stage heart failure. Circulation 75:331–339

9. Levine TB, Francis GS, Goldsmith SR, Simon AB, Cohn JN (1982) Activity of the sympathetic nervous system and renin-angiotensin system assessed by plasma hormone levels and their relation to hemodynamic abnormalities in congestive heart failure. Am J Cardiol 49:1659–1666

10. von der Leyen H, Mende U, Meyer W, Neumann J, Nose M, Schmitz W, Scholz H, Starbatty J, Stein B, Wenzlaff H, Döring V, Kalmár P, Haveric A (1991) Mechanism underlying the reduced positive inotropic effects of the phosphodiesterase III inhibitors pimobendan, adibendan and saterinone in failing as compared to nonfailing human cardiac muscle preparations. Naunyn-Schmiedeberg's Arch Pharmacol 344:90–100

11. Neumann J, Schmitz W, Scholz H, Meyerinck L von, Döring V, Kalmár P (1988) Increase in myocardial G_i-proteins in heart failure. Lancet II:936–937

12. Reeves ML, Leigh BK, England PJ (1987) The identification of a new cyclic nucleotide phosphodiesterase activity in human and guinea-pig cardiac ventricle. Biochem J 241:535–541

13. Schmitz W, Scholz H, Erdmann E (1987) Effects of α- and β-adrenergic agonists, phosphodiesterase inhibitors and adenosine on isolated human heart muscle preparations. Trends Pharmacol Sci 8:447–450

14. Steinfath M, Danielsen W, von der Leyen H, Mende U, Meyer W, Neumann J, Nose M, Reich T, Schmitz W, Scholz H, Starbatty J, Stein B, Döring V, Kalmar P, Haverich A (1992) Reduced α_1- and β_2-adrenoceptor mediated positive inotropic effects in human end-stage heart failure. Brit J Pharmacol (in press)

15. Thomas JA, Marks BH (1978) Plasma norepinephrine in congestive heart failure. Am J Cardiol 41:233–243

16. Thompson WJ, Appleman MM (1971) Multiple cyclic nucleotide phosphodiesterase activities from rat brain. Biochemistry 10:311–316

17. Thompson WJ, Terasaki WL, Epstein PM, Strada SJ (1979) Assay of cyclic nucleotide phosphodiesterase and resolution of multiple molecular forms of the enzyme. Adv Cyclic Nucleotide Res 10:69–92

18. Wilmshurst PT, Walker JM, Fry CH, Mounsey JP, Twort CHC, Williams BT, Davies MJ, Webb-Peploe MM (1984) Inotropic and vasodilator effects of amrinone on isolated human tissue. Cardiovasc Res 18:302–309

Author's address:
Prof. Dr. W. Schmitz
Abteilung Allgemeine Pharmakologie
Universitäts-Krankenhaus
Eppendorf
Martinistraße 52
W-2000 Hamburg 20, FRG

Cardiovascular cyclic nucleotide phosphodiesterases and their role in regulating cardiovascular function

E.D. Pagani, R.A. Buchholz, P.J. Silver

Department of Cardiovascular Pharmacology, Sterling Winthrop Pharmaceuticals Research Division, Rensselaer, New York, USA

Summary: We have described five phosphodiesterase (PDE) isozymes that can be found in cardiac and vascular smooth muscle of animals and humans. Much of the evidence for the role that these isozymes have in the regulation of cellular processes has been generated through, or awaits, the identification of selective and potent PDE inhibitors. While selective inhibitors of the cGMP-inhibitable (cGi)-PDE isozyme have been approved for use in the acute treatment of heart failure, selective inhibitors of the cGMP-PDE have not been extensively explored as potential candidates for the treatment of cardiovascular diseases. More potent selective inhibitors of the cGMP-PDE isozyme are needed to determine whether these pharmacological potentiators of EDRF and ANP will be useful in the therapy of angina, hypertension or heart failure.

Key words: PDE isozymes; cardiac and vascular smooth muscle; PDE inhibitors; EDRF; ANP

Introduction

The mechanism of action of a number of endogenous hormones or neurotransmitters (e.g., vasopressin, atrial natriuretic factor, norepinephrine) and drugs (nitroglycerin, milrinone) that affect hemodynamics and cardiac performance directly involve regulation of the intracellular concentration of cyclic adenosine monophosphate (cAMP) and cyclic guanosine monophosphate (cGMP). Cyclic nucleotides represent a class of second messengers whose role it is to trigger a cascade of intracellular events that regulate cellular function. The intracellular concentration of cyclic nucleotides is determined through their synthesis via cyclases and their degradation via phosphodiesterases (PDE). Cyclic AMP and cGMP are hydrolyzed to the 5' monophosphate by diverse families of distinct PDE isozymes which have been identified and partially characterized from cardiac, smooth muscle, renal cells, platelets, lung, rod and cone cells from the retina and from other cell types (for review, see (3)). Abnormalities in cell function could result from defects in either cyclase or phosphodiesterase activities that give rise to either a deficiency or excess of cyclic nucleotides. For example, cholera toxin affects the G_s regulatory protein in the microvillus membrane of the intestines, causing an overstimulation of adenylate cyclase activity and cAMP synthesis (23) while studies (11) have shown that in human heart failure there is an overexpression of the G_i regulatory protein which suppresses cardiac adenylate cyclase activity and reduces the concentration of cAMP. Little information exists, however, regarding alterations in the cAMP and cGMP PDEs as a cause or consequence of cardiovascular disease. In this paper, we describe some of the work we have performed to evaluate the types of PDE isozymes present in cardiac and smooth muscle, the characteristics of a cGMP-inhibitable, cAMP PDE isozyme in cardiac tissue from patients and animals with heart failure, and we give some examples of how selective inhibition of the cGMP-inhibitable, cAMP-PDE isozyme and the zaprinast-inhibitable, cGMP-PDE isozyme enhance cardiac performance and cardiovascular hemodynamics.

Phosphodiesterase isozymes: isolation, enzymatic characteristics and regulation of activity

Nomenclature and isolation of PDE isozymes

The emergence of different nomenclatures for the identification of PDE isozymes has created some confusion. One widely used nomenclature identifies PDE isozymes based on their order of elution from an anion-exchange resin and refers to the isozymes as peaks of PDE activity, ie., Peak I through Peak V. However, since peaks of PDE activity often contain multiple PDE isozymes, a more definitive nomenclature has been proposed which is based on each isozyme's substrate specificity and/or susceptibility to selective chemical inhibition (3). Eventually, as the PDE isozymes become purified and their amino acid sequence determined, a nomenclature based on primary sequence would appear to be the most accurate (4). However, only a small fraction of the PDE isozymes have been sequenced to date, too small a number to use as a framework to create an identification system. In this paper we will refer to PDE isozymes using both nomenclatures since in certain types of experiments ample amounts of tissue were not available for further separation and purification of PDE isozymes as described below.

In our studies, we have evaluated both human and animal tissues from the cardiovascular system and kidney, and have identified a 1) zaprinast-inhibitable, cGMP-PDE (cGMP-PDE), 2) a calcium-calmodulin stimulated PDE (CaCM-PDE), 3) a cGMP-stimulated, cAMP-PDE (cGs-PDE), 4) a rolipram-inhibitable, cGMP-insensitive, cAMP-PDE (rolipram-PDE) and 5) a cGMP-inhibitable, cAMP-PDE (cGi-PDE). These isozymes have been isolated via anion-exchange chromatography and further purified with the use of affinity resins prepared from chemical analogs of selective inhibitors (Table 1).

Anion-exchange chromatography has been used to isolate four PDE isozymes in the soluble fractions of cardiac and smooth muscle tissue, however, the isozymes cannot be resolved into four single peaks of activity. In most instances the isozymes elute in three major peaks of PDE activity, although their relative proportions vary across animal species and cell types. Present in the first peak of activity are the cGMP-PDE and the CaCM-PDE. The second peak of activity mostly contains the cGs-PDE, while the third peak contains both the rolipram-inhibitable and the cGi-PDE isozymes.

We have further purified a cGMP-PDE isozyme from the first peak of an anion-exchange resin from dog aorta and human saphenous vein by using a combination of calmodulin-affinity chromatography and an affinity resin prepared from a zaprinast-analog (Fig. 1). The cGMP-PDE is released from the zaprinast-analog resin by elution with 50 mM cGMP. The cGMP-PDE isozyme preferentially hydrolyzes cGMP over cAMP. Substrate hydrolysis is non-cooperative and follows simple Michaelis-Menten kinetics with an apparent Km value of about 1 μM for smooth muscle from dog aorta and human saphenous vein. Zaprinast (M&B 22,948) is a

Table 1. Cyclic nucleotide phosphodiesterase isozymes isolated from cardiac and vascular smooth muscle tissue

Isozyme isolation	PDE family	Substrate Km (μM)	Selective Inhibitor Ki (nM)	Resin used in isolation
Peak I				
cGMP-PDE	V	cGMP (1)	Zaprinast (260 nM)	Zaprinast-analog sepharose resin
Calcium-Calmodulin stimulated PDE	I	cAMP (1–2)	Calmodulin-antagonists	Calmodulin-sepharose resin
Peak II				
cGMP-stimulated	II	cAMP (20–30)	None reported	cGMP-sepharose resin
Peak III				
cGMP-inhibited	III	cAMP (0.25)	Milrinone (300 nM)	Milrinone-analog sepharose resin
Rolipram-inhibitable	IV	cAMP (4–7)	Rolipram (300 nM)	Rolipram-analog sepharose resin

competitive inhibitor of the cGMP-PDE with an apparent Ki value of 260 nM (Table 1). Studies are ongoing to determine if the cGMP-PDE isozyme isolated from dog and human vascular smooth muscle contains a non-catalytic, cGMP binding site similar to that described for the cGMP binding PDE isolated from platelets (7) and lung (13).

The CaCM-PDE also appears in the first peak of PDE activity from an anion-exchange resin and can be further enriched with the use of calmodulin-affinity chromatography. CaCM-PDE hydrolyzes cAMP and cGMP and the rate of hydrolysis is stimulated by calcium and calmodulin (Table 1). The observed effect of calcium/calmodulin on activation of PDE activity is to increase Vmax without affecting Km for substrate. The concentration of calmodulin required for half maximal activation appears to vary slightly with the cell type from which the CaCM-PDE is isolated. Calmodulin antagonists, like calmidazolium and trifluoperazine, can attenuate the stimulatory effect of calcium/calmodulin on CaCM-PDE activity (Table 1). We have also synthesized submicromolar potent, selective inhibitors of the basal CaCM-PDE activity (unpublished observation) which are not calmodulin antagonists.

The second peak of PDE activity from an anion-exchange resin contains the cGs-PDE isozyme whose activity for cAMP hydrolysis can be stimulated by micromolar concentrations of cGMP. We have obtained enriched fractions of this PDE with the use of a cGMP-sepharose resin. In the absence of cGMP, the kinetics for cAMP hydrolysis are cooperative, but follows Michaelis-Menten kinetics in the presence of cGMP. The role of cGs-PDE isozyme in the regulation of cardiovascular function remains elusive, primarily due to the absence of a potent and selective inhibitor of this isozyme.

The third peak of PDE activity isolated by anion-exchange chromatography routinely contains both a rolipram-inhibitable, cGMP-insensitive, cAMP-PDE and a cGMP-inhibitable (cGi), low Km cAMP-PDE. We have further purified the rolipram-PDE using a rolipram-analog affinity resin. The enzyme is easily released

VASCULAR SMOOTH MUSCLE cGMP-PDE

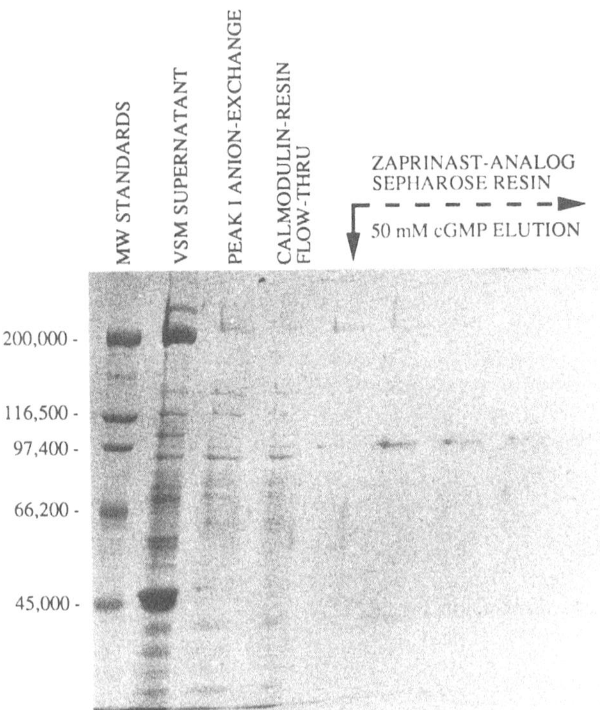

Fig. 1. SDS polyacrylamide (7.5%) gel of fractions leading to purification of the cGMP-PDE isozyme. Dog aorta was homogenized in 10 mM Tris (pH 7.5) containing proteolytic enzyme inhibitors and centrifuged at 20 000 x g. The vascular smooth muscle (VSM) supernatant fraction (lane 2) was applied to an anion-exchange resin, washed and eluted with 0.2 M sodium acetate, to yield Peak I PDE isozymes (lane 3). Peak I was applied to a calmodulin-sepharose resin, and the flow through fraction (lane 4) was applied to the zaprinast-analog sepharose resin. The column was washed and the cGMP-PDE activity was eluted with 50 mM cGMP (lanes 5–9). The majority of the cGMP activity was associated with the lanes 6, 7 and 8, where a major protein band of about 10, 000 daltons was observed. Molecular weight standards are seen in lane 1.

from the affinity resin by eluting with 50 mM cAMP. The apparent Km for cAMP is about 4–7 µM and hydrolysis of cAMP follows Michaelis-Menten kinetics. cAMP hydrolysis is competitively inhibited by rolipram with an apparent Ki of 300 nM (Table 1). We have also further purified the cGi-PDE from heart and vascular smooth muscle using a milrinone-analog affinity resin. The enzyme is also easily released from the affinity resin by elution with 5 mM cAMP. The cGi-PDE has an apparent Km for cAMP of 0.25 µM and follows Michaelis-Menten kinetics. The apparent Ki for milrinone is 300 nM (Table 1).

PDE Isozymes in human and rat cardiac tissue with heart failure

We have evaluated the PDE isozymes in normal and diseased human (20) and rat hearts (14). Several in vitro studies of ventricular tissue obtained from patients in various stages of heart failure have shown that inhibitors of the cGi-PDE isozyme become less efficacious as positive inotropes as the stage of heart failure progresses (12, 6). The cardiac inotropic potency and efficacy of the cGi-PDE inhibitors are also significantly reduced or absent in the rat relative to other animals, suggesting the possibility of species differences. In contrast, the vasodilatory activity of cGi-PDE inhibitors appears to be somewhat equiefficacious across species. To determine if the loss of cardiotonic efficacy of these agents could be attributed to diseased-related alterations or species-dependent differences in the cGi-PDE, we evaluated the enzymatic properties and the effect of reference inhibitors on PDE isozymes isolated from normal and failed human (20) and rat cardiac tissue (14).

Evaluation of the cGi-PDE isozyme in human cardiac tissue from patients with end-stage heart failure

Segments of the left ventricle were obtained from Brigham and Women's Hospital, Boston, in collaboration with Dr. Paul Allen, Department of Anesthesia. Diseased heart tissue was harvested from patients undergoing orthotopic heart transplantation because of end-stage failure due to idiopathic cardiomyopathy. Control heart tissue was obtained from organ donors whose hearts were not considered acceptable for transplantation for reasons other than cardiovascular disease. The PDE isozymes were isolated by either DEAE-Sephacel or Mono-Q High Performance Liquid Chromatography (HPLC). In both normal and failed hearts, four major peaks of PDE activity were present in the soluble fraction using either chromatographic technique, however, only one peak of activity was present in the particulate fraction from the same hearts (Fig. 2). The first peak of activity of the soluble fraction contained a zaprinast-inhibitable, cGMP-PDE and a CaCM-PDE isozyme. Peak II contained predominantly the cGs-PDE isozyme. Peak IIIa contained mostly cGi-PDE isozyme whereas peak IIIb contained both the rolipram-inhibitable and the cGi-PDE isozyme. The evidence for the presence of these isozymes was obtained using a pharmacological method in which an IC_{50} value for a specific inhibitor is generated for one of the isozymes while the other isozyme in the peak is inactivated with a selective inhibitor. For peak IIIa from normal and failed hearts, the IC_{50} value for rolipram was essentially unchanged in the presence of 3 μM CI-930, a selective inhibitor of the cGi-PDE isozyme; Peak IIIb, however, contained a mixture of the rolipram-inhibitable and the cGi-PDE isozymes. In normal hearts, the IC_{50} value for rolipram alone in peak IIIb was about 200 μM, but decreased to 8 μM in the presence of 3 μM CI-930. The peak IIIb fraction from myopathic hearts appeared to contain slightly more cGi-PDE, as suggested by the lower IC_{50} value for rolipram (1 μM) in the presence of 3 μM CI-930. Similar results were obtained when the PDE isozymes in peaks IIIa and IIIb from some of the hearts were isolated using affinity resins.

The particulate fraction of both normal and failed human ventricular tissue predominantly contained the cGi-PDE isozyme, as suggested by the IC_{50} value for CI-930, which was essentially unaffected by 10 μM rolipram. Moreover, very little rolipram-inhibitable activity could be recovered from the particulate fraction with

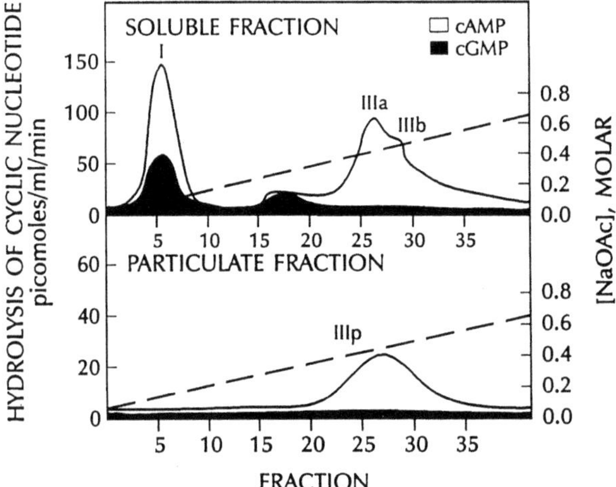

Fig. 2. Representative anion-exchange (DEAE-Sephacel) chromatogram of phosphodiesterase (PDE) activity obtained from the soluble (top panel) and particulate (bottom panel) fractions of human cardiac ventricular tissue as previously described (22). PDE activity (pmol/ml/min) was measured using either 1 μM cAMP or cGMP as substrate. The peaks of activity are designated I, IIIa and IIIb as described in the text. The dashed line indicates the concentration (M) of sodium acetate (NaOAc).

the use of rolipram-analog affinity resin. The specific activity, apparent Km, Vmax, and IC_{50} values for milrinone for the cGi-PDE isozyme from the soluble and particulate fractions of normal and failed cardiac muscle were compared (Table 2). While the apparent Km values for cAMP hydrolysis by the cGi-PDE were not significantly different between normal and failed hearts, we found a 50–75% reduction in the maximal activity of the cGi-PDE isolated from diseased hearts relative to normal hearts. The relative potencies and efficacies of milrinone and CI-930 for the cGi-PDE from normal and failed human heart tissue were unchanged; the Ki values of milrinone obtained for inhibition of the cGi-PDE from the particulate fraction of normal (300 nM) and failed human hearts (330 nM) were not different (Table 2). The results from our study suggest that the lack of positive inotropic efficacy of cGi-PDE isozyme inhibitors in failed human heart tissue in vitro is probably not related to an alteration in pharmacological potency of these inhibitors. In intact tissue, the magnitude of the effect that a cGi-PDE inhibitor has is intimately related to the activity of adenylate cyclase. Thus the major cause of the lack of efficacy of these agents in failed heart muscle in vitro could be the attenuation of cAMP production by adenylate cyclase (12, 6, 8).

cGMP-Inhibitable PDE isozyme in ventricular tissue of rats
with myocardial infarction

The rat model of myocardial infarction is an experimental model of heart failure (19, 9). Rats with 10-week-old, healed myocardial infarcts exhibit depressed indices of cardiac performance, elevations in total peripheral vascular resistance (TPR) and

Table 2. Comparison of specific activity, apparent K_m, V_{max} and IC_{50} value for Milrinone of particulate and soluble human cardiac PDE isozymes

Peak	Substrate	Specific Activity[a]	$K_m{}^b$	$V_{max}{}^b$	Milrinone	
					IC_{50} (μM)	Ki (nM)
Normal						
Soluble I	cGMP	293 ± 26	0.47 ± 0.03	554 ± 138		
	cAMP	492 ± 63	0.38 ± 0.05	891 ± 264		
Soluble IIIa	cAMP	5080 ± 285	0.21 ± 0.01	2371 ± 204	0.44 ± 0.07	
Soluble IIIb	cAMP	5596 ± 974	0.25 ± 0.03	1994 ± 477	0.86 ± 0.07	
Particulate III	cAMP	7905 ± 2325	0.27 ± 0.03	2262 ± 535	0.40 ± 0.06	300
Myopathic						
Soluble I	cGMP	325 ± 83	0.46 ± 0.05	505 ± 129		
	cAMP	482 ± 91	0.36 ± 0.06	663 ± 115		
Soluble IIIa	cAMP	3560* ± 504	0.211 ± 0.01	808* ± 106	0.53 ± 0.11	
Soluble IIIb	cAMP	7737 ± 1752	0.29 ± 0.05	2825 ± 484	1.50 ± 0.37	
Particulate III	cAMP	2347* ± 488	0.29 ± 0.04	1153* ± 305	0.86 ± 0.18	330

Results are the mean ± SEM for normal (n = 5) and myopathic (n = 5) samples per group. The notation (*) indicates statistically different (p ≤ 0.05) from the corresponding PDE isozymes from normal hearts.

[a]Specific activity is expressed as pmol cyclic nucleotide hydrolyzed/min/mg protein applied to the column measured at 1 μM substrate.

[b]Apparent Km (μM) and V_{max} (pmol/mg protein/min) were determined with a Lundon-1 computer program, which uses the method of Hofstee.

pulmonary pressure and a significant reduction in their expected survival time. Daily administration of the cGi-PDE inhibitor milrinone and the angiotensin converting enzyme (ACE) inhibitors captopril and enalapril to rats with myocardial infarction have been shown to prolong survival (19, 21). The reason for the decrease in mortality in this model is not entirely clear, but may be attributable to a reduction in TPR and a concomitant unloading of the heart. While the vasodilatory potency/efficacy of cGi-PDE inhibitors appear to be similar across species, the cardiotonic potency/efficacies do vary across species and also with the extent of cardiac disease. For example, the effect of milrinone on increasing force development in isolated heart tissue from different species is rabbit = guinea pig = cat ≫ rat (1); in studies of isolated human heart tissue, milrinone appears to be a more efficacious inotrope in normal tissue compared to failed heart tissue (6). The cardiotonic efficacy of cGi-PDE inhibitors could be influenced by numerous factors, some of which include the tissue content and cellular location (cytosolic vs particulate) of the cGi-PDE and the synthesis of cAMP via adenylate cyclase.

To gain a better understanding of the cardiotonic efficacy of cGi-PDE inhibitors in rats, we isolated the cardiac PDE isozymes from normal rats and from rats with healed myocardial infarcts. Ten weeks after ligation (infarcted) or sham-ligation (sham-operated) of the left coronory artery in age-matched, adult Sprague-Dawley rats (9), rats were euthanized and unscarred segments of ventricular tissue were collected for the isolation of PDE isozymes using anion-exchange chromatography. The first two peaks of ventricular PDE activity of the soluble fraction obtained from control rats and from rats with cardiac infarcts contained a cGMP-PDE and a CaCM-PDE isozyme (peak I) and a cGs-PDE isozyme (peak II); these isozymes were not present in the particulate fraction of either groups. The cGi-PDE isozyme and the rolipram-inhibitable PDE isozymes were present in the third peak of activity (peak III) of the soluble fraction and also in the corresponding peak of activity from the particulate fraction of control and infarcted hearts. The presence of these PDE isozymes in peak III of the cytosolic and particulate fractions from normal and infarcted cardiac tissue was confirmed in experiments using a rolipram-analog and a milrinone-analog affinity resin and also by pharmacologically separating the isozymes with the use of selective PDE inhibitors, as we previously described for experiments with human cardiac tissue. There were no significant differences between the respective apparent Km values for cAMP hydrolysis by the cGi-PDE isolated from the soluble and particulate fractions of control and infarcted hearts which ranged between 0.28 to 0.87 μM.

The isozymes in peaks I, II, and III were evaluated for sensitivity to inhibition by standard PDE inhibitors (Table 3). In general, no major differences in inhibitor sensitivity were observed between corresponding peaks of PDE activity from the soluble and particulate fractions of control and infarcted hearts. However, milrinone and CI-930 tended to be more potent for inhibiting peak III PDE activity in the soluble fraction compared to the particulate fraction. The pharmacological significance of this finding is not apparent. The conclusion from this study is that the cGi-PDE isozyme is present in the soluble and particulate fraction of rat heart and the enzyme characteristics do not appear to change 10 weeks after myocardial infarction, when overt heart failure is apparent. This is somewhat similar to our findings with human hearts in failure. Moreover, the decreased level of positive inotropy of cGi-PDE inhibitors in the rat does not appear to be related to the absence of a cGi-PDE

Table 3. The effect of PDE inhibitors on peak I and peak III PDE activity isolated from hearts of sham-operated rats and from rats with myocardial infarct

| Agent | Sham-operated | | | Myocardial infarct | | |
| | Peak I | Peak III | | Peak I | Peak III | |
		Soluble	Particulate		Soluble	Particulate
IBMX	9	5	6	9	4	7
	(7–12)	(4–6)	(4–9)	(7–12)	(3–6)	(4–11)
zaprinast	8	131	103	15	135	74
	(4–15)	(119–143)	(97–110)	(8.9–23)	(113–162)	(35–154)
Milrinone	165	1	3	168	0.9	6
	(138–198)	(0.9–2)	(2–4)	(142–204)	(0.7–1)	(1–11)
CI-930	186	0.25	0.3	371	0.3	NA
	(107–324)	(0.2–0.4)	(0.1–0.8)	(169–814)	(0.2–0.6)	
Rolipram	527	47	27	668	78	31
	(411–675)	(19–116)	(9–79)	(434–1027)	(61 100)	(8–114)

Results are mean IC_{50} values with corresponding 95% confidence limits for isozyme preparations from hearts of sham-operated (n = 4) rats and from rats with myocardial infarction (n = 4). NA indicates that an IC_{50} values was not attained.

isozyme in the soluble or particulate fraction or lack of efficacy of cGi-PDE inhibitors against the isolated enzyme.

In a parallel study to the one described above, we evaluated the effect of milrinone and other types of positive inotropes working via other mechanisms on the contractile performance of segments of cardiac tissue from sham-operated rats and from rats 10 weeks after myocardial infarction was induce (14). These studies (Fig. 3) showed that: 1) milrinone had little to no effect on contractile force development by ventricular tissue from normal and infarcted rat hearts, 2) isoproterenol increased contractile force development by ventricular tissue from sham-operated hearts, but not from hearts of rats 10 weeks post myocardial infarction, and 3) forskolin increased force development in both normal and infarcted rat ventricular tissue. As discussed above, the lack of effectiveness of a beta-agonist in increasing contractile force of failed hearts has been explained by an uncoupling of the beta-receptor via an increase in the Gi (11) or a decrease in the Gs regulatory proteins. Milrinone did display functional PDE inhibition as it potentiated contractile force development in normal rat ventricular tissue in response to isoproterenol (data not shown) and forskolin (Fig. 3), and in response to forskolin in tissue from infarcted rat hearts (Fig. 3). Thus, the ineffectiveness of cGi-PDE inhibitors in increasing cardiac contractile force in the rat appears to be related to the basal turnover of cAMP rather than the absence or lack of pharmacological potency for the cGi-PDE isozyme. Moreover, these data suggest that the cGi-PDE in rat heart may regulate contractility under conditions when cAMP synthesis is active.

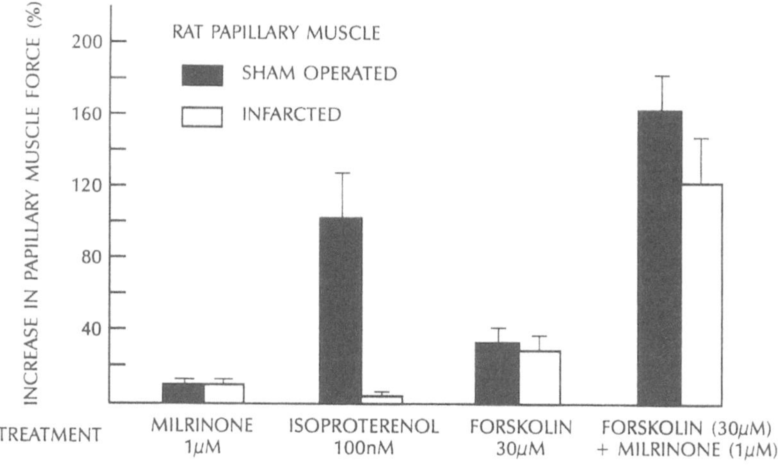

Fig. 3. The effects of milrinone (1 µM), isoproterenol (100 nM), forskolin (30 µM), and milrinone (1 µM) plus forskolin (30 µM) on force development by right ventricular papillary muscles isolated from hearts of sham-operated rats (filled bars) and from hearts of rats 10 weeks after myocardial infarction (open bars). Papillary muscles were paced at 60 beats per minute at preload tension of 0.5 g for 60 min prior to the addition of the test compounds. In experiments in which forskolin was used, the calcium concentration of the Tyrode's solution was reduced from 1.8 mM to 0.25 mM in order to eliminate the incidence of irregularities in the rate of contraction. Data represent the mean of the percent change ± SEM from baseline force (n = 8–12).

Hemodynamic responses to inhibition of the cGi cAMP-PDE and cGMP-PDE isozymes

The hemodynamic responses to selective inhibition of the cGi-PDE isozyme in both animals (2, 18) and man (17) have been well documented. As a consequence of inhibition of cGi-PDE in cardiac muscle (Fig. 4), increased cAMP levels produce an increase in the rate and magnitude of developed force, as well as an enhancement in the rate of muscle relaxation. There is also a direct effect on increasing heart rate, although this usually occurs at higher concentrations of the cGi-PDE inhibitor. Concurrently, in vascular smooth muscle, elevated cAMP causes a reduction in total peripheral resistance, enhanced coronary blood flow, reduced pulmonary vascular resistance and reduced right atrial pressure, indicative of venodilation (16). The hemodynamic responses to selective inhibitors of the cGMP-PDE isozyme have not been widely reported or well documented, since there are, as yet, no clinically used selective inhibitors of this isozyme. However, the activity of many vasorelaxant agents, such as nitrate vasodilators and endothelium derived relaxing factor (EDRF), and the natriuretic activity of atrial natriuretic peptide (ANP) are known to involve stimulation of cGMP synthesis via guanylate cyclase in vascular smooth muscle and renal tubule cells. Hydrolysis of cGMP occurs via two distinct PDE isozymes, one of which is selectively and competitively inhibited by zaprinast. We have identified that

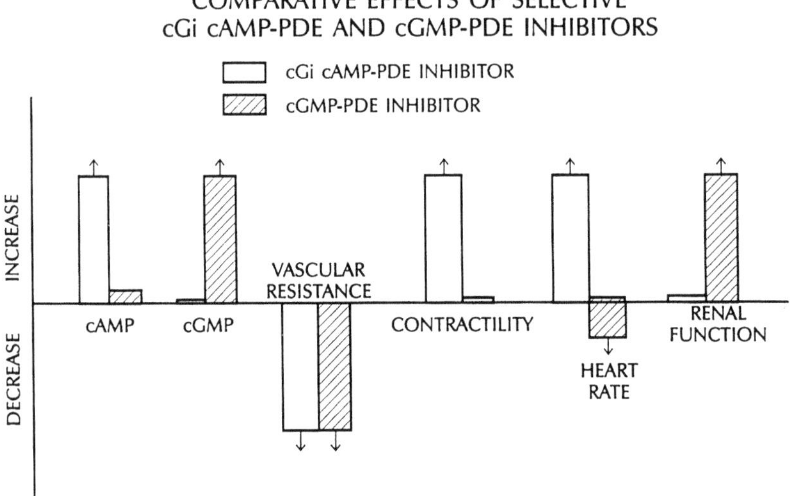

Fig. 4. Comparative cardiovascular effects of selective inhibitors of the cGi cAMP-PDE isozyme (e.g., milrinone, enoximone,) and of the cGMP-PDE isozyme (e.g., zaprinast). Inhibitors of the cGi-PDE increase intracellular cAMP levels in target tissues, reduce vascular bed resistances, increase cardiac contractility and heart rate and have little effect on renal function (i.e., natriuresis and diuresis). In contrast, inhibitors of the cGMP-PDE increase cGMP levels in target tissues (with a possible small increase in cAMP levels through cGMP-induced inhibition of the cGi-PDE), decrease vascular bed resistances with no effect on cardiac contractility, tend to reduce heart rate and enhance renal function by stimulating diuresis and natriuresis.

vascular smooth muscle from human saphenous vein (unpublished observation) and renal cell lines from pig (5) and human (unpublished observation) contain a zaprinast-inhibitable, GMP-PDE isozyme. Thus, inhibitors of the cGMP-PDE isozyme may have therapeutic utility in the treatment of cardiovascular disease in conditions where the need for enhanced cardiac inotropy is not required, but where a reduction in cardiac afterload and the stimulation of natriuresis and diuresis are desirable (Fig. 4).

Just as the efficacy of cGi cAMP-PDE inhibitors is dependent upon the synthesis of cAMP via adenylate cyclase, the vasorelaxant efficacy of cGMP-PDE inhibitors is dependent upon the synthesis of cGMP via guanylate cyclase. One way to show that the vasorelaxant and hypotensive activities of a cGMP-PDE inhibitor such as zaprinast are linked to the accumulation of cGMP in vascular smooth muscle is to measure vasorelaxation or hypotensive activity in the presence of a stimulator (e.g., sodium nitroprusside (SNP), ANP) or an attenuator (e.g., N^w-nitro-L-arginine (NNA), methylene blue) of cGMP synthesis with the concomitant measurement of intracellular cGMP in target tissues. The vasorelaxant activity of zaprinast in the presence and absence of SNP, a stimulator of soluble guanylate cyclase, was evaluated in rat aortic vascular smooth muscle contracted with phenylephrine (Fig. 5, top panel). In the presence of 10 μM zaprinast, which is a nonvasorelaxant concentration in denuded rat aortic smooth muscle, the EC_{50} value for SNP was significantly reduced from 15 nM to 4 nM, suggesting functional inhibition of the cGMP-PDE activity in vascular smooth muscle (Fig. 5, top panel). Similarly, both zaprinast and SNP alone slightly increased cGMP levels in VSM, however, the combined effect of zaprinast and SNP on cGMP levels was greater than that expected through an additive effect, thus demonstrating the functional inhibition of the cGMP-PDE by a PDE inhibitor (Fig. 5, middle panel).

The natriuretic effect of ANP is linked to the stimulation of particulate guanylate cyclase with the concomitant increase in cGMP levels in renal tubule cells (22). The LLC-PK$_1$ cell is a cell line derived from the proximal tubule of the pig kidney. These cells express functional ANP receptors as evidenced by the stimulation of cGMP synthesis by exposure to ANP. This cell line also contains a zaprinast-inhibitable cGMP-PDE that elutes in the first peak of PDE activity from a Mono-Q HPLC resin (5). The effect of ANP on cGMP stimulation in this cell line is potentiated by zaprinast, as shown in Figure 6. Exposing LLC-PK$_1$ to 1 μM ANP or to 100 μM zaprinast increased intracellular cGMP levels from a control value of 150 fmoles/dish to 550 fmoles/dish for ANP and to 300 fmoles/dish for zaprinast. Pre-treatment of the cells with zaprinast potentiated the response to ANP which increased to 1650 fmole/dish, about twice the level of cGMP if the effects of ANP and zaprinast were simply additive. This study demonstrates the functional inhibition of a cGMP-PDE inhibitor in renal cells and suggests the possibility that such agents may have utility as a natriuretic agent through potentiation of endogenous ANP and also may circumvent the observed effect of an uncoupling of the ANP receptor from the natriuretic response in heart failure.

As described above, both EDRF and zaprinast promote vasorelaxation through different mechanisms, but through a common pathway that leads to elevations in intracellular cGMP. To further elucidate the hypotensive effect of zaprinast, Dundore et al. (10) determined if NNA, an inhibitor of the synthesis and/or release of nitric oxide (or EDRF) (15), alters the hypotensive response to zaprinast. Zaprinast or vehicle was given to conscious SHR in a cumulative i.v. dose-dependent manner 3C min after a 30-min pretreatment with NNA (3 mg/kg) or saline (1 ml/kg). Mean

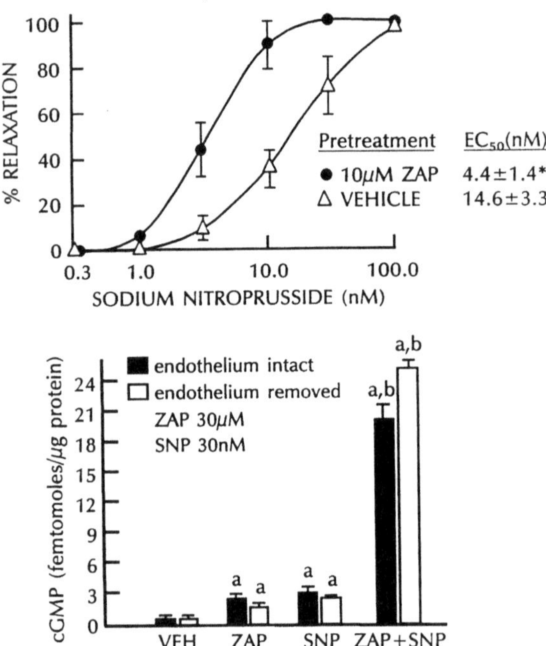

Fig. 5. The potentiating effect of zaprinast (ZAP) on the vasorelaxant activity of sodium nitroprusside (SNP) in denuded rat aortic smooth muscle (top panel; n = 6) and on SNP-induced elevation in intracellular cGMP levels (bottom panel; n = 8) in rat aortic smooth muscle with and without endothelium. Data in the top panel are expressed as the mean percent relaxation ± SEM of smooth muscle tissue contracted to 2 grams of tension by 1 uM phenylephrine; data in the bottom panel are expressed as femtomoles cGMP/µg smooth muscle tissue. The notations (a) and (*) indicate significantly different from vehicle, the notation (b) indicates significantly different from zaprinast or SNP alone, respectively. Statistical significance was achieved at a p value ≤ 0.05.

arterial pressure was measured 5 min after each dose of zaprinast. Five minutes after the last dose of zaprinast (30 mg/kg), the rats were anesthetized with pentobarbital. A segment of the abdominal aorta was freeze-clamped in situ and removed for the determination of cGMP levels. NNA (3 mg/kg) significantly decreased basal aortic cGMP levels by 54% and increased mean arterial pressure by 37 ± 2 mm Hg. Zaprinast (30 mg/kg) increased aortic cGMP by 187% and decreased mean arterial pressure by 49 ± mm Hg. NNA (3 mg/kg) reduced the accumulation of cGMP in aortic tissue and attenuated the depressor response produced by zaprinast. These data are consistent with the hypothesis that NNA inhibits the tonic release of EDRF and that the depressor effects of zaprinast are due to potentiation of the vasodilator effects of EDRF in vivo. Moreover, since the changes in mean arterial pressure produced by NNA and zaprinast were significantly correlated with cGMP levels in aortic tissue, the concentration of cGMP in vascular smooth muscle may be one of several factors that determine blood pressure in SHR.

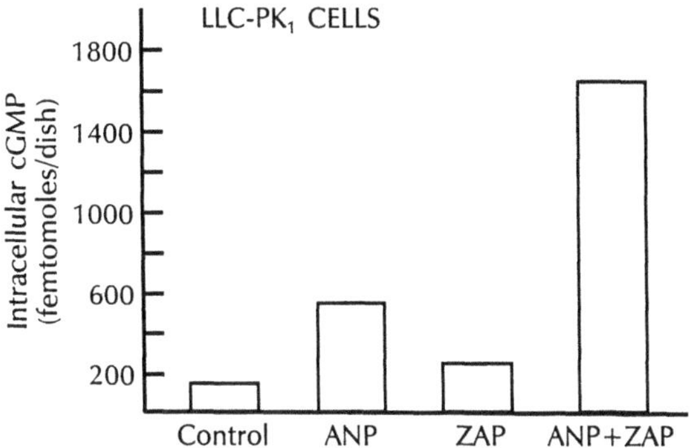

Fig. 6. The effect of 1 μM atrial natriuretic peptide (ANP), 100 μM zaprinast (ZAP) and ANP plus ZAP on intracellular levels of cGMP in LLC-PK₁ (pig renal tubule) cells. Cells were exposed to vehicle or zaprinast 60 min prior to the addition of ANP. After 60 min, cells were rinsed in fresh medium and cGMP levels in cell extracts were determined by RIA. Data are mean values from a representative experiment.

References

1. Alousi AA, Stankus GP, Stuart JC, Walton LH (1983) Characterization of the cardiotonic effects of milrinone, a new and potent cardiac bipyridine, on isolated tissues from several animal species. J Cardiovasc Pharmacol 5:804–811
2. Alousi AA, Canter JM, Cicero F, Fort DJ, Helstosky A, Lesher GY, Montenaro MJ, Stankus GP, Stuart JC, Walton LH (1984) Pharmacology of milrinone. In: Braunwald E, Sonnenblick EH, Chakrin LW, Schwarz Jr. RP (eds) Milrinone Investigation of New Inotropic Therapy for Congestive Heart Failure. Raven Press, New York, pp 21–48
3. Beavo JA (1988) Multiple isozymes of cyclic nucleotide phosphodiesterases. In: Greengard P, Robinson GA (eds) Advances in Second Messenger and Phosphoprotein Research, Vol 22. Raven Press, New York, pp 1–38
4. Beavo JA, Reifsnyder DH (1990) Primary sequence of cyclic nucleotide phosphodiesterase isozymes and the design of selective inhibitors. Trends Pharmacol Sci 11:150–155
5. Bode DC, Pagani ED, O'Connor B, Silver PJ (1991) Potentiation of cGMP accumulation in renal cells by inhibition of cGMP phosphodiesterase (PDE). FASEB J 5(6) :A1592
6. Böhm D, Diet F, Erdman E (1988) Enhancement of the effectiveness of milrinone to increase force of contraction by stimulation of cardiac beta adrenoceptors in the failing human heart. Klin Wochenschr 66:957–962
7. Coquil J-F, Franks DJ, Wells JN, Dupuis M, Hamet P (1980) Characteristics of a new binding protein distinct from the kinase for guanosine 3':5'-monophosphate in rat platelets. Biochim Biophy Acta 631:148–165
8. Danielson W, V der Leyen H, Meyer W, Neumann J, Schmitz W, Scholz H, Starbatty J. Stein B, Doring V, Kalmer P (1989) Basal and isoprenaline-stimulated cAMP content in failing versus nonfailing human cardiac preparations. J Cardiovasc Pharmacol 14:171–173
9. DeFelice AF, Harris AL, Frering R, Horan P (1989) Beneficial hemodynamic effects of milrinone and enalapril in conscious rats with healed myocardial infarction. Eur J Pharmacol 167:211–220
10. Dundore RL, Pratt PF, O'Connor B, Buchholz RA, Pagani ED (1991) N^W-nitro-L-arginine attenuates the accumulation of aortic cyclic GMP and the hypotension produced by zaprinast. Eur J Pharmacol 200:83–87

11. Feldman AM, Cates AE, Veazcy WB, Hershberger RE, Bristow MR, Baughman KL, Baumgartner WA, Dop CV (1988) Increase of the 40, 000-mol wt pertussis toxin substrate (G protein) in the failing human heart. J Clin Invest 82:189–197
12. Feldman MD, Copelas L, Gwathmey JK, Philips P, Warren SE, Schoen FJ, Grossman W, Morgan JP (1987) Deficient production of cyclic AMP: pharmacologic evidence of an important cause of contractile dysfunction in patients with end-stage heart failure. Circ 75:311–339
13. Francis S, Lincoln TM, Corbin JD (1980) Characterization of a novel·cGMP binding protein from rat lung. J Biol Chem 255:620–626
14. Harris AL, VanAller G, DeFelice AF, Horan PJ, Frering R, Pagani E, Silver P (1989) Effect of milrinone on isolated papillary muscles from 10 week myocardially-infarcted (MI) and sham-operated rats. FASEB J 3:A1040
15. Ishii K, Chang B, Kerwin JF, Huang ZJ, Murad F (1990) N^w-nitro-L-arginine: a potent inhibitor of endothelium-derived relaxing factor formation. Eur J Pharmacol 176:219–223
16. Lee KC, Canniff PC, Hamel DW, Pagani ED, Ezrin AM (1991) Cardiovascular and renal effects of milrinone in beta-adrenoreceptor blocked and non-blocked anesthetized dogs. Drugs Under Exp Clin Res 3:145–158
17. LeJemtel TH, Maskin CS, Chadwick B, Sonnenblick EH (1984) Hemodynamic effects of intravenous and oral milrinone in patients with chronic heart failure. In: Braunwald E, Sonnenblick EH, Chakrin LW, Schwarz Jr. RP (eds) Milrinone Investigation of New Inotropic Therapy for Congestive Heart Failure. Raven Press, New York, pp 133–141
18. Pagani E, Fort DJ, Ezrin AM, Silver PJ (1988) In vitro cardiovascular phosphodiesterase inhibition and in vivo cardiovascular activity of milrinone in dogs. Fed Proc 2 (4) : A366
19. Pfeffer MA, Pfeffer JM, Steinberg BS, Finn P (1985) Survival after an experimental myocardial infarction: beneficial effects of long-term captopril therapy. Circ 72:406–412
20. Silver PJ, Allen P, Etzler J, Hamel L, Bentley RG, Pagani ED (1990) Cellular distribution and pharmacological sensitivity of low Km cyclic nucleotide phosphodiesterase isozymes in human cardiac muscle from normal and cardiomyopathic subjects. Sec Mes Phospho 13:13–25
21. Sweet CS, Ludden CT, Stabilito II, Emmert SE, Heyse IF (1988) Beneficial effects of milrinone and enalapril on long-term survival of rats with healed myocardial infarction. Eur J Pharmacol 147:29–37
22. Wong XR, Xie MH, Shi LB, Liu FY, Huang CL, Gardner DG, Cogan MG (1988) Urinary cGMP as biological marker of the renal activity of atrial natriuretic factor. Amer J Physiol 255: F1220–F1224
23. Zemelman BV, Chu SW, Walker WA (1989) Host response to escherichia coli heat-liable enterotoxin via two microvilli membrane receptors in the rat intestine. Infection and Immunity 57:2947–2952

Author's address:
Dr. Edward D. Pagani
Department of Cardiovascular Pharmacology
Sterling Winthrop Pharmaceuticals Division
Rensselaer, New York 12144, USA

Na,K-ATPase expression in normal and failing human left ventricle

P.D. Allen[1], T.A. Schmidt[2], J.D. Marsh[3], K. Kjeldsen[2]

Department of Anesthesia, Brigham and Women's Hospital[1], Boston,
Massachusetts, USA
Department of Medicine B, Division of Cardiology,
Rigshospitalet, Copenhagen University School of Medicine[2], Denmark
Department of Medicine, Cardiovascular Division, Brigham and Women's Hospital[3]

Summary: The expression of the Na,K-ATPase was studied in both normal and failing human myocardium which was collected within 5 min of cardiac explantation in preparation for orthotopic transplantation or at the time of organ harvest. Abundance of mRNA for all three catalytic α subunits of the Na,K-ATPase was analyzed in samples from patients with end-stage heart failure due to either ischemic or dilated cardiomyopathy, as well as from normal controls. Vanadate facilitated ^{3}H-oubain binding before and after a Digibind wash was analyzed on tissue from a subset of these patients. mRNA analysis demonstrated that all three catalytic Na,K-ATPase α subunits were expressed in human heart and that there was no evidence for change in relative expression or abundance induced by disease. The specific digitalis receptor concentration was 760 ± 58 and 614 ± 47 pmol/g wet weight in the samples from normal and failing hearts, respectively (p = NS). From these studies it can be concluded that, whereas there is a tendency for a decrease in the number of oubain receptors in heart failure, there is no significant alteration in the expression of Na,K-ATPase message or protein caused by chronic heart failure.

Key words: Heart failure; myocardium; Na,K-ATPase; 3H-oubain binding; heart; mRNA

Introduction

The Na,K-ATPase catalyzes the active transport of Na^+ and K^+ across the plasma membrane of the heart and other cells and thus maintains the electrochemical gradient across the cell membrane which allows depolarization to occur. The Na,K-ATPase is generally accepted to be the receptor for digitalis glycosides which increase cardiac contractility by inhibiting the Na,K-ATPase and allow intracellular Na^+ to rise. This in turn leads to an augmentation in intracellular Ca^{2+} concentration via the Na^+, Ca^{2+} exchanger and increases Ca^{2+} loading of the sarcoplasmic reticulum (31).

It has been suggested that the Na,K-ATPase may be upregulated after chronic digitalis administration, leading to tolerance to the inotropic effect of the drug, but molecular mechanisms that may underlie this effect have not been examined in detail in human myocardium. Upregulation of the Na,K-ATPase has been demonstrated in HeLa cells, chick heart cells, rat skeletal myotubes, and guinea pig myocardium after digitalis administration (3, 5, 7, 19, 27). This has also been shown in human

This work was supported in part by grants from the Danish Heart Foundation, Novo's Foundation, and the Anesthesia Foundation, Inc.

eythrocytes from patients on chronic digoxin therapy (13, 14). It is possible to measure the number of digitalis binding sites very accurately using vanadate-facilitated ^{3}H-oubain binding (24), and it is possible to correct for any receptors bound by circulating digitalis by preincubating the samples with an excess of digitalis antibody fragments (F_{ab}) (28).

The Na,K-ATPase has two subunits, α catalytic subunit (α) which spans the membrane seven times, and α glycoprotein subunit (β) which is required for assembly and insertion in the plasma membrane. Each of these subunits has a multi-gene family with three known α isoforms ($\alpha1$, $\alpha2$, $\alpha3$) derived from different genes (21). In the rat, the isoforms have markedly different Ki's for inhibition by oubain (32). Their cardiac glycoside affinities are controlled by a limited number of amino acid substitutions that have been carefully defined (21). In other animal species the difference in oubain affinity is controversial, but, if present, is much less obvious than the difference seen in the rat (6, 10, 11, 12, 25). Specific Na,K-ATPase isoform switching has been demonstrated in the rat in response to cardiac hypertrophy and hypertension. (17). The effect of this switch is that, in response to disease, there are more high affinity sites.

The population of Na,K-ATPase isoforms has not yet been determined in human myocardium. Furthermore, any change in oubain binding affinity as a result of cardiac disease in humans is controversial (6, 11). The purpose of the current investigation is to ascertain expression of the three α isoform messages in human heart and to determine whether each message or protein is regulated in advanced heart failure or in response to chronic digoxin administration. To date, all human studies have been conducted either on small endomyocardial biopsies or on post mortem tissue collected several hours after death. The current studies were performed on pieces of left-ventricular free wall that were frozen in liquid nitrogen within 5 min after cardiac explantation on hearts from patients who were undergoing orthotopic cardiac transplantation or from organ donors with no cardiac disease whose heart was not needed or not suitable for transplantation.

Materials and methods

Tissue:

Na,K-ATPase isoform expression and vanadate-facilitated oubain binding was examined from failing and normal human left ventricle. Tissue was obtained at the time of orthotopic cardiac transplantation from patients with severe heart failure (n = 6 oubain binding; n = 12 mRNA expression) and at the time of organ harvest from patients whose hearts were not acceptable or not needed for transplantation (n = 5 oubain binding; n = 6 mRNA expression). All tissues were obtained using protocols approved by the Brigham and Women's Hospital Institutional Committee for the Protection of Human Subjects. Left-ventricular free wall was excised and frozen in liquid nitrogen within 5 min of excision of the heart, and then stored in liquid nitrogen. All patients for the oubain binding study had end-stage ischemic myocardiopathy and had been receiving chronic digitalis therapy for a mean duration of 22 months. The patients for the mRNA studies were equally distributed between the same six patients with ischemic cardiomyopathy as were studied for oubain binding and six patients with dilated idiopathic cardiomyopathy. The normal samples which were used for both mRNA and oubain binding were obtained from organ donors who had their tissue harvested a mean of 2 days after the institution of life support and the only cardio-active drug administered during this period was dopamine. None of these patients had received digoxin.

^{3}H-oubain binding:

All procedures were carried out using a freshly prepared vanadate buffer (24). Clearance of exogenous digoxin from its "receptor" was performed by washing the samples in excess digoxin F_{ab} for 16 h (28). Na,K-ATPase activity was quantified using standard ^{3}H-oubain binding (9). Specific uptake was calculated by subtracting nonspecific uptake from total uptake and is expressed per gram wet weight of tissue. All reagents were as described previously. Specific digoxin antibody (ovine), Digibind was a gift of the Welcome Foundation LTD (Kent, UK). Purity of the ^{3}H-oubain was checked by the Na,K-ATPase extraction method as the percentage of ^{3}H activity that could bind to purified Na,K-ATPase (16).

mRNA analysis:

Total RNA was extracted from human heart using the guanidium isothiocyanate-cesium chloride method (8). After determining the specificity of the probes and stringency of wash conditions on Northern blots, 2.5 and 5 µg of total RNA were directly applied to nitrocellulose paper using a "slot blot" apparatus, and the RNA was bonded to the nitrocellulose, both with ultraviolet crosslinking and by baking for 2 h in a vacuum oven. One of five identical strips was then hybridized using as probe a random primed ^{32}PdCTP labeled cDNA specific for the coding region of each of the rat α1, α2, or α3 cDNAs (800 bp, 1200 bp, 750 bp respectively). The blots were then washed at the appropriate stringency for each probe, and allowed to expose x-ray film for a time necessary to produce a signal within the linear range. All blots were stripped of their signal by washing in 0.1% sodium dodecal laural sulfate at 95 °C for 5 min, re-exposed to x-ray film to confirm the absence of signal, and then hybridized with a 800bp probe specific for human GAPDH (a "housekeeping gene," whose expression has previously been shown to be unchanged by these disease states (21)) to quantitate the amounts of mRNA on each strip. The x-ray film was scanned using a Joyce Lobel densitometric scanner, and the area under the curve for all Na,K-ATPase isoform measurements was normalized for the area under the curve for GAPDH. (Specific rat Na,K-ATPase α isoform cDNAs were the kind gift of Dr. Charles Simmons, CHMC, Boston, MA).

Statistics:

Results are given as means $\pm$ SEM. mRNA quantitation was done by normalizing each sample to the signal produced by hybridization with GAPDH and by arbitrarily setting the mean normalized α1 value for the normal subjects to 1. Differences in oubain binding between and among groups were ascertained using Students two tailed t-test for paired and unpaired data, respectively.

Results

^{3}H-oubain binding:

The control group's mean age was 28 years (four males, one female). The heart failure group's mean age was 45 years and all six were males. The mean ejection fraction of the heart failure group was 20%. Their serum digoxin level was within the therapeutic range (0.9–1.8 ng/ml).

^{3}H-oubain binding for the heart failure patients was 614 ± 47 pmol/g wet weight and from individuals without heart disease it was 760 ± 58 pmol/g wet weight. (p = 0.08) (Fig. 1).

mRNA:

In these human samples the message for all three α subunit isoforms was present. The relative abundance of the three α isoforms was 1/0.3/0.5 (α1, α2, α3 respectively; Fig. 2) There was no difference in the relative abundance of any of the three α isoforms among the three groups.

Discussion

In human myocardium, all previously described isoforms of the α Na,K-ATPase subunits are present in significant amounts, at least at the level of gene transcription.

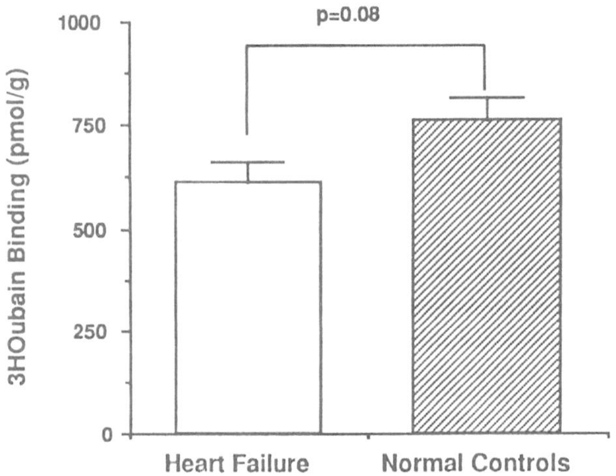

Fig. 1. Mean values for ^{3}H-oubain binding after clearance of previously bound digoxin in samples of left ventricle from six patients (heart recipients) treated with digoxin, and from five controls (potential organ donors) not exposed to digoxin. ^{3}H-oubain binding was determined using vanadate-facilitated ^{3}H-oubain binding to intact myocardial specimens. Clearance of previously bound digoxin was performed by washing samples in buffer containing excess fragments of digoxin antibody, and was followed by vanadate-facilitated ^{3}H-oubain binding (mean ± SEM).

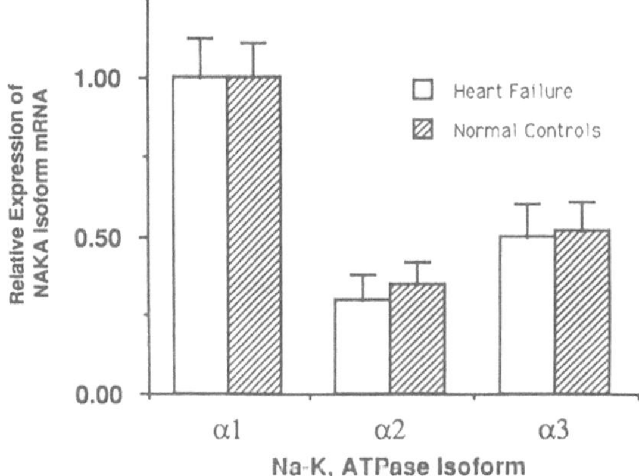

Fig. 2. Mean relative mRNA expression levels of the three catalytic α subunits of the Na,K-ATPase in left-ventricular tissue from normal and failing hearts. Values were obtained from densitometric scans of exposed X-ray film corrected relative to the expression of GAPDH. The GAPDH normalized level of α1 signal in normal hearts was arbitrarily assigned a relative value of 1 (mean ± SEM).

From these data it appears that there is no change in the expression of Na,K-ATPase in human heart as a result of either end-stage heart failure caused by ischemic or dilated cardiomyopathy or by chronic digitalis administration. The relative amounts of mRNA's of the three catalytic α isoforms in all groups is similar to the amounts of the corresponding proteins seen in human hearts (Sweadner, K; personal communication). It is possible that a small change in protein expression might be missed by studies of mRNA abundance, because to accurately discern a change using the second method a 1.5–2 fold difference is required. It is also possible that a small change in protein abundance is due to post-transcriptional factors and would not be present at the level of mRNA abundance.

The possible functional differences or specific roles for the three different isoforms in human heart is yet unknown. The existence of more than one α isoform in the human heart (which might have a different affinity for oubain) and the apparent difference in Na,K-ATPase sensitivity to oubain after chronic digitalis administration seen in human erythrocytes (13, 14) and in cardiac tissue in other species (3, 5, 7, 19, 27) raises the question of whether or not there is altered sensitivity of diseased hearts. Both the inotropic and toxic effects of cardiac glycosides might be due to a differential expression of one of the α isoforms, or be caused by a upregulation of one or all of the α isoforms. (15, 30) In contrast to what has been seen in human erythrocytes and lymphocytes (13, 14), and in cardiac tissue of other species (3, 5, 7, 19, 27), there is no evidence for upregulation of myocardial digitalis receptor concentration found in human hearts which had been exposed to chronic digitalis administration. From the data in this study, it seems unlikely that there is neither a significant switch in any single α isoform expression, nor in expression of the pump as a whole in response to chronic congestive heart failure.

From our data on oubain binding and wash-out (24), a single, high-affinity digitalis glycoside binding site is apparent, but the presence of a slightly lower affinity site cannot be ruled out by these data. It is not yet entirely clear whether the human Na,-K-ATPase isoforms have different affinities for the cardiac glycosides. A single-affinity site has been described in human myocardium assayed either as vanadate supported oubain binding to biopsy or necropsy specimens, or as Mg^{2+}-P_i supported binding in isolated membrane preparation (12, 29). On the other hand, when binding was studied under conditions of Na^+, ATP, and Mg^{2+} two classes of oubain binding sites were seen (10). Interestingly, the $\alpha1$ isoform in the human is attributed to being the high-affinity binding protein and the $\alpha2/\alpha3$ forms are postulated to have the lower affinity, in direct contrast to the rat. In human brain samples oubain-binding in the presence of Na^+, ATP and Mg^{2+} gave linear Scatchard plots, but complex displacement curves. There was little difference between human brain and heart, but that is not surprising as both tissues have all three isoforms present. There is no question that, even if there are two different affinities in humans, the difference between the two affinities is not nearly as large as that found in the rat (5.6 mM, and 56 mM in the rat and 1.2×10^{-8} M and 1.8×10^{-7} M for human) (10, 11).

There is some evidence that patients with some forms of advanced heart disease are more sensitive to cardiac glycosides and may be more prone to toxicity (31). However, it is uncertain whether that is a pharmacodynamic or pharmicokinetic effect. Grupp (15) has shown that the sensitivity and effectiveness to cardiac glycosides is directly related to ejection fraction. Norgaard (26) has shown independently that the number of available pumps is also related to ejection fraction. Böhm (6) has not shown this difference in membrane fractions from patients with dilated cardiomyopathy. It is not easy to reconcile these two discordant findings, however, it may be due to prior receptor occupancy by digoxin administered to some of these patients or a problem with recovery of the Na,K-ATPase in the membrane fractions. Although our data show a trend for a reduced Na,K-ATPase activity as evidenced by a modest reduction in vanadate-facilitated 3H-oubain binding in the failing hearts after F_{ab} pretreatment (p = 0.08) there is not a significant change from the non failure "controls".

Studies in animals have shown a selective reduced expression of $\alpha2$ and $\alpha3$ mRNAs in DOCA-salt and angiotensin hypertensive rat hearts (17), as well as a decrease in all isoform mRNAs in Milan hypertensive rat hearts (9), a reduced expression of $\alpha2$ mRNA in hypokalemic rat hearts (2), no change in the high/low affinity in spontaneously hypertensive rats (1, 20), and a decrease in the high/low affinity in dog hearts (22). Recent studies have demonstrated a genetic alteration in the Dahl sodium sensitive rat (18) and spontaneous hypertensive rat (23) which could modify the transport properties of the Na,K-ATPase in these hypertensive animals. Thus, there is no real consensus that can be drawn from these animal data. However, it is clear from the data in this study that we were not able to detect a significant difference in the relative expression of any of the isoforms of the human Na,K-ATPase caused by disease in humans.

In conclusion, normal human left ventricle has been found to have a Na,K-ATPase concentration of 760 ± 58 pmol/g wet weight. mRNAs of all three α catalytic subunits were detected in both normal and failing hearts. This correlates well with the amounts of Na,K-ATPase isoform proteins detected in human hearts by immunoblots (Sweadner K; personal communication). No evidence was found for a adaptation to digitalization in the failing human heart. There was a trend for a lower number of pumps in the failure group, but there was no significant decrease in abundance of oubain binding sites in failing heart. This is confirmed by the finding of no significant decrease in the expression of the mRNAs in any of the three α catalytic subunits in response to disease.

Acknowledgements. The authors thank Dr. Frederick Schoen, Division of Cardiac Pathology, Dr. Verdi Disesa and Dr. Lawrence Cohen, Division of Cardiothoracic Surgery, and Dr. Rosemarie Maddi, Dr. John Fox, and Mrs. Santawana Muckherjee, Department of Anesthesia, Brigham and Women's Hospital, and the New England Organ Bank, Boston, Massachusetts, for their help in procuring the tissue used in this study. We also thank Grete Simonsen for skilled technical assistance, Stig Haunso for valuable discussions, and the Wellcome Foundation for a gift of digoxin F_{ab} antibody fragments (Digibind), and Ms. Alyce Russo for secretarial support.

References

1. Aalkjaer C, Kjeldsen K, Norgaard A, Clausen T, Mulvany MJ (1985) Oubain binding and Na + content in resistance vessels and skeletal muscles of SHR and K + -depleted rats. Hypertension 7:277–286
2. Azuma KK, Hensley CB, Putnam DS, McDonough AA (1990) Differential regulation of Na,K-ATPase $\alpha 1$, $\alpha 2$, and β in heart, muscle and brain of hypokalemic rats. J Gen Physiol 96:87a
3. Bluschke V, Bonn R, Greeff K (1976) Increase in the $(Na^+ + K^+)$-ATPase activity in heart muscle after chronic treatment with digitoxin or potassium deficient diet. Eur J Pharmacol 37:189–191
4. Boheler KR, Carrier L, Bastie D, Allen PD, Komajda M, Mercadier JJ, Schwartz K (1991) Skeletal actin mRNA increases in the human heart during ontogenic development and is the major isoform of control and failing adult hearts. J. Clin Invest 88:323–330
5. Bonn R, Greeff K (1978) The effect of chronic administration of digitoxin on the activity of the myocardial (na + K)-ATPase in guinea-pigs. Arch Int Pharmacodyn Ther 233:53–64
6. Böhm M, Ungerer M, Erdmann E (1990) Beta adrenoceptors and m-cholinoceptors in myocardium of hearts with coronary artery disease or idiopathic dilated cardiomyopathy removed at cardiac transplantation. Am J Cardiol 66:880–882
7. Brodie C, Sampson SR (1985) Effects of chronic oubain treatment on [^{3}H] oubain binding sites and electrogenic component of membrane potential in cultured rat myotubes. Brain Res 247:121–123
8. Chirgwin JM, Przybyla AE, MacDonald RJ, Rutter WJ (1979). Isolation of biologically active ribonucleic acid from sources enriched in ribonuclease. Biochemistry (Wash) 18:5294–5299.
9. Cutler CP, Cramb G (1990) The isoforms of Na,K-ATPase in myocardial tissues of the Milan hypertensive rat. J Gen Physiol 96:63a
10. DePover A, Godfraind T: Interaction of oubain with $(Na^+ + K^+)$ ATPase from human heart and from guinea-pig heart. Biochem Pharmacol 28:3051–3056
11. DePover A, Grupp G, Schwartz A, Grupp IL (1991) Coupling of contraction through effects on Na,K-ATPase: changes in Na,K-ATPase isoforms in heart disease? Heart Failure 1:201–258
12. Erdmann E, Hasse W (1975) Quantitative aspects of oubain binding to human erythrocyte and cardiac membranes. J Physiol (Lond) 251:617–682
13. Erdmann E, Werdan K, Krawietz W (1984) Influence of digitalis and diuretics on oubain binding sites on human erythrocytes. Klin Wochenschr 62:87–92
14. Ford AR, Aronson JK, Grahame-Smith DG, Carver JG (1979) Changes in cardiac glycoside receptor sites, [86] rubidium uptake and intracellular sodium concentrations in the erythrocytes of patients receiving digoxin during the early phases of treatment of cardiac failure in regular rhythm and of atrial fibrillation. Br J Clin Pharmacol 8:125–134
15. Grupp G, Grupp IL, Melvin DB, Schwartz A (1988) Functional evidence in diseased human heart fibers for multiple sensitivities of the inotropic oubain receptor Na,K-

ATPase, in Membrane Biophysics III: Biological Transport. New York, Alan R Liss, pp. 215–222.

16. Hansen O, Skou JC (1973) A study on the influence of the concentration of Mg^{2+}, P_i, K_1^+, Na^+, and Tris on (Mg^{2+} + P_i)-supported g-strophanthin binding to ($Na^+ - K^+$) activated ATPase from ox brain. Biochim Biophys Acta 311:51–66

17. Herrera VL, Chobanian AV, Opazo NR: Isoform specific modulation of Na^+, K^+-ATPase alpha-subunit gene expression in hypertension. Science 241:221–223

18. Herrera VLM, Ruiz-Opazo N (1990) Alteration of a1 Na,K-ATPase $^{86}Rb^+$ influx by a single amino acid substitution. Science 249:1023

19. Kim D, Marsh JD, Barry WH, Smith TW (1984) Effects of growth in low potassium medium or oubain on membrane Na,K-ATPase, cation transport, and contractility in cultured chick heart cells. Circ Res 55:39–48

20. Lee SW, Schwartz A, Adams RJ et al (1983) Decrease in Na,K-ATPase activity and [^{3}H]oubain binding sites in sarcolemma from hearts of spontaneously hypertensive rats. Hypertension 5:683

21. Ligrel JB, Orlowski J, Shull MM, Price EM (1990) Molecular genetics of Na,K-ATPase. Prog Nucleic Acid Res Mol Biol 38:37–89

22. Maixent JM, Lelievre L (1987) Differential inactivation of inotropic and toxic digitalisreceptors in ischemic dog hearts. Molecular basis for the deleterious effects of digitalis. J Biol Chem 262:12458

23. Nojima H, Yagawa Y, Kawakami K (1989) The Na,K-ATPase alpha 2 subunit gene displays restriction fragment length polymorphisms between the genomes of normotensive and hypertensive rats. J Hypertension 7:937

24. Norgaard A, Kjeldsen K, Hansen O, Clausen T, Larsen CG, Larsen FG (1986) Quantification of the 3H-oubain binding site concentration in human myocardium: a postmortem study. Cardiovasc Res 30:428–435

25. Norgaard A, Bagger JP, Bjerregaard P, Baandrup U, Kjeldsen K, Thomsen PEB (1988) Relation of left ventricular function and Na,K-pump concentration in suspected idiopathic dilated cardiomyopathy. Am J Cardiol 61:1312–1315

26. Norgaard A, Kjeldsen K (1989) Human myocardial Na,K-pumps in relation to heart disease. J Appl Cardiol 4:239–245

27. Rayson BM (1989) Rates of synthesis and degradation of $Na^+ - K^+$-ATPase during chronic oubain treatment. Am J Physiol 356:C75–C80

28. Schmidt TA, Kjeldsen K (1991) Enhanced clearance of specifically bound digoxin from human myocardial and skeletal muscle samples by specific digoxin antibody fragments. Subsequent complete digitalis glycoside receptor (Na,K-ATPase) quantification. J Cardiovasc Pharmacol 17:670–677

29. Schwinger RH, Böhm M, Erdmann E (1990) Effectivness of cardiac glycosides in human myocardium with the without "downregulated" beta-adrenoceptors. J Cardiovas. Pharmacol 15:692–697

30. Smith TW, Antman EM, Friedman PL (1984) Digitalis glycosides: mechanisms and manifestations of toxicity. Progr Cardiovasc Dis 1984; 26–413

31. Smith TW (1988) Digitalis. Mechanisms of action and clinical use. N Engl J Med 318:358–365

32. Sweadner KJ (1979) Two molecular forms of ($Na^+ + K^+$) simulated ATPase in brain separation, and difference in affinity for strophanthidin. J Biol Chem 254:6060–6067

Author's address:
Paul D. Allen, M.D., Ph.D.
Department of Anesthesia
Brigham and Women's Hospital
75 Francis Street
Boston, MA 02115
USA

*Excitation-contraction coupling
and contractile proteins*

Structural and functional diversity of human ventricular myosin

H. Rupp, R. Jacob

Institute of Physiology II, University of Tübingen, FRG

Summary: The role of subcellular alterations in the process of heart failure remains ill-defined. Because contractile performance of failing heart muscle is depressed, possible alterations in the myosin molecule could be of particular relevance. There is increasing evidence that myofibrillar ATPase activity is reduced in congestive heart failure, whereas the findings on myosin ATPase are still controversial. The molecular causes of the reduced activity are currently not known. Because α-MHC is present only in small amounts in normal ventricles, a shift in favor of β-MHC is of minor importance. Also immunohistochemical data on subspecies of β-MHC seem not to provide an explanation. A new type of myosin heterogeneity was found by optimizing native polyacrylamide gel electrophoresis in the presence of pyrophosphate. Two bands (V_A and V_B) were observed in ventricles of patients with valvular disease. Because the two bands were detected also in normal hearts of large mammals, the existence of V_A/V_B cannot be diagnostic of diseased heart. However, the V_A/V_B ratio was influenced by the hemodynamic load, whereby the fast migrating band (V_A) increased with the diastolic and systolic load. Because a relationship with the hemodynamic load was observed only in surgical muscle specimens, it appears that this heterogeneity is prone to post mortem modification. Further work is required to identify the molecular nature of this heterogeneity and to examine the therapeutic potential of a pharmacological modification of the V_A/V_B ratio.

Key words: Subcellular alterations; heart failure; myosin ATPase; V_A/V_B ratio

Introduction

A hallmark of congestive heart failure is the depressed mechanical performance of heart muscle. Although the pump function can be improved by various pharmacological approaches, it is still not possible to increase markedly the overall survival, and no intervention has until now stopped or reversed the progression of congestive heart failure (14, 29). The depressed heart performance arises from a number of unfavorable changes involving chamber geometry, coronary blood supply and molecular structures of the heart muscle cell (26, 27, 59). The elucidation of subcellular changes in the heart has been a goal for cardiologists of various disciplines. Because myosin represents, besides actin, the major component of the contractile machinery, it has attracted special interest. Despite great efforts, changes in the structure or enzymatic activity of myosin remain controversial. One of the problems of identifying molecular defects in myosin of diseased hearts arises from the fact that the sampling conditions of diseased and normal hearts are not identical. As a consequence, post-mortem modifications are likely to be different for diseased and control hearts. Furthermore, because the purification of myosin, actomyosin and myofibrils is continuously improved, apparently controversial findings could arise from technical shortcomings which varied over the years. One has also to take into account that no in vitro assay exists which mimics the complex process of force generation and shortening of intact sarcomeres. Thus, subtle changes in myosin structure relevant to in vivo muscle function could occur, which do not show up in ATPase assays. In such a case, one would underestimate the impact of structural alterations. One has also to take into account that a certain defect may arise only in pre-end stage heart failure or end-stage heart failure but not in apparently compensated hearts.

In the present approach it was attempted to critically evaluate published findings on human myosin and to focus on those properties which do not seem to be influenced by post mortem modifications. Furthermore, a novel technique involving an improvement of the native pyrophosphate gel electrophoresis is presented which demonstrates a myosin heterogeneity not observed in small mammalian ventricles. Since the pump function of the heart is influenced only to a minor extent by atria and because acquired congestive heart failure rarely becomes manifest in children, only ventricular myosin of adults is considered in detail.

Autoptic specimens

ATPase assays. In a pioneering approach, Alpert and Gordon (2) showed that myofibrillar ATPase was decreased in ventricular specimens (3–12 h post mortem) of seven hypertensive subjects with congestive heart failure compared with specimens obtained following traumatic death in nine normal subjects, and it was suggested that a lesion in congestive heart failure may reside in the contractile protein itself. This finding was later confirmed, and it was shown that the depression of myofibrillar ATPase activity (4–11 h post mortem) was a function of failure, but not of hypertrophy per se (16). Myofibrillar ATPase was also depressed in hearts (autopsies, < 36 h post mortem) with a weight range of 500–700 g compared with hearts weighing less than 300 g (32).

When the ATPase activity of isolated myosin was studied, no unifying picture emerged. In autoptic specimens (< 24 h post mortem) from four patients with hypertension (one with congestive heart failure), no differences were found in K^+-EDTA, Ca^{2+}, Mg^{2+} and actin-activated myosin ATPase activity compared with autoptic specimens of four adults with no history of heart disease (51). Also in autoptic (< 12 h post mortem) samples of 10 patients with hypertrophic hearts (one with pulmonary edema) the early phosphate burst size and steady-state ATPase activities stimulated by K^+-EDTA, Ca^{2+}, Mg^{2+} and actin were unaltered compared with five control hearts (36, 43). In 12 patients with the genetically determined asymmetric septal hypertrophy (surgical specimens) the K^+-EDTA and Ca^{2+} activated myosin ATPase activity was also unaltered compared with specimens obtained from four subjects without heart disease 6–20 h after death (35). When a cross-linked complex between myosin subfragment 1 and actin was studied, the ATPase activity was 700-fold higher than that of myosin S1 (31). There was, however, no difference in this activity between two normal and three hypertrophied hearts of hypertensive patients (two with heart failure, one without signs of failure) (all hearts frozen within 20 h of death) (31).

Electrophoretic techniques. A major progress in the identification of myosin isoforms was achieved by the separation of native myosin on polyacrylamide gels in the presence of pyrophosphate (25). Using this technique, only a single band was observed for autoptic (< 24 h post mortem) hearts from four subjects with hypertension (one with congestive heart failure) and four subjects without heart disease (51). A single band was observed also for ventricles obtained within 24 h after death (8). This finding was confirmed for patients with hypertrophic hearts and subjects with no heart disease (< 12 h post mortem) (36). If the post mortem period was shorter (4–6 h; disease not specified), also only a single band was observed (9). Because this band migrated comparable to rat V_3, it was identified as V_3 of small mammals.

Using a different approach involving separation of α-myosin heavy chains (MHC) and β-MHC on dissociating sodium dodecyl sulphate (SDS) polyacrylamide gels, only a single β-MHC was detectable in 11 ventricles of subjects (< 2–24 h post mortem) with no apparent cardiovascular disease; based on the sensitivity of the method it was concluded that α-MHC did not exceed 5% (10).

An additional band which amounts up to 25% was observed on native pyrophosphate gels for ventricular samples (< 12–24 h post mortem) of nine patients with myocardial infarction; this band comigrated with one (HA-1) of the two atrial isoforms (22–24). Because in ventricles (< 8–12 h post mortem) of 10 subjects without heart disease, only one band was observed under identical electrophoretic conditions, it was concluded that the occurrence of the second band is typical for infarcted ventricles (22–24).

Immunohistochemical techniques. Based on an immunohistochemical approach using autoptic ventricular specimens (< 12 h post mortem) the majority of muscle fibers was stained by an antibody directed against rat V_3 (36). The amount of a rat V_1-like myosin (HV_1) was estimated to range from almost 0 to 15% (13 subjects with less than 2%, five subjects with 4–10%, and one subject with 14%) (36). In samples from patients with valvular disease, HV_1 was absent (52). The amount of HV_1 was correlated with the Ca^{2+}-stimulated myosin ATPase activity (36) in accordance with findings on V_1 of rat ventricles (37, 42). In an independent study (3–24 h post mortem samples), a level of 14–24% ventricular α-MHC was observed in one specimen, but the other samples were estimated to contain less than 10% α-MHC (17). The human ventricle would thus have only a very limited range for reacting to an overload which results in an increased β-MHC expression and thus a reduction of myosin ATPase activity. Because a correlation of α,β-MHC fiber percentage with the cardiac index was found, it was suggested that small amounts of α-MHC isoform may be important for the contractile performance of the heart (7).

In favor of a more complex myosin heterogeneity in ventricles was the suggestive finding of two subpopulations of myosin which were postulated to be V3-type myosins (6). It was tentatively concluded that the variant (β′-myosin) not present in the rat was relatively rare, whereas the other represented 90–95% of total myosin (6). This heterogeneity could be resolved further by demonstrating ventricular fibers containing either αβ-, ββ- or ββ′-myosins. The ββ′-fibers varied in three normal ventricles from < 1% to 5% of total fibers and were practically absent in diseased hearts (two patients with ischemic heart disease and four with primary dilated cardiomyopathy; post mortem time not specified) (7). Superimposed on the α,β-MHC heterogeneity is thus another heterogeneity which could, however, not be correlated with hemodynamic parameters (7). The range of adaptation of the heterogeneity arising from β′ would be small.

Peptide mapping and peptide sequencing. Peptide maps of myosin of adults (autoptic specimens, < 24 h post mortem) did not differ from those of patients (surgical specimens) with obstructive hypertrophic cardiomyopathy (51). Furthermore, partial sequence data of myosin light chains exhibited no difference in greatly hypertrophied hearts (30).

Surgical or cardiac transplant specimens

ATPase activities. Because subtle changes in myosin might be lost during the post mortem period, it is essential to examine also muscle specimens obtained during heart surgery. In the early experiments of Gordon and Brown (16) the myofibrillar ATPase activity of failing papillary muscles (surgical specimens of two patients with long-standing mitral insufficiency and congestive failure) was depressed compared with the activity of hypertrophic nonfailing pulmonary infundibulum (surgical specimens of five patients with tetralogy of Fallot); the failing papillary muscles exhibited a myofibrillar ATPase activity which was in the range of failing post mortem ventricular ATPase activity (16). When the myofibrillar ATPase activity was assayed in the ventricular homogenate, it was found reduced in congestive cardiomyopathy, similar to that of patients with impaired ventricular function secondary to valvular disease (41). In a recent study it was shown that in ventricular specimens of patients with end-stage heart failure (obtained during cardiac transplant surgery) caused by coronary artery disease or idiopathic cardiomyopathy, the myofibrillar Mg-ATPase activity was decreased compared with that of normal human ventricular muscle of accident victims whose hearts were unsuitable for cardiac transplantation (40). This decrease in ATPase activity was not causally linked to a reduced amount of myofibrillar protein because in ventricular (surgical) specimens of patients with mitral valve insufficiency, the myofibrillar ATPase was reduced to the same extent, although the myofibrillar protein content was unchanged. Because the Ca^{2+}-ATPase of myosin of patients with end-stage heart failure was not reduced (stated by the authors as unpublished results), it was suggested that the interaction between the thick and thin filaments is abnormal (40). The myofibrillar Mg-ATPase activity was found reduced to the same extent in heart failure irrespective of etiology (immunological rejection, coronary heart disease, idiopathic cardiomyopathy); control hearts were from accident victims (1). There is, however, no general agreement that myofibrillar ATPase is decreased in failing hearts. Thus, in a study using explanted hearts with dilated cardiomyopathy and normal hearts from renal transplant donors, a significant change in basal and maximal myofibrillar ATPase could not be observed (61).

Less clear again, are, studies on the ATPase activity of isolated myosin. The Ca^{2+} activated myosin ATPase activity was reduced in failing hearts from patients with coronary artery disease, but not in failing hearts with idiopathic cardiomyopathy or immunological rejection (1). The K^+-EDTA, Ca^{2+} and actin-activated myosin ATPase activity was found identical in explanted hearts with dilated cardiomyopathy compared with normal hearts from renal transplant donors (61). However, in another study the Ca^{2+}-activated myosin ATPase was decreased in failing hearts with idiopathic dilated cardiomyopathy (eight explanted hearts) compared with the activity of a normal heart which could not be transplanted for technical reasons (11).

Electrophoretic techniques. In accordance with the findings using autoptic muscle, ventricular myosin of subjects with coronary heart disease appeared as a single band on native pyrophosphate gels (38). Based on the separation by SDS polyacrylamide gel electrophoresis, it was concluded that no gross changes occur in the subunit structure of myosin from subjects with assymmetric septal hypertrophy (35). In biopsies of nine patients with primary and 27 patients with secondary ventricular hypertrophy, an atrial-like light chain 1 was detected which was correlated with peak circumferential wall stress (20). The mean content was highest in dilated cardiomyopathy (12%), less in cases with pressure overload (6%), volume overload (3%), and infarction (2%). No such increase was observed in hypertrophic cardiomyopathy or in coronary heart disease without infarction (20, 49). The total light chain 1 content, when referred to tropomyosin, remained constant (20). An atrial light chain 2 was not detected in ventricles (20). In ventricular infundibular muscle of patients with tetralogy of Fallot the atrial light chain 1 was also detected (average 3%, one subject with 34%) with an unchanged ratio of light chain 1 and 2 (3). The appearance of the atrial light chain 1 seems to be restricted to ventricular myocardium in the early stages of cardiomyopathy or during the development of hypertrophy, because it was only barely discernible in failing ventricles of patients (eight explanted hearts) with idiopathic dilated cardiomyopathy (11). Based on these data, it appears that an increased atrial light chain 1 cannot be the cause of a reduced ATPase activity of failing hearts.

Immunohistochemical techniques. Also in surgical specimens the early findings demonstrated that muscle fibers of ventricular samples of patients with ischemic heart disease or valvular dysfunction (56) and patients with valvular disease (52) are stained by antibodies cross-reacting with β-MHC of small mammals. The total score of α-MHC decreased in ventricles of patients with valvular disease as the mean left ventricular pressure increased, and a transmural gradient was observed, whereby in the epicardial regions α-MHC was present in up to 10–15% of myofibers and almost 0% in subendocardial regions (57).

Also in surgical specimens, evidence could be obtained for a heterogeneity of β-MHC. Two β_1 and β_2 myosins coexisted in all ventricular myofibers. The antibody stained the myofibers of pressure overloaded ventricles homogeneously and an isozymic shift between β_1 and β_2 could not be elucidated (57). The two molecular variants β_1- and β_2-MHC were isolated and differences in the primary structure were deduced from the analysis of peptides produced by chymotryptic digestion (58). The finding that on the basis of over 200 resolved peptides, no differences could be found between MHC from normal and hypertrophied surgical specimens (total 57 samples) (50) would also suggest that no pronounced shifts in the β_1/β_2 ratio occurred. It remains to be shown whether alterations occur in congestive heart failure.

Mechanical and energetic performance of heart muscle. In volume-overloaded ventricular myocardium (surgical specimens of five patients with severe mitral regurgitation), peak twitch tension was reduced by 55% and maximum rate of tension rise by 65% compared with specimens from subjects with normal left-ventricular function (one with mitral stenosis and three with coronary artery disease) (18). The average crossbridge force-time integral was increased by 85% in volume-overloaded myocardium, which could arise from a reduced ATPase activity due to a higher actomyosin binding constant (18). Noteworthy is the finding that this change was observed, not only in volume-overloaded human, but also in pressure overloaded nonfailing rabbit myocardium. It was concluded that the changes in crossbridge behavior are dominated by alterations in factors other than α-MHC/β-MHC shifts (18). Such undefinded influences could also be responsible for the intriguing finding that the mechanical behavior of the failing heart muscle is not consistently depressed. Thus, the Ca^{2+} response curve of papillary muscle force generation (six explanted hearts with dilated cardiomyopathy or severe coronary heart disease) was not different from that of control hearts (5).

Improved native electrophoresis. Although the native electrophoresis using the conditions of Hoh (25) proved to be a successful tool for identifying myosin isozymes of small mammals, it typically results in a single band of large mammalian myosin. In our first study on myosin of the cat and dog, also only a single band was observed (45). During optimization of this technique, it was found that large mammalian myosin could be separated into two bands which under identical electrophoretic conditions were not observed for the rabbit with homogeneous V3 (47, 48). In keeping with the original nomenclature of Hoh (25) of classifying myosin bands according to their migration velocity on the native gel, but avoiding confusion with the V1, V2, V3 nomenclature, the two bands were named V_A and V_B (Fig. 1) (47, 48). Because these bands were observed in large mammals which evolved independently of each other, it appears that the bands are characteristic of large mammals but do

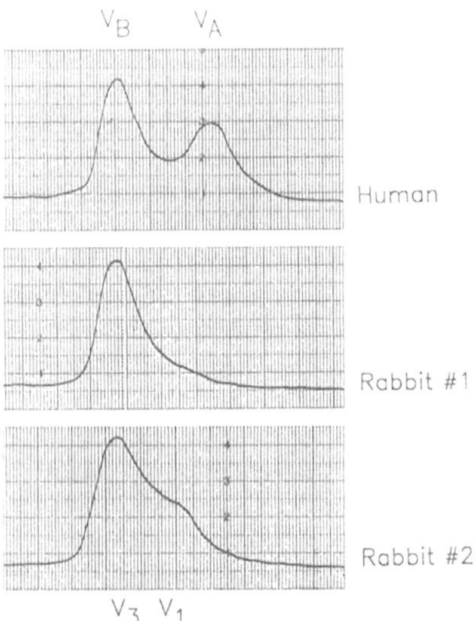

Fig. 1. Representative densitometric tracings of ventricular myosins separated on a polyacrylamide gel in the presence of pyrophosphate. Myosins were extracted from ventricular specimens of a subject with mitral regurgitation and two rabbits. Note, the higher proportion of V_1 in rabbit No. 2 compared with rabbit No. 1, and the different electrophoretic migration behavior of V_1 compared with V_A. The gel was stained with Coomassie blue. Left-ventricular specimens were obtained during the time of valve replacement in accordance with the local ethical research procedures and were provided by Dr. G. Fenchel, Department of Thoracic and Cardiovascular Surgery, University of Tübingen.

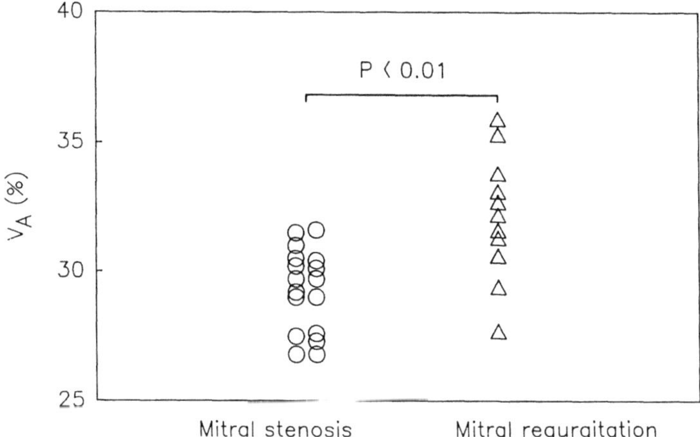

Fig. 2. The proportion of V_A of ventricular specimens from patients with mitral stenosis or mitral regurgitation irrespective of secondary valvular diseases. The statistical comparisons were made by two-tailed Student's t-test.

not occur in the evolutionary distinct group of small mammals of high heart rates (Zak and Rupp, unpublished). The proportion of V_A seems to range within 25–40%, as deduced in a preliminary study on surgical specimens of patients with valvular disease (46, 54). Because V_A/V_B was observed not only in domesticated but also in wild-living species (unpublished), the existence of V_A/V_B cannot be diagnostic of diseased hearts. However, in surgical specimens of patients with a volume overload due to mitral regurgitation, the V_A proportion was significantly greater than in patients with mitral stenosis (Fig. 2). Based on a multiple regression analysis, it was found that in patients with mitral regurgitation V_A correlated primarily with enddiastolic pressure, whereas in patients with mitral stenosis a significant correlation was observed with the systolic work load (product of aortic pressure and cardiac index) (unpublished). The existence of V_A and V_B was recently confirmed by Alousi et al. (1). Although no quantitative data were given, it was reported that the magnitude of the fast migrating band varied considerably among the hearts. However, no measurable differences in myosin patterns between the normal and diseased hearts were found (1). Several attempts at measuring ATPase activity of the two bands on the gel by the Ca^{2+} precipitation technique failed to give reproducible results (1). These findings would not necessarily be inconsistent with our data, because a relationship between work load and V_A/V_B heterogeneity was observed only for surgical specimens of hearts with a defined hemodynamic performance, but not in explanted hearts or autoptic material (unpublished). This could indicate that the V_A/V_B heterogeneity is influenced by the acute load of the heart and that the V_A/V_B ratio is subject to post mortem changes.

Conclusions

Although a unifying picture has not emerged yet, there is increasing evidence that human ventricular myosin is a target for modifications in normal and diseased heart.

The major controversial point relates to alterations in ATPase activity of diseased hearts. If only patients are considered with congestive heart failure, then the main evidence is in favor of a depressed myofibrillar ATPase activity, whereas there is less evidence for a reduction of myosin ATPase activity. The situation is unclear for patients with various heart diseases in the absence of congestive heart failure. Because the transition from an apparently compensated to a failing heart is an ill-defined and unpredictable process, the question can currently not be answered of whether, in pre-end stage heart failure, a depressed activity is already present which could be a factor contributing to heart failure. The fact that a depressed ATPase activity can be detected more consistently for myofibrils than for myosin could indicate that the defect is related to the interaction of myosin with the thin filament involving actin or regulatory proteins. Such a change would not necessarily result in an altered in vitro myosin ATPase activity.

There is now general agreement that the proportion of α-MHC is present only in small amounts in human ventricle which is replaced by β-MHC already by an overload not leading to congestive heart failure. This mechanism can thus not be responsible for the depressed myofibrillar ATPase activity of failing hearts.

Superimposed on the α-/β-MHC heterogeneity is a β-MHC heterogeneity which is ill-defined. This heterogeneity has been demonstrated by independent groups using immunohistochemical techniques (7, 58). Because the pattern of antibody staining of β_1 and β_2-MHC is different from the antibody staining of β and β'-myosin, it was concluded that the two antibodies do not stain the same subspecies of β-MHC (57). Because the β,β'-heterogeneity disappeared in ischemic heart disease (7), it is also not likely to be responsible for the depressed myofibrillar ATPase activity of failing hearts.

A new heterogeneity was demonstrated by separation of myosin into two bands using an improved native gel electrophoresis. In contrast to the apparent defect of myosin of failing hearts, this heterogeneity is present in normal hearts and is influenced by the systolic and diastolic load of the heart. A relationship between this heterogeneity and the mechanical performance of heart muscle can be observed only if the muscle specimens are sampled under well defined conditions. Because the fast-migrating band (V_A) was found in the range of 25–40% and increased with the hemodynamic load, it is not likely to arise from α-MHC. It is not known how this heterogeneity changes in patients with congestive heart failure and the possible link with the depressed ATPase activity of failing hearts remains to be established.

As regards the molecular nature of the various myosin heterogeneities, one should a priori take into account all possible mechanisms involving different myosin genes, alternative splicing or post-translational modifications. The present evidence is not in favor of an expression of genes in addition to the well characterized small mammalian α-MHC and β-MHC genes (4, 12, 21, 28, 33, 34, 60). Although in Drosophila an alternative splicing mechanism has been demonstrated, giving rise to three developmentally regulated transcripts (44), there is currently no evidence for alternative splicing in the human heart. Thus, post-translational modifications could be important involving probably only few amino acid residues.

Of particular interest is the possible relationship of the myosin heterogeneities in human heart with the β-MHC subspecies observed in normal and chronically hypertensive baboons (19). The two β-MHC were reported to have 210 and 200 kD and to share several antigenic determinants. A fivefold increase in the level of the 200-kD β-MHC was observed by electrophoresis in gradient pore polyacrylamide gels in hypertrophied ventricles of baboons with longstanding renal hypertension (19). The Ca^{2+} activated myosin ATPase was decreased by 35% in the hypertrophied ventricles (19). However, only a single myosin band was observed on native pyrophosphate gels both for normotensive and hypertensive baboons (19). Also the observation of two myosin β-type heavy chains in bovine heart (13) should be examined further.

The heterogeneity of the myosin molecule is increased further by taking into account the phosphorylated species of myosin light chain 2 (39). Although the current data are not in favor of a major role of myosin phosphorylation in ventricular muscle, and a dependency on hemodynamic parameters has not been observed for the ventricle (39), the caveat remains that post mortem modifications could wipe out differences present in vivo.

Superimposed on the various alterations in the myosin molecule are genetic defects. It was recently demonstrated that an αβ-MHC hybrid and a point mutation can occur in familial hypertrophic cardiomyopathy (15, 53, 55). The finding that the modification of a single amino acid is associated with cardiomyopathy suggests that structural changes in the myosin molecule must not be extensive in order to affect heart function.

Clearly, further work is required to strengthen the concept that human myosin represents a unique form which is not functionally comparable with small mammalian myosin V3. The elucidation of these heterogeneities should prove helpful in tracing the pathogenesis of various heart diseases and could open new strategies of therapy and prevention of heart failure by controlling the manifestation of certain heterogeneities.

Acknowledgements. The experimental study was supported by the Deutsche Forschungsgemeinschaft and the Alfred-Teufel-Stiftung.

References

1. Alousi AA, Grant AM, Etzler JR, Cofer BR, Van der Bel-Kahn J, Melvin D (1990) Reduced cardiac myofibrillar Mg-ATPase activity without changes in myosin isozymes in patients with end-stage heart failure. Mol Cell Biochem 96:79–88
2. Alpert NR, Gordon MS (1962) Myofibrillar adenosine triphosphatase activity in congestive heart failure. Am J Physiol 202:940–946
3. Auckland LM, Lambert SJ, Cummins P (1986) Cardiac myosin light and heavy chain isotypes in tetralogy of Fallot. Cardiovasc Res 20:828–836
4. Bober E, Buchberger-Seidl A, Braun T, Singh S, Goedde HW, Arnold HH (1990) Identification of three developmentally controlled isoforms of human myosin heavy chains. Eur J Biochem 189:55–65
5. Böhm M, Morano I, Pieske B, Rüegg JC, Wankerl M, Zimmermann R, Erdmann E (1991) Contribution of cAMP-phosphodiesterase inhibition and sensitization of the contractile proteins for calcium to the inotropic effect of pimobendan in the failing human myocardium. Circ Res 68:689–701
6. Bouvagnet P, Leger J, Pons F, Dechesne C, Leger JJ (1984) Fiber types and myosin types in human atrial and ventricular myocardium: An anatomical description. Circ Res 55:794–804
7. Bouvagnet P, Mairhofer H, Leger JOC, Puech P, Leger JJ (1989) Distribution pattern of α and β myosin in normal and diseased human ventricular myocardium. Basic Res Cardiol 84:91–102

8. Clark WA, Chizzonite RA, Everett AW, Rabinowitz M, Zak R (1982) Species correlations between cardiac isomyosins. A comparison of electrophoretic and immunological properties. J Biol Chem 257:5449–5454
9. Cummins P (1984) Contractile protein transitions in human cardiac overload: reality and limitations. Eur Heart J 5 (suppl F):119–127
10. Cummins P, Lambert SJ (1986) Myosin transitions in the bovine and human heart. A developmental and anatomical study of heavy and light chain subunits in the atrium and ventricle. Circ Res 58:846–858
11. Dalla Libera L, Pauletto P, Piccolo D, Scannapieco G, Vescovo G (1991) The idiopathic dilated cardiomyopathy in man. A biochemical and molecular study on myosin. Basic Res Cardiol 86:70–78
12. Feghali R, Leinwand LA (1989) Molecular genetic characterization of a developmentally regulated human perinatal myosin heavy chain. J Cell Biol 108:1791–1797
13. Flink IL, Morkin E (1984) Sequence of the 20-kilodalton heavy chain peptide from the carboxyl-terminus of bovine cardiac myosin subfragment-1. J Clin Invest 74:639–646
14. Francis GS, Cohn JN (1990) Heart failure: mechanisms of cardiac and vascular dysfunction and the rationale for pharmacologic intervention. FASEB J 4:3068–3075
15. Geisterfer-Lowrance AAT, Kass S, Tanigawa G, Vosberg HP, McKenna W, Seidman CE, Seidman JG (1990) A molecular basis for familial hypertrophic cardiomyopathy: a β cardiac myosin heavy chain gene missense mutation. Cell 62:999–1006
16. Gordon MS, Brown AL (1966) Myofibrillar adenosine triphosphatase activity of human heart tissue in congestive failure: effects of ouabain and calcium. Circ Res 18:534–542
17. Gorza L, Mercadier JJ, Schwartz K, Thornell LE, Saverio S, Schiaffino S (1984) Myosin types in the human heart: An immunohistofluorescence study of normal and hypertrophied atrial and ventricular myocardium. Circ Res 54:694–702
18. Hasenfuss G, Mulieri LA, Blanchard EM, Holubarsch C, Leavitt BJ, Ittleman F, Alpert NR (1991) Energetics of isometric force development in control and volume-overload human myocardium. Comparison with animal species. Circ Res 68:836–846
19. Henkel RD, VandeBerg JL, Shade RE, Leger JJ, Walsh RA (1989) Cardiac beta myosin heavy chain diversity in normal and chronically hypertensive baboons. J Clin Invest 83:1487–1493
20. Hirzel HO, Tuchschmid CR, Schneider, J, Krayenbuehl HP, Schaub MC (1985) Relationship between myosin isoenzyme composition, hemodynamics, and myocardial structure in various forms of human cardiac hypertrophy. Circ Res 57:729–740
21. Hixson JE, Henkel RD, Britten ML, Vernier DT, deLemos RA, VandeBerg JL, Walsh RA (1989) Alpha-myosin heavy chain cDNA structure and gene expression in adult, fetal, and premature baboon myocardium. J Mol Cell Cardiol 21: 1073–1086
22. Hoffmann U, Axmann C, Grisk A (1986) Myosin isoenzymes in normal and hypertrophied human hearts. Biochim Biomed Acta 45:985–996
23. Hoffmann U, Siegert E (1987) Atrial and ventricular myosins from human hearts. I. Isoenzyme distribution during development and in the adult. Basic Res Cardiol 82: 348–358
24. Hoffmann U, Axmann C, Palm N (1987) Atrial and ventricular myosins from human hearts. II. Isoenzyme distribution after myocardial infarction. Basic Res Cardiol 82:359–369
25. Hoh JFY, McGrath PA, Hale PT (1978) Electrophoretic analysis of multiple forms of rat cardiac myosin: Effects of hypophysectomy and thyroxine replacement. J Mol Cell Cardiol 10:1053–1076
26. Jacob R, Ebrecht G, Kissling G, Rupp H, Takeda N (1986) Functional consequences of cardiac myosin isoenzyme redistribution. In: Rupp H (ed) The Regulation of Heart Function. Thieme, New York, pp 305–326
27. Jacob R, Kissling G (1989) Ventricular pressure-volume relations as the primary basis for evaluation of cardiac mechanics. Return to Frank's diagram. Basic Res Cardiol 84:227–246

28. Jaenicke T, Diederich KW, Haas W, Schleich J, Lichter P, Pfordt M, Bach A, Vosberg HP (1990) The complete sequence of the human beta-myosin heavy chain gene and a comparative analysis of its product. Genomics 8:194–206
29. Klamerus KJ (1986) Current concepts in clinical therapeutics: congestive heart failure. Clin Pharm 5:481–498
30. Klotz C, Leger JJ, Elzinga M (1982) Comparative sequence of myosin light chains from normal and hypertrophied human hearts. Circ Res 50:201–209
31. Lauer B, van Thiem N, Swynghedauw B (1989) ATPase activity of the cross-linked complex between cardiac myosin subfragment 1 and actin in several models of chronic overloading. A new approach to the biochemistry of contractility. Circ Res 64:1106–1115
32. Leclercq JF, Swynghedauw B, Bouveret P, Faucomprez C, Piguet V (1976) Myofibrillar ATPase, DNA and hydroxyproline content of human hypertrophied heart. Eur J Clin Invest 6:27–33
33. Liew CC, Sole MJ, Yamauchi-Takihara K, Kellam B, Anderson DH, Lin LP, Liew JC (1990) Complete sequence and organization of the human cardiac beta-myosin heavy chain gene. Nucleic Acids Res 18:3647–3651
34. Mahdavi V, Izumo S, Nadal-Ginard B (1987) Developmental and hormonal regulation of sarcomeric myosin heavy chain gene family. Circ Res 60:804–814
35. Maron BJ, Ferrans VJ, Adelstein RS (1977) Isolation and characterization of myosin from subjects with asymmetric septal hypertrophy. Circ Res 40:468–473
36. Mercadier JJ, Bouveret P, Gorza L, Schiaffino S, Clark WA, Zak R, Swynghedauw B, Schwartz K (1983) Myosin isoenzymes in normal and hypertrophied human ventricular myocardium. Circ Res 53:52–62
37. Mercadier JJ, Lompre AM, Wisnewsky C, Samuel JL, Bercovici J, Swynghedauw B, Schwartz K (1981) Myosin isoenzymic changes in several models of rat cardiac hypertrophy. Circ Res 49:525–532
38. Morano I, Arndt H, Gärtner C, Rüegg JC (1988) Skinned fibers of human atrium and ventricle: Myosin isoenzymes and contractility. Circ Res 62:632–639
39. Morano I, Wankerl M, Böhm M, Erdmann E, Rüegg JC (1989) Myosin P-light chain isoenzymes in the human heart: evidence for diphosphorylation of the atrial P-LC form. Basic Res Cardiol 84:298–305
40. Pagani ED, Alousi AA, Grant AM, Older TM, Dziuban SW, Allen PD (1988) Changes in myofibrillar content and Mg-ATPase activity in ventricular tissues from patients with heart failure caused by coronary artery disease, cardiomyopathy, or mitral valve insufficiency. Circ Res 63:380–385
41. Peters TJ, Wells G, Oakley CM, Brooksby IAB, Jenkins BS, Webb-Peploe MM, Coltart DJ (1977) Enzymic analysis of endomyocardial biopsy specimens from patients with cardiomyopathies. Brit Heart J 39:1333–1339
42. Pope B, Hoh JFY, Weeds A (1980) The ATPase activities of rat cardiac myosin isoenzymes. FEBS Lett 118:205–208
43. Rappaport L, Swynghedauw B, Mercadier JJ, Lompre AM, de la Bastie D, Samuel JL, Schwartz K (1986) Physiological adaptation of the heart to pathological overloading. Fed Proc 45:2573–2579
44. Rozek CE, Davidson N (1983) Drosophila has one myosin heavy-chain gene with three developmentally regulated transcripts. Cell 32:23–34
45. Rupp H (1982) Polymorphic myosin as the determinant of myofibrillar ATPase in different haemodynamic and thyroid states. Basic Res Cardiol 77:34–46
46. Rupp H, Fenchel G (1988) Interrelationship between catecholamine stores and myosin structure in human heart. Z Kardiol 77 (Suppl 1): 267
47. Rupp H, Jacob R (1986) Myocardial transitions between fast- and slow-type muscle as monitored by the population of myosin isoenzymes. In: Rupp H (ed) The Regulation of Heart Function. Thieme, New York, pp 271–291
48. Rupp H, Jacob R (1985) Ventricular myocardium as a fast- or slow-type muscle. The influence of stressors and the preventive action of intense exercise. In: Beamish R, Panagia

V, Dhalla NS (eds) Pathogenesis of Stress-induced Heart Disease. Martinus Nijhoff, Boston, pp 147–158
49. Schaub MC, Hirzel HO (1987) Atrial and ventricular isomyosin composition in patients with different forms of cardiac hypertrophy. Basic Res Cardiol 82 (suppl 2):357–367
50. Schaub MC, Tuchschmid CR, Srihari T, Hirzel HO (1984) Myosin isoenzymes in human hypertrophic hearts. Shift in atrial myosin heavy chains and in ventricular myosin light chains. Eur Heart J 5 (suppl F):85–93
51. Schier JJ, Adelstein RS (1982) Structural and enzymatic comparison of human cardiac muscle myosins isolated from infants, adults, and patients with hypertrophic cardiomyopathy. J Clin Invest 69:816–825
52. Schwartz K, Apstein C, Mercadier JJ, Lecarpentier Y, de la Bastie D, Bouveret P, Wisnewsky C, Swynghedauw B (1984) Left ventricular isomyosins in normal and hypertrophied rat and human hearts. Eur Heart J 5(suppl F):77–83
53. Solomon SD, Geisterfer-Lowrance AA, Vosberg HP, Hiller G, Jarcho JA, Morton CC, McBride WO, Mitchell AL, Bale AE, McKenna WJ et al. (1990) A locus for familial hypertrophic cardiomyopathy is closely linked to the cardiac myosin heavy chain genes, CRI-L436, and CRI-L329 on chromosome 14 at q11–q12. Am J Hum Genet 47:389–394
54. Takeda N, Rupp H, Fenchel G, Hoffmeister HE, Jacob R (1985) Relationship between the myofibrillar ATPase activity of human biopsy material and hemodynamic parameters. Jap Heart J 26:909–922
55. Tanigawa G, Jarcho JA, Kass S, Solomon SD, Vosberg HP, Seidman JG, Seidman CE (1990) A molecular basis for familial hypertrophic cardiomyopathy: an α/β cardiac myosin heavy chain hybrid gene. Cell 62:991–998
56. Tsuchimochi H, Sugi M, Kuro-o M, Ueda S, Takaku F, Furuta S, Shirai T, Yazaki Y (1984) Isozymic changes in myosin of human atrial myocardium induced by overload. Immunohistochemical study using monoclonal antibodies. J Clin Invest 74:662–665
57. Tsuchimochi H, Kuro-o M, Takaku F, Yoshida K, Kawana M, Kimata S, Yazaki Y (1986) Expression of myosin isozymes during the developmental stage and their redistribution induced by pressure overload. Jap Circ J 50:1044–1052
58. Tsuchimochi H, Kuro-o M, Koyama H, Kurabayashi M, Sugi M, Takaku F, Furuta S, Yazaki Y (1988) Heterogeneity of β-type myosin isozymes in the human heart and regulational mechanisms in their expression. J Clin Invest 81:110–118
59. Vogt M, Jacob R, Kissling G, Rupp H (1987) Chronic cardiac reactions. II. Mechanical and energetic consequences of myocardial transformation versus ventricular dilatation in the chronically pressure-loaded heart. Basic Res Cardiol 82 (Suppl 2):147–159
60. Vosberg HP, Horstmann-Herold U, Jaenicke T, Morano I, Rüegg JC, Eldin P, Leger JJ (1990) Regulation of the human β-myosin heavy chain gene and an approach to a functional analysis of recombinant protein subregions of the β-chain. In: Pette D (ed) The Dynamic State of Muscle Fibers. de Gruyter, Berlin, pp 45–60
61. Wiegand V, Ebecke M, Figulla H, Schuler S, Kreuzer H (1989) Structure and function of contractile proteins in human dilated cardiomyopathy. Clin Cardiol 12:656–660

Author's address:

Dr. H. Rupp
Physiologisches Institut II, Universität Tübingen
Gmelinstrasse 5
7400 Tübingen, FRG

Contractile protein function in failing and nonfailing human myocardium

G. Hasenfuss[1], L.A. Mulieri[2], B.J. Leavitt[2], P.D. Allen[3], C. Holubarsch[1], H. Just[1], N.R. Alpert[2]

[1] Department of Medicine, Cardiology, University of Freiburg, FRG
[2] Department of Physiology and Biophysics and Department of Cardiothoracic Surgery, University of Vermont, Burlington, Vermont, USA
[3] Department of Anesthesiology, Brigham and Women's Hospital, Boston, Massachusetts, USA

Summary: Isometric heat and force measurements were used to relate mechanical performance to function of contractile proteins in muscle strips from failing and nonfailing human hearts (37°C, 60 beats per minute). Compared to control myocardium, crossbridge behavior was altered in myocardium from hearts with end-stage failing dilated and ischemic cardiomyopathy, resulting in increased crossbridge force-time integral by 33% and 36%, respectively.

Peak isometric twitch tension was reduced significantly by 46% in muscle strips from hearts with dilated cardiomyopathy. In myocardium from hearts with ischemic cardiomyopathy peak isometric twitch tension was comparable to values from nonfailing hearts. Including all three types of myocardium, there was a close correlation between the number of crossbridge interactions during the isometric twitch (tension-dependent heat) and peak twitch tension ($r = 0.88$; $p < 0.001$). Compared to control, in failing myocardium from dilated cardiomyopathic hearts, tension-independent heat (calcium cycling) was significantly reduced. This indicates that in dilated cardiomyopathy reduced peak twitch tension results from decreased calcium activation of contractile proteins with reduced number of crossbridge interactions during the isometric twitch.

In ischemic cardiomyopathy mechanisms different from those observed in dilated cardiomyopathy seem to be involved in the development of heart failure.

Key words: Myothermal measurements; myocardial failure; crossbridge cycle; crossbridge force-time integral; calcium cycling

Introduction

Systolic and diastolic performance of the heart, both of which are altered in heart failure, critically depend upon the characteristics of the myosin actin crossbridge cycle and the course of activation and deactivation of the acto-myosin system. The key element in myocardial force development or shortening is the crossbridge cycle. During each crossbridge interaction, the myosin crossbridge head attaches to actin, rotates, in a manner that develops force or causes shortening with thick and thin filaments sliding past each other, and then detaches from the actin filament to start another cycle. There is an obligatory hydrolysis of one high energy phosphate bond with each crossbridge cycle (19). Contractile force of the myocardium depends on the number of crossbridges activated during each contraction-relaxation cycle and on the force development and maintenance of the individual crossbridge interaction (crossbridge force-time integral). The number of crossbridges activated can be increased by a recruitment of additional units of crossbridges or by a decrease in the "off-time" during which the crossbridge is in the weakly bound, non force-generating state. The latter is consistent with an increase in the rate constant "f", promoting the transition from the weakly bound into the strongly bound crossbridge state (6). Crossbridge force-time integral can be increased by a higher force development of the individual crossbridge cycle or by prolonged crossbridge attachment-time. The latter is consistent with a decrease in the rate constant of crossbridge detachment "g" (6). While the number of crossbridges activated mainly depends on excitation-contraction coupling processes, crossbridge force-time integral seems to be regulated by factors such as myosin heavy chain or light chain isoform shifts, myosin light chain phosphorylation or changes in thin filament regulatory systems.

In the present study, we used the myothermal method to investigate in the intact muscle how alterations in crossbridge behavior and number of crossbridge inter-actions may be involved in the development of myocardial failure in dilated and ischemic cardiomypathy.

Methods

Muscle strip preparations

Simultaneous heat and force measurements were performed in thin muscle strips which were dissected from larger pieces of human myocardium. The dissection procedure was described in detail by Mulieri et al. (27, 29). Experiments were performed in different types of human myocardium: 1) Nonfailing human myo-cardium was obtained from subepicardial left-ventricular biopsies dissected during coronary artery bypass surgery in patients with coronary artery disease (29). All patients exhibited normal left-ventricular function. 2) Failing human myocardium was dissected from the left ventricle of explanted hearts with endstage failing dilated or ischemic cardiomyopathy (13).

The study was reviewed and approved by the Committee on Human Research of the University of Vermont. Patients gave written informed consent before participat-ing in the study.

Myothermal measurements

Changes in muscle temperature were measured with 14-junctions, Hill-type thermo-piles fabricated by vacuum deposition of bismuth and antimony junctions on mica substrates (26). To perform the heat and mechanical measurements, the muscles were mounted in contact with the active region of a thermopile and connected to the force gauge. The muscles were stretched gradually to the length at which maximum steady-state isometric twitch force was reached (l_{max}). Experiments were performed under isometric conditions at 37°C with a stimulation frequency of 60 beats per minute (Ca^{++}-content of the Krebs-Ringer solution: 2.5 mM). Under steady-state isometric conditions all of the energy turned over by the muscle is liberated as heat by the end of the twitch. The total activity related heat is divisable into initial and recovery components. The latter arises from the metabolic resynthesis of high-energy phos-phates. Initial heat is composed of the tension-dependent heat and tension-independent heat. Tension-dependent heat results from high-energy phosphate hydrolysis by cycling crossbridges. Assuming that one high-energy bond is hydrolyzed during each crossbridge cycle, tension-dependent heat directly reflects the number of crossbridge interactions during the isometric twitch (15). Tension-independent heat reflects high-energy phosphate hydrolysis of excitation-contraction coupling processes. This includes high-energy phosphate hydrolysis by sarcoplasmic reticulum and sarcolemmal calcium pumps, and high-energy phosphate hydrolysis by sarcolemmal sodium-potassium ATPases and other ATP utilizing pumps. The major portion of tension-independent heat results from high-energy phosphate hydrolysis by sarcoplasmic reticulum calcium pumps to remove calcium from the cytosol (2). Assuming a constant stoichiometry of calcium transport by the sarcoplasmic reticulum Ca^{++}-ATPase (two Ca^{++}-ions being transported per molecule of ATP hydrolyzed), tension-independent heat mainly reflects the amount of calcium removed and thus, during steady-state conditions, the amount of calcium released during the isometric twitch.

From mechanical and myothermal measurements, the average force-time integral of the individual crossbridge cycle can be calculated (15).

Measurements were performed in 14 muscle strip preparations from seven hearts with normal left-ventricular function, in 13 muscle strips from six hearts with end-stage failing dilated cardiomyopathy, and in seven muscle strips from three hearts with end-stage failing ischemic cardiomyopathy.

Myothermal measurements and analysis were performed as described recently (15). Data are expressed as mean $\pm$ SEM. To evaluate the statistical significance of differences between groups the unpaired t-test was applied. A p value < 0.05 was accepted as statistically significant.

Results

Mechanical parameters in failing and nonfailing myocardium

In nonfailing human myocardium, peak isometric twitch tension was 26 $\pm$ 4 mN/mm^2, maximum rate of tension rise was 193 $\pm$ 26 mN/mm^2·s and maximum rate of tension fall was 148 $\pm$ 23 mN/mm^2·s. In myocardium from dilated cardiomyopathic hearts, peak twitch tension, rate of tension rise, and rate of tension fall were significantly reduced by 46%, 51%, and 46%, respectively (Fig. 1). In myocardium from ischemic cardiomyopathic hearts, peak twitch tension, as well as rate of tension rise and fall were not significantly different from values obtained in nonfailing myocardium (Fig. 1).

Crossbridge function in failing and nonfailing myocardium

Average force-time integral of the individual crossbridge cycle was calculated from isometric force-time integral of the twitch and tension-dependent heat (15). Compared to nonfailing myocardium, in failing myocardium from both dilated and ischemic cardiomyopathic hearts crossbridge force-time integral was significantly increased by 33% and 36%, respectively (Fig. 2).

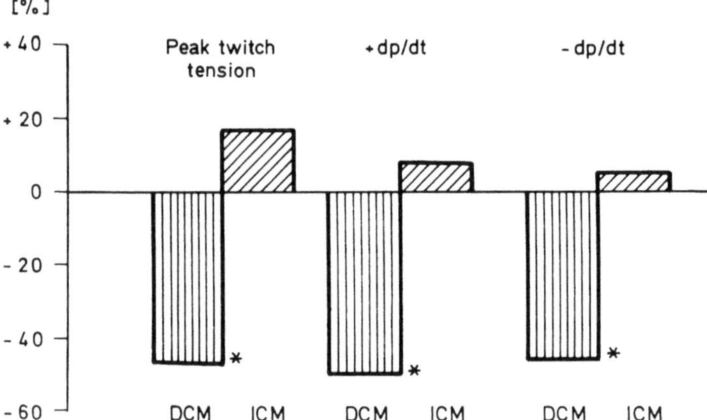

Fig. 1. Percentage change of peak twitch tension, maximum rate of tension rise (+ dp/dt) and maximum rate of tension fall (− dp/dt) compared to control myocardium. DCM = dilated cardiomyopathy; ICM = ischemic cardiomyopathy; * = p < 0.05 versus control.

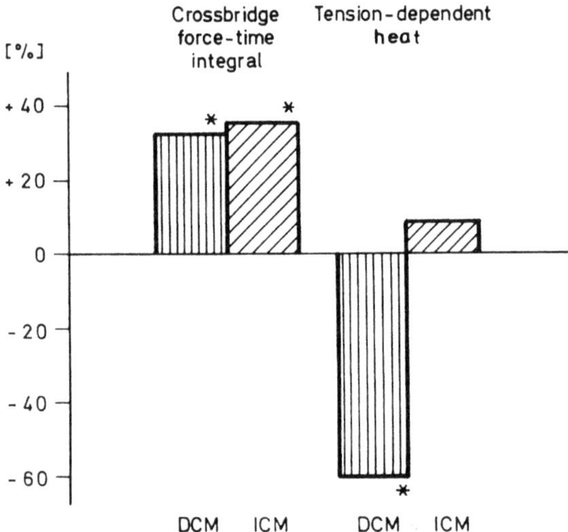

Fig. 2. Percentage change of crossbridge force-time integral and tension-dependent heat compared to control myocardium. DCM = dilated cardiomyopathy; ICM = ischemic cardiomyopathy; * = p < 0.05 versus control.

Number of crossbridge interactions in failing and nonfailing myocardium

Assuming that one high-energy phosphate bond is hydrolyzed during each crossbridge cycle in both failing and nonfailing human myocardium, tension-dependent heat reflects the number of crossbridge interactions during the isometric twitch. Compared to control myocardium, in myocardium from dilated cardiomyopathic hearts, tension-dependent heat was significantly reduced by 61%, whereas in myocardium from ischemic cardiomyopathic hearts, tension-dependent heat was not significantly different from values in control myocardium (Fig. 2).

Including nonfailing myocardium and failing myocardium from both dilated and ischemic cardiomyopathic hearts, there was a close linear correlation between tension-dependent heat liberated by the muscle strips and isometric peak twitch tension developed (r = 0.88; p < 0.001).

Calcium cycling in failing and nonfailing myocardium

To further evaluate the reduced force development and the reduced number of crossbridge interactions in failing myocardium from dilated cardiomyopathic hearts, tension-independent heat was analyzed. Compared to nonfailing myocardium, in failing dilated cardiomyopathy tension-independent heat was significantly reduced by 69% (p < 0.05), indicating considerable reduction in the number of calcium ions cycling during the isometric twitch (Fig. 3).

Discussion

In the present study, we used the myothermal method to relate mechanical parameters of nonfailing and failing human myocardium to function of subcellular systems. The average peak twitch tension value of $26\,mN/mm^2$ in the isolated nonfailing human myocardium falls within the range of clinical wall stress values obtained from patients with normal left-ventricular function measured during cardiac catheterization (14, 18). Mechanical parameters of the isometric twitch were considerably altered in myocardium from dilated cardiomyopathic hearts. Compared

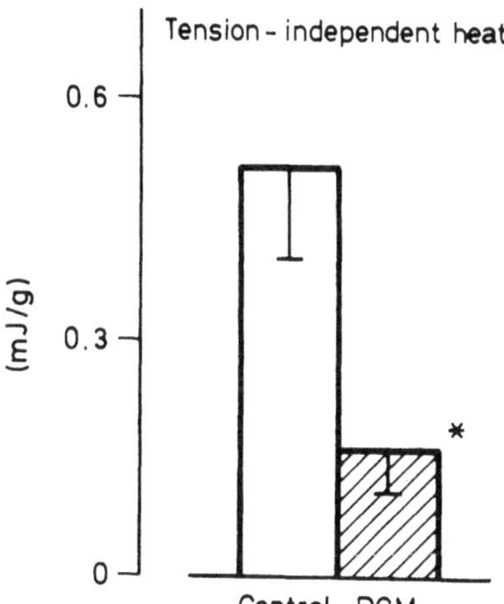

Fig. 3. Tension-independent heat in control myocardium and myocardium from hearts with dilated cardiomyopathy (DCM). * = p < 0.05.

to control, in failing dilated cardiomyopathy peak twitch tension was reduced by 46%, maximum rate of tension rise was reduced by 51%, and maximum rate of tension fall was reduced by 46%. Interestingly, in myocardium from ischemic cardiomyopathic hearts, mechanical parameters were not significantly different from values in nonfailing myocardium.

On a subcellular level, reduced contractile force in the failing myocardium may result from alteration of the function of the individual crossbridge cycle or from a reduced number of crossbridge interactions during the isometric twitch. Crossbridge function was evaluated by calculating average crossbridge force-time integral. As is obvious from Fig. 2, crossbridge force-time integral was increased significantly in failing myocardium from dilated cardiomyopathic hearts. However, since similar alterations in crossbridge behavior occur in myocardium from hearts with dilated cardiomyopathy (reduced peak twitch tension) and ischemic cardiomyopathy (normal peak twitch tension) the alterations of the individual crossbridge cycle may not be the cause of reduced systolic performance observed in dilated cardiomyopathy. Furthermore, increased crossbridge force-time integral compared to control myocardium has recently been observed in volume-overload human myocardium (mitral regurgitation), as well as in nonfailing pressure-overload rabbit myocardium (1, 15). In the rabbit myocardium, alterations of the crossbridge cycle which go parallel with reduced myosin ATPase activity and reduced unloaded shortening velocity, have been considered to reflect adaptational changes of the contractile proteins to increased hemodynamic stress (1, 15).

Increased crossbridge force-time integral may result from increased crossbridge force, increased attachment-time, or both. If the increased crossbridge force-time integral is caused by prolonged attachment time, this would be consistent with a reduction in the rate constant of crossbridge detachment "g" (6).

Increased crossbridge force-time integral in the hypertrophied and failing myocardium may have two different consequences with respect to myocardial function. On the one hand, increased crossbridge force-time integral may be favorable from an energy economy point of view, since more force or force-time integral is produced per unit of high-energy phosphate hydrolyzed. On the other hand, reduced rate constant of crossbridge detachment may – by reducing shortening velocity – prevent the myocardium from developing high power output or may impair relaxation.

It is important to note that in pressure-overload rabbit myocardium and in other animal models of myocardial hypertrophy, changes in crossbridge behavior have been closely related to shifts in myosin heavy chain isoforms (17, 34). This is different in the human ventricular myocardium in which no significant isoform shifts have been observed (10, 21). Therefore, in the human myocardium, crossbridge force-time integral may be regulated by alterations in myosin light chains (16, 23) or thin filament regulatory systems (3). Alternatively, myosin heavy chain modifications, not detectable by previous techniques, may occur in the human heart (33).

Since reduced force development in myocardium from failing dilated cardiomyopathic hearts cannot be explained by alterations within the crossbridge cycle, the number of crossbridge interactions during the isometric twitch was investigated by analyzing tension-dependent heat. Tension-dependent heat which results from high-energy phosphate hydrolysis of crossbridges was reduced by 61% in failing myocardium from dilated cardiomyopathic hearts, and was not different from control in myocardium from ischemic cardiomyopathic hearts. Furthermore, including all types of myocardium, there was a close correlation between tension-dependent heat and peak twitch tension. This indicates that reduced peak twitch tension in failing dilated cardiomyopathy results from reduced number of crossbridge interactions during the isometric twitch. In contrast, in myocardium from ischemic cardiomyopathic hearts both force development and number of crossbridge interactions were not different from nonfailing control myocardium. It is tempting to speculate from this preliminary data that heart failure in dilated cardiomyopathy may primarily result from altered function at the level of the myocyte. In contrast, in ischemic cardiomyopathy heart failure may mainly result from the replacement of myocardium by scar tissue. In addition, ischemia or hypoxia, not present in the well oxygenized myocardial strip under in vitro conditions, may cause disturbed myocardial function under in vivo conditions in the ischemic cardiomyopathic heart.

The decreased number of crossbridge interactions observed in failing dilated cardiomyopathy may result from a reduced amount of myofibrillar proteins per cross-section or from disturbed excitation-contraction coupling processes. The former possibility seems to be unlikely, since we were able to show that myosin content was reduced by only about 20% in myocardium dissected from failing dilated cardiomyopathic hearts as compared to myocardium dissected from nonfailing hearts (13).

To evaluate excitation-contraction coupling, tension-independent heat measurements which reflect high-energy phosphate hydrolysis of excitation-contraction coupling processes (see methods) were performed. As shown in Fig. 3 tension-independent heat was significantly reduced by 69% in failing compared to nonfailing myocardium, indicating a considerable decrease in the amount of calcium cycling during the isometric twitch in failing dilated cardiomyopathy. This finding suggests that reduced isometric force development in dilated cardiomyopathy results from a decreased number of crossbridge interactions which, in turn, is due to decreased activation, i.e., decreased number of calcium ions released. From the present data it cannot be differentiated whether the decreased number of crossbridge interactions during the isometric twitch is caused by a decreased recruitment of crossbridges or by reduced cycling rate of the individual crossbridge. According to Brenner (6), calcium activation increases the rate constant "f" promoting the transition from the weakly bound crossbridge state into the strongly bound, force-generating step of the crossbridge. Thus, reduced calcium release in dilated cardiomyopathy may reduce crossbridge cycling rate by prolongation of the weakly bound period (off-time) of the individual crossbridge cycle.

It is important to note that the present finding of reduced isometric force development in myocardium from dilated cardiomyopathy contrasts to previous studies suggesting similar isometric tension development in normal and myopathic myocardium (5, 9, 11, 30). However, the apparent discrepancy can most likely be explained by different experimental conditions. Lower stimulation frequency used in other studies seems to be of major relevance since the force-frequency relationship is altered in failing dilated cardiomyopathy (8, 28–31). At low frequencies (30 beats per minute) peak twitch tension is similar in failing and nonfailing human myocardium. Increasing frequency from 30 to 60 beats per minute increases peak twitch tension in nonfailing and decreases peak twitch tension in failing myocardium from dilated cardiomyopathic hearts. This results in significant lower tension values in failing compared to nonfailing myocardium at 60 beats per minute. In other words, differences in contractile force between failing dilated cardiomyopathy and nonfailing myocardium are present at physiological frequencies, but absent at low frequencies.

Likewise, previous aequorin studies have been interpreted to indicate that there is no reduction in the amount of activation in dilated cardiomyopathy (24). Again, this discrepancy from the present interpretation seems to be related to differences in experimental conditions. While the present study was performed on left-ventricular strip preparations at 37 °C with a stimulation rate of 60 beats per minute, most aequorin studies were performed predominantly on right-ventricular trabeculae at 30 °C with a stimulation rate of 20 beats per minute, at which twitch tension of failing human myocardium is reported to be not significantly different from normal (8, 24). Furthermore, in accordance with our findings, recent measurements using the fluorescent Ca^{++}-indicator fura-2 showed that peak Ca^{++} transients were significantly reduced in isolated myocytes from failing human hearts (4).

The reduced amount of calcium for activation of contractile proteins could result from altered transsarcolemmal calcium fluxes to trigger calcium release from the sarcoplasmic reticulum (SR) or from disturbed SR function. Whereas there is no evidence of altered transsarcolemmal calcium fluxes in heart failure (4, 32), there is considerable evidence of disturbed SR calcium handling. Mercardier et al. reported that the messenger RNA of SR calcium ATPase is reduced in the failing human heart (22). This may indicate that the number of SR calcium pumps is reduced and would be consistent with findings of Limas et al. who demonstrated that SR calcium uptake is depressed in failing human myocardium (20); the latter, however, has not been observed by Movsesian et al. (25). Reduced calcium transport capacity of the SR may cause SR calcium depletion and thus may be the cause of reduced calcium release. This hypothesis may also explain why the force-frequency relation is altered in the failing myocardium from dilated cardiomyopathic hearts. An increase in frequency reduces the time available for SR calcium reuptake and, thus, in presence of depressed SR calcium transport capacity, may reduce the availability of calcium to be released during the following twitch. Low cytosolic cyclic AMP concentrations may be additionally involved in disturbed sarcoplasmic reticulum calcium handling (7).

In summary, disturbed calcium cycling with reduced activation of contractile proteins provides a potential subcellular explanation for reduced systolic performance in end-stage failing dilated cardiomyopathy. In ischemic cardiomyopathy mechanisms different from those observed in dilated cardiomyopathy seem to be involved in the development of heart failure.

References

1. Alpert NR, Mulieri LA (1982) Increased myothermal economy of isometric force development in compensated cardiac hypertrophy induced by pulmonary artery constriction in the rabbit. Circ Res 50:491–500
2. Alpert NR, Blanchard EM, Mulieri LA (1989) Tension-independent heat in rabbit papillary muscle. J Physiol (Lond) 414:433–453
3. Anderson PAW, Malouf NN, Oakeley AE, Pagani ED, Allen PD (1992) Troponin T isoform expression in the normal and failing human left ventricle: a correlation with myofibrillar ATPase activity. In: Hasenfuss G, Holubarsch Ch, Just H, Alpert NR (eds): Cellular and molecular alterations in the failing human heart. Steinkopff Verlag, Darmstadt
4. Beuckelmann DJ, Erdmann E (1992) Calcium currents and intracellular calcium transients in single ventricular myocytes isolated from terminally failing human myocardium. In:

Hasenfuss G, Holubarsch Ch, Just H, Alpert NR (eds) Cellular and molecular alterations in the failing human heart. Steinkopff Verlag, Darmstadt

5. Böhm M, Beuckelmann D, Brown L, Feiler L, Lorenz B, Näbauer M, Kemkes B, Erdmann E (1988) Reduction of beta-adrenoceptor density and evaluation of positive inotropic responses in isolated, diseased human myocardium. Eur Heart J 9:844–852

6. Brenner B (1988) Effect of Ca^{2+} on crossbridge turnover kinetics in skinned single rabbit psoas fibers: Implications for regulation of muscle contraction. Proc Natl Acad Sci 85:3265–3269

7. Feldman MD, Copelas L, Gwathmey JK, Phillips P, Warren SE, Schoen FJ, Grossman W, Morgan JP (1987) Deficient production of cyclic AMP: pharmacologic evidence of an important cause of contractile dysfunction in patients with end-stage heart failure. Circulation 75:331–339

8. Feldman MD, Gwathmey JK, Phillips P, Schoen F, Morgan JP (1988) Reversal of the force-frequency relationship in working myocardium from patients with end-stage heart failure. J Applied Cardiol 3:273–283

9. Ginsburg R, Bristow MR, Billingham ME, Stinson EB, Schroeder JS, Harrison DC (1983) Study of the normal and failing isolated human heart: Decreased response of failing heart to isoproterenol. Am Heart J 106:535–540

10. Gorza L, Mercadier JJ, Schwartz K, Thornell LE, Sartore S, Schiaffino S (1984) Myosin types in the human heart. Circ Res 54:694–702

11. Gwathmey JK, Copelas L, MacKinnon R, Schoen FJ, Feldman MD, Grossman W, Morgan JP (1987) Abnormal intracellular calcium handling in myocardium from patients with end-stage heart failure. Circ Res 61: 70–76

12. Gwathmey JK, Slawsky MT, Hajjar RJ, Briggs GM, Morgan JP (1990) Role of intracellular calcium handling in force-interval relationship of human ventricular myocardium. J Clin Invest 85: 1599–1613

13. Hasenfuss G, Mulieri LA, Leavitt BJ, Allen PD, Haeberle JR, Alpert NR. Alteration of contractile function and excitation-contraction coupling in dilated cardiomyopathy. Circ Res (in press)

14. Hasenfuss G, Holubarsch Ch, Blanchard EM, Mulieri LA, Alpert NR, Just H (1989) Influence of isoproterenol on myocardial energetics. Experimental and clinical investigations. Basic Res Cardiol 84 (suppl 1):147–155

15. Hasenfuss G, Mulieri LA, Blanchard EM, Holubarsch Ch, Leavitt BJ, Ittleman F, Alpert NR (1991) Energetics of isometric force development in control and volume-overload human myocardium: comparison with animal species. Circ Res 68:836–846

16. Hirzel HO, Tuchschmid CR, Schneider J, Krayenbühl HP, Schaub MC (1985) Relationship between myosin isoenzyme composition, hemodynamics and myocardial structure in various forms of human cardiac hypertrophy. Circ Res 57: 729–740

17. Holubarsch Ch, Litten RZ, Mulieri LA, Alpert NR (1985) Energetic changes of myocardium as an adaptation to chronic hemodynamic overload and thyroid gland activity. Basic Res Cardiol 80:582–593

18. Hood WP, Rackley CE, Rolett EL (1968) Wall Stress in the normal and hypertrophied human left ventricle. Am J Cardiol 22:550–558

19. Huxley AF (1957) Muscle structure and theories of contraction. Prog Biophys Biophys Chem 7:255–318

20. Limas CJ, Olivari M, Goldenberg IF, Levine TB, Bendit DG, Simon A (1987) Calcium uptake by sarcoplasmic reticulum in human dilated cardiomyopathy. Cardiovasc Res 21:601–605

21. Mercadier JJ, Bouveret P, Gorza L, Schiaffino S, Clark WA, Zak R, Swynghedauw B, Schwartz K (1983) Myosin isoenzymes in normal and hypertrophied human ventricular myocardium. Circ Res 53:52–62

22. Mercardier JJ, Lompré AM, Duc P, Boheler KR, Fraysse JB, Wisnewski C, Allen PD,

Komajda M, Schwartz K (1990). Altered sarcoplasmic reticulum calcium-ATPase gene expression in the human ventricle during end-stage heart failure. J Clin Invest 85:305–309
23. Morano I (1992) Effects of different expression and posttranslational modifications of myosin light chains on contractility of skinned human cardiac fibers. In: Hasenfuss G, Holubarsch Ch, Just H, Alpert NR (eds) Cellular and Molecular alterations in the failing human heart. Steinkopff Verlag, Darmstadt
24. Morgan JP (1991) Abnormal intracellular modulation of calcium as a major cause of cardiac contractile dysfunction. N Engl J Med 325:625–632
25. Movsesian MA, Bristow MR, Krall J (1989) Calcium uptake by cardiac sarcoplasmic reticulum from patients with idiopathic dilated cardiomyopathy. Circ Res 65:1141–1144
26. Mulieri LA, Luhr G, Trefry J, Alpert NR (1977) Metal-film thermopiles for use with rabbit right ventricular papillary muscles. Am J Physiol 233:C146–C156
27. Mulieri LA, Hasenfuss G, Ittleman F, Blanchard EM, Alpert NR (1989) Protection of human left ventricular myocardium from cutting injury with 2,3-butanedione monoxime. Circ Res 65:1441–1444
28. Mulieri LA, Hasenfuss G, Leavitt B, Ittleman F, Allen PD, Alpert NR (1990) Altered tension and force-frequency relation in failing human myocardium. J Mol Cell Cardiol 22(1):32
29. Mulieri LA, Leavitt BJ, Hasenfuss G, Allen PD, Alpert NR (1992) Contraction frequency dependence of twitch and diastolic tension in human dilated cardiomyopathy. In: Hasenfuss G, Holubarsch Ch, Just H, Alpert NR (eds) Cellular and molecular alterations in the failing human heart. Steinkopff Verlag, Darmstadt
30. Phillips PJ, Gwathmey JK, Feldman MD, Schoen FJ, Grossman W, Morgan JP (1990) Post-extrasystolic potentiation and the force-frequency relationship: Differential augmentation of myocardial contractility in working myocardium from patients with end-stage heart failure. J Mol Cell Cardiol 22: 99–110
31. Pieske B, Hasenfuss G, Holubarsch Ch, Just H (1992) Alterations of the force-frequency relationship in the failing human heart depend on the underlying cardiac disease. In: Hasenfuss G, Holubarsch Ch, Just H, Alpert NR (eds) Cellular and molecular alterations in the failing human heart. Steinkopff Verlag, Darmstadt
32. Rasmussen PR, Minobe W, Bristow MR (1990) Calcium antagonist binding sites in failing and nonfailing human ventricular myocardium. Biochemical Pharmacol 39:691–696
33. Rupp H, Jacob R (1992) Structural and functional diversity of human ventricular myosin. In: Hasenfuss G, Holubarsch Ch, Just H, Alpert NR (eds) Cellular and molecular alterations in the failing human heart. Steinkopff Verlag, Darmstadt
34. Swynghedauw B (1986) Developmental and functional adaptation of contractile proteins in cardiac and skeletal muscle. Physiol Rev 66:710–771

Author's address:
Dr. G. Hasenfuss
Dept. of Medicine, Cardiology
University of Freiburg
Hugstetter Strasse 55
7800 Freiburg, FRG

Troponin T isoform expression in the normal and failing human left ventricle: a correlation with myofibrillar ATPase activity

P.A.W. Anderson[1,2], N.N. Malouf[3], A.E. Oakeley[1], E.D. Pagani[4], P.D. Allen

Departments of Pediatrics[1] and Cell Biology[2], Duke University,
Department of Pathology[3], University of North Carolina at Chapel Hill,
Department of Cardiovascular Pharmacology[4], Sterling Research Group,
Rensselaer, New York,
Department of Anesthesiology, Brigham and Women's Hospital, Boston,
Massachusetts, USA

Summary: The expression of troponin T, a thin filament regulatory protein, was examined in normal and failing left ventricles. The samples were obtained from the hearts of patients with severe heart failure who were undergoing cardiac transplantation, and from normal adult hearts that could not be used for transplantation. Western blots of the myofibrillar proteins demonstrated two isoforms, troponin T 1 (TnT_1) and troponin T 2 (TnT_2). TnT_2 is expressed at significantly higher levels in failing hearts (p < 0.004). Western blots of two-dimension SDS-PAGE gels resolved two dominant spots of TnT_1 and of TnT_2 and several minor troponin T species. Alkaline phosphatase treatment markedly decreased the sizes of the two acidic spots while increasing the two more basic spots by a comparable amount. Myofibrillar ATPase activity had an inverse and negative linear relationship (r = 0.7, p < 0.02) with the myofibrillar percentage of total troponin T comprised of TnT_2. In that heart failure in these transplant patients had multiple bases, we propose that rather than a cause of heart failure, the disease-associated changes in troponin T isoform expression are an adaptation to abnormal myocardial function.

Key words: Heart failure; myocardium; contractile proteins; monoclonal antibody; adult heart; thin filament regulatory proteins

Introduction

Abnormalities of many aspects of myocardial structure and function have been found in the failing heart. Myofibrillar ATPase activity, a measure of a process that is basic to myocardial contraction, was first characterized as being depressed in the failing heart by Alpert and Gordon (1). A recent study (13) found that in failing human myocardium myofibrillar ATPase activity is lower than that of the normal human, while in the normal and failing left ventricle myosin ATPase activity does not differ. The latter finding is consistent with beta myosin heavy chain being the dominant isozyme expressed in normal and failing human myocardium (11,14). The presence of changes in myofibrillar ATPase activity and the absence of changes in myosin ATPase activities suggested to us that the expression of the thin filament proteins, which control actin-myosin interaction, is altered by cardiac disease processes.

We selected as the candidate protein, troponin T, which binds troponin I and C to tropomyosin and is required for the calcium sensitivity of myofibrillar ATPase

---

*This work was supported in part by the following grants from the NIH: HL 42250, HL 20749, HL 37358.

activity. This selection was based in part on the presence in rabbit fast skeletal muscle of a relationship between troponin T isoforms and the sensitivity of the myofilaments to calcium (16). In addition, in cardiac muscle, multiple troponin T isoforms have been found in other species (2,3), while in a reconstituted *in vitro* system myofibrillar ATPase activity is affected by cardiac troponin T isoforms.

We have found that in adult human myocardium one dominant isoform, troponin T1 (TnT_1), is expressed and that in the failing left ventricle the expression of a second isoform, troponin T2 (TnT_2) is upregulated. The relative amount of TnT_2 had a negative inverse relationship with myofibrillar ATPase activity. These data suggest that the heart failure-associated changes in troponin T expression represent an adaptive response to a pathophysiological state.

Methods

Tissue: Troponin T isoform expression was examined in tissue from failing and normal human left ventricle. Tissue was obtained at the time of orthotopic cardiac transplantation from patients with severe heart failure (n = 10) and at the time of organ harvest from patients whose hearts were unacceptable for transplantation (n = 12). The tissues from all patients in both the severe heart failure and the non-heart failure groups were obtained using protocols approved by the Brigham and Women's Hospital institutional committee for the protection of human subjects. Three samples in the non-failure group were obtained from trauma patients harvested 2-5 h after their sudden death (a gift of Dr. Valdur Saks, Moscow, USSR). Three others were from organ donors who were on life support for less than 2 h (a gift of Dr. Frank Sreter, Budapest, Hungary). The remaining six non-failure samples were obtained from patients (Brigham and Womens's Hospital, Boston, USA) after variable periods of life support.

Left-ventricular free wall was cut into small transmural pieces and frozen in liquid nitrogen within 5 min of excision of the heart. Specimens were transported on dry ice and kept in liquid nitrogen until the specimens were prepared for electrophoresis and measurement of myofibrillar ATPase activity.

Materials: The reagents were the same as described previously (3). Monoclonal antibody 13-11, raised against a rabbit cardiac troponin T, was obtained from a hybridoma clone (10), purified on Fractogel CM, and used at 50 micrograms/ml. A monoclonal antibody (JLT-12), that recognizes an epitope common to mammalian cardiac and fast skeletal muscle troponin T, was purchased from Amersham.

Gel electrophoresis and Western blots: The samples were prepared for SDS-PAGE and for two-dimension electrophoresis using previously published methods (2,6,12). SDS-PAGE was performed using the method of Laemmli (9), with these modifications: 30% acrylamide and 1.1% bis-acrylamide stock solutions were used in 7.5% running gels and a 3.3% stacking gel (2,3). Silver staining of the gels was performed following the method of Schachat et al. (15). Western blots were performed (3) using an alkaline phosphatase labeled secondary antibody and NBT and BCIP (Promega) as the color reagent. The Western blots were scanned with an LKB laser densitometric scanner, and the areas under the troponin T densitometric waveform integrated. The amount of TnT_2 (see Results) was described as a percentage of the total amount of troponin T.

The effects of proteolysis were tested for by thawing myocardial samples, leaving them at room temperature in the absence of protease inhibitors, placing portions of the preparation in sample buffer after one-half, one, two, and four hours, and using Western blots to examine the relative amounts of the isoforms. Dephosphorylation of myofibrillar preparations was performed following the method of Holroyde et al. (8) using bacterial alkaline phosphatase (Gibco BRL).

The maximal myofibrillar ATPase activity, using pCa5, was measured previously by Pagani et al. (13) in left-ventricular samples from four of the organ donors and six of the heart failure patients.

Statistics: The percentage of total left-ventricular troponin T comprised of TnT_2 in the patient groups were compared by one way analysis of variance and Kruskal-Wallis analysis. Linear regression analysis of the relationship between percentage of total troponin T comprised of TnT_2 and myofibrillar ATPase activity of individual hearts was performed. A $p < 0.05$ was considered significant.

Results

The Heart Failure and Organ Donor Population: The control group (n = 12) contained 10 males and two females. Their ages at the time of organ harvest ranged from 18 to 58 years (29 $\pm$ 11, mean $\pm$ SD). Patients in the control group who were cared for at Brigham and Women's Hospital were on a life support system for 24 to 72 h, while those from Hungary were on a life support system for less than 2 hs. The heart failure group (n = 10) contained nine males and one female whose ages at the time of orthotopic transplantation ranged from 14 to 54 years (38 $\pm$ 16). The causes of heart failure were multiple. They included coronary artery disease, two patients (pts.); idiopathic dilated cardiomyopathy, six pts.; idiopathic hypertrophic subaortic stenosis, one pt.; and ventricular septal defect, one pt.

One-dimensional SDS-PAGE demonstrated that human myocardium contains proteins with electrophoretic mobilities similar to those of cardiac troponin T isoforms of other species (Fig. 1). Western blots of control myofibrillar proteins, probed with the cardiac specific MAb 13-11, demonstrated that one of these proteins is a cardiac troponin T isoform, TnT_1 (Fig. 1), while Western blots of failing heart myofibrillar preparations contained two polypeptides recognized by MAb 13-11. One migrated with TnT_1, while the other protein, TnT_2, had a faster electrophoretic mobility. Densitometric scans of Western blots of each control heart preparation demonstrated, in general, either no TnT_2 or a very small amount of this polypeptide (Fig. 2). The heart-failure group preparations had a greater percentage of TnT_2 relative to total troponin T than those from the normal adult hearts, $p < 0.001$ (see Fig. 2). Interestingly, the hearts of patients from Hungary and Russia had a lower percentage of TnT_2 than did the hearts of organ donors from the United States, who were maintained for longer times on life support system (Fig. 2). The control patient with the highest percentage of TnT_2, 9.1%, was maintained on life support for the longest period of time, 72 h. However, the percentage of TnT_2 did not differ significantly between the six patients from Russia and Hungary and the six patients from the United States.

SDS-PAGE and Western blots of myocardium left at room temperature for up to 4 h in the absence of protease inhibitors demonstrated TnT_2 did not arise from

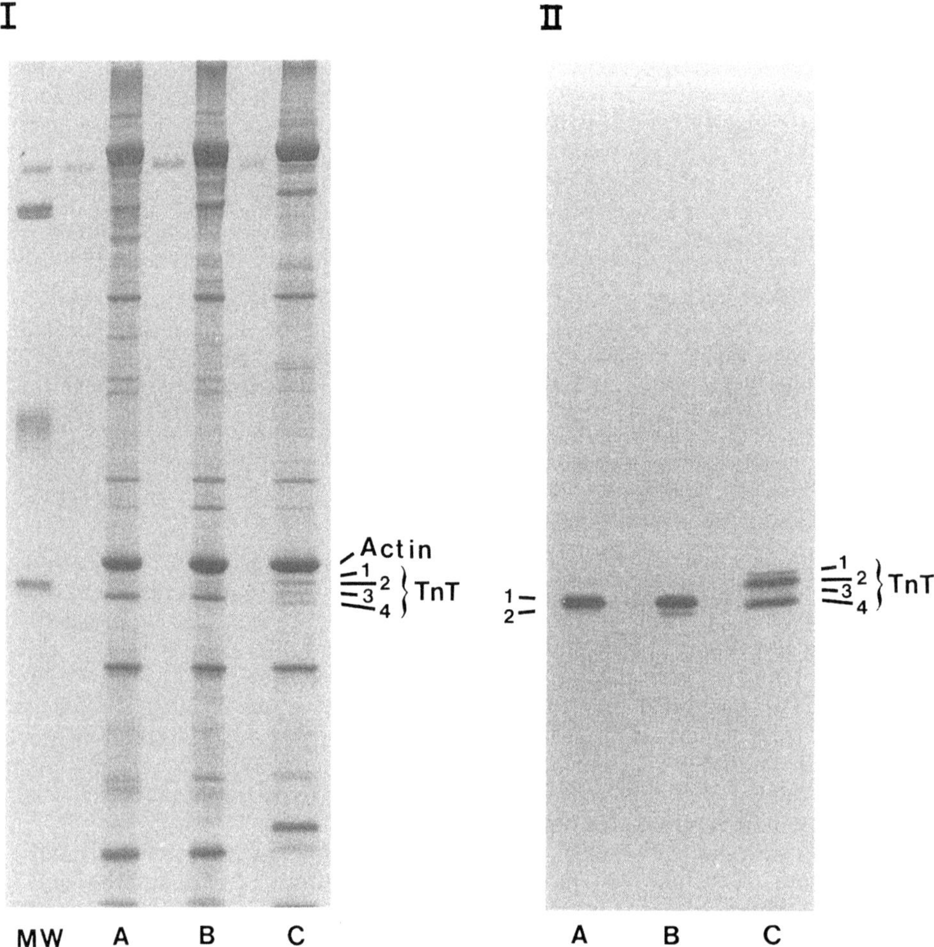

Fig. 1. Panel I: Silver stained one-dimensional 7.5% polyacrylamide gel loaded with preparations from (A) normal human left ventricle, (B) failing human left ventricle, and (C) two-day postnatal rabbit heart. The region in the gel to which cardiac troponin T (TnT) isoforms migrate is bracketed. The four rabbit cardiac troponin T isoforms are numbered based on their electrophoretic mobilities. The molecular weight standards (MW) are 165 000; 155 000; 68 000; and 39 000 Daltons. Panel II: Western blot of proteins transblotted from the same one-dimensional 7.5% polyacrylamide gel loaded with the same myofibrillar preparations as the gel in Panel I. Monoclonal antibody 13-11, which was used as the primary antibody, was raised against the rabbit cardiac troponin T isoform numbered 4. The two human cardiac troponin T isoforms are numbered, based on their electrophoretic mobilities.

proteolysis (see Fig. 3). The relative amount of TnT_2 did not increase in normal myocardium left at room temperature, remaining the same as found in either frozen myocardium placed directly into sample buffer or myocardium subjected to the protocols to make myofibrillar preparations. Similarly, myocardium from failing hearts left at room temperature demonstrated no change in the relative amounts of

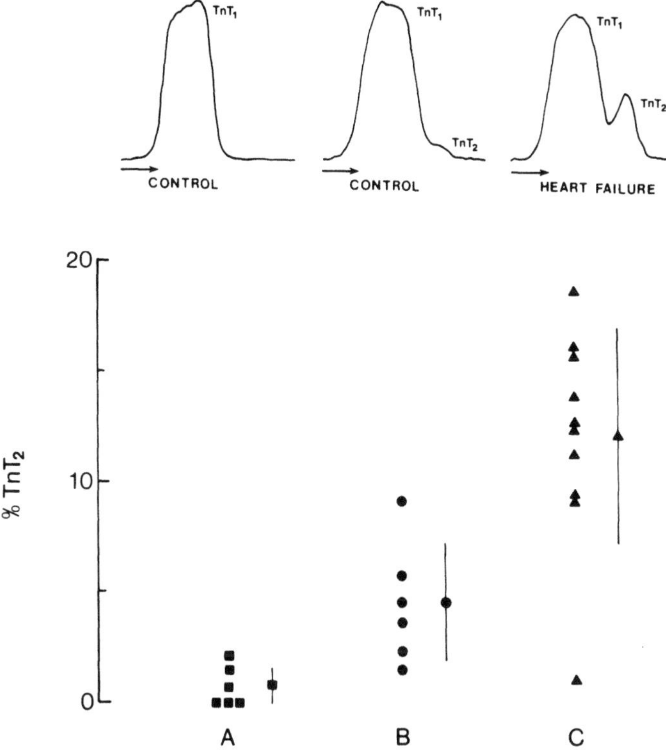

Fig. 2. Upper panel: Examples of densitometric waveforms obtained from Western blots of myofibrillar preparations from control and failing myocardium. The waveform on the left is typical of patients who were maintained on life support systems for less than 2 h. The arrow represents the direction of electrophoresis. Lower panel: The percentage of the total troponin T comprised of troponin T_2 (TnT_2) for each of the left ventricles and the mean value for each group $\pm$ one standard deviation. The values were lowest for those patients maintained on life support for less than 2 h or who died traumatically (A, squares). The highest values were for patients with severe heart failure (C, triangles) while the intermediate group (B, circles) was made up of the patients with no heart disease who were maintained on life-support systems for 24-72 h. Groups A and B are both significantly different from group C, while groups A and B do not differ significantly from each other.

TnT_1 and TnT_2 (Fig. 3). Troponin T was sensitive to proteolysis in that proteins with electrophoretic mobilities faster than TnT_1 and TnT_2 were identified in Western blots of myocardium kept at room temperature for less than 1 h. The longer that myocardium was kept at room temperature, the greater was the amount of these proteolytic fragments. The failure of TnT_2 to increase in amount with the appearance of the proteolytic fragments confirms our conclusion that TnT_2 is not a proteolytic fragment of TnT_1.

MAb JLT-12 recognized in normal and failing heart myofibrillar proteins the same two polypeptides, TnT_1 and TnT_2, recognised by MAb 13-11. JLT-12 did not

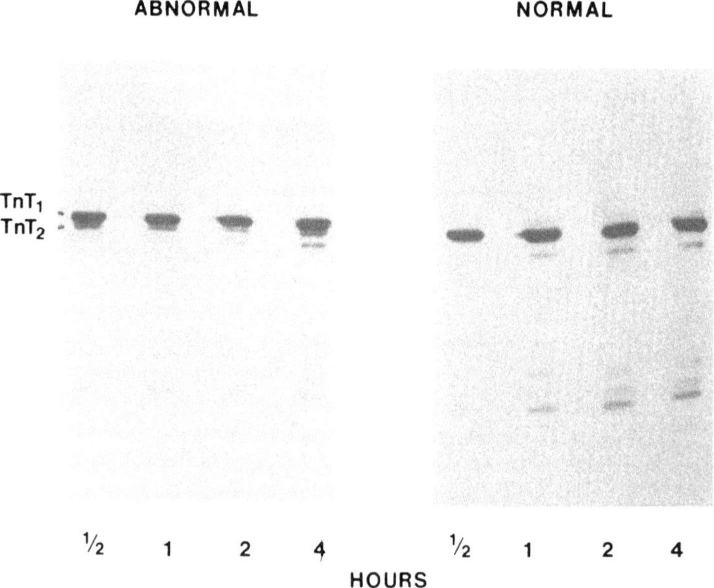

Fig. 3. Western blots of myocardium from a failing (abnormal) and from a normal left ventricle that were allowed to stay at room temperature for .5 to 4h. MAb 13-11 was used as the primary antibody. No increase in TnT_2 occurred with increasing time in conditions that enhance proteolysis. Protein degradation did occur as is evidenced by the presence of proteins with faster electrophoretic mobilities than those of TnT_1 and TnT_2 with incubations at room temperature for periods as short as .5 and 1 h. Note the increase in the amounts of these proteins with longer incubations at room temperature.

recognize in these preparations any proteins with electrophoretic mobilities similar to those of fast skeletal muscle troponin T isoforms.

The troponin T isoforms, identified in the normal and the failing heart, were characterized more thoroughly using two-dimensional gel electrophoresis (Fig. 4). Western blots of proteins transblotted from two-dimensional gels demonstrated that TnT_1 was separated into three spots while TnT_2 was resolved into two spots (Figs. 4b and 5). MAb 13-11 also recognized its epitope in minor troponin T isoforms that were not identified in the silver-stained gels and that had electrophoretic mobilities and pIs that differed from those of TnT_1 and TnT_2 (Fig. 4). Myocardial preparations that were thawed and placed directly in IEF buffer or those that were processed to obtain myofibrillar preparations demonstrated the same pattern of spots. Phosphatase treatment markedly decreased the two dominant acidic spots of TnT_1 and TnT_2 while the more basic spots were comparably increased (Fig. 5).

The myofibrillar ATPase activities of heart failure patients (n = 6) and of organ donors (n = 4) were significantly different (p < 0.018), the ATPase activities of the normal myocardium being larger (Fig. 6). A significant, linear, and inverse relationship was found between ATPase activity and the percentage of TnT_2, p < 0.02, r = 0.7 (Fig. 6).

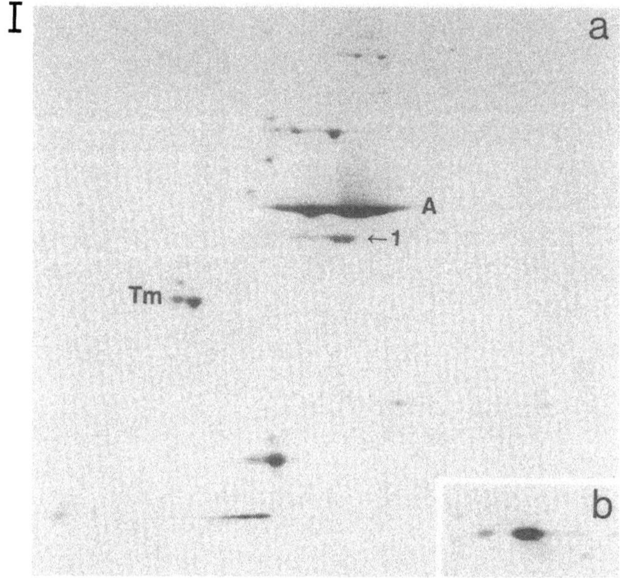
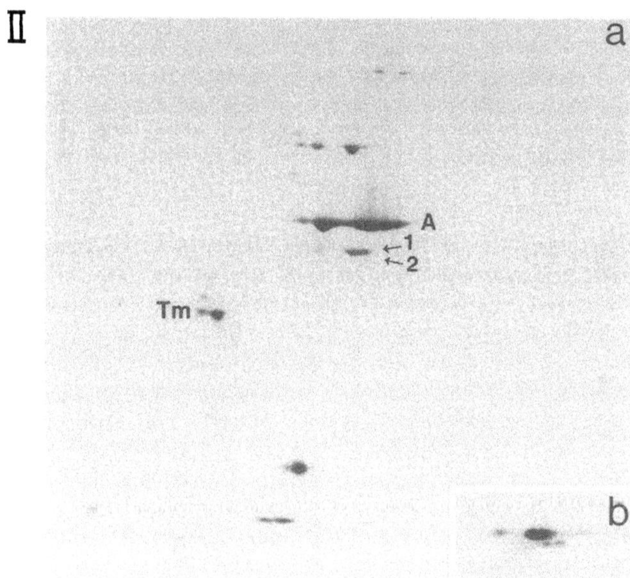

Fig. 4. Silver stained two-dimensional 7.5% polyacrylamide gels of myofibrillar preparations from a normal (Ia) and a failing (IIa) left ventricle. The insets (b) illustrate Western blots of the proteins resolved on two-dimensional gels and probed with MAb 13-11 (see also Fig. 5). The troponin T isoforms TnT_1 and TnT_2 are labeled 1 and 2, respectively. A is actin, Tm is tropomyosin.

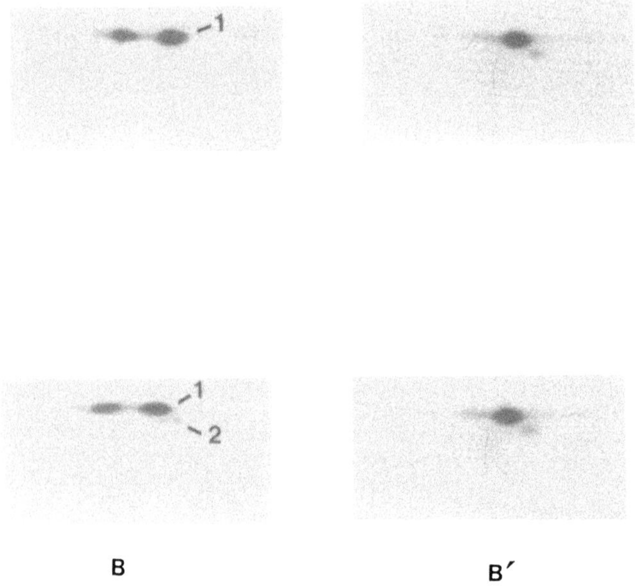

Fig. 5. Western blots of two-dimensional gels, probed with MAb 13-11, illustrate the differences between the relative amounts of the troponin T isoforms of a normal left ventricle (A) and those of a failing left ventricle (B). In A and B TnT_1 and TnT_2 are labeled as 1 and 2, respectively. A' and B' illustrate Western blots of preparations of the same proteins used in A and B following dephosphorylation. The two major spots of TnT_1 and TnT_2 are collapsed into single spots by the phosphatase treatment, while one of the minor isoforms of TnT_1 is not affected.

Discussion

Previous studies have demonstrated that multiple cardiac troponin T isoforms (2, 3, 7, 17) are expressed in several mammalian species in a developmentally and regionally regulated manner and have suggested that these isoforms may be functionally important (19). Those studies coupled with the results of Pagani et al. (13) led us to investigate in man the effects of disease on the cardiac expression of troponin T, a thin filament regulatory protein. We found that in man disease alters myocardial troponin T expression, and that myofibrillar ATPase activity and troponin T isoforms are correlated.

We hypothesize that the two human cardiac troponin T isoforms, TnT_1 and TnT_2, arise from alternative RNA splicing and not from either post-translational modifications or proteolysis. Alternative splicing of a primary transcript is the basis of troponin T isoform expression in other species, e.g., in chicken heart (7) and rat fast skeletal muscle (4). Isolation and sequencing of genomic and cDNA will help resolve whether this is the basis of TnT_1 and TnT_2.

Our finding that TnT_1 and TnT_2 have different electophoretic mobilities is consistent with these isoforms being products of different mRNA: post-translational modifications of fast skeletal muscle troponin T result in proteins with the same electrophoretic mobilities and different isoelectric points (5). The dephosphorylation

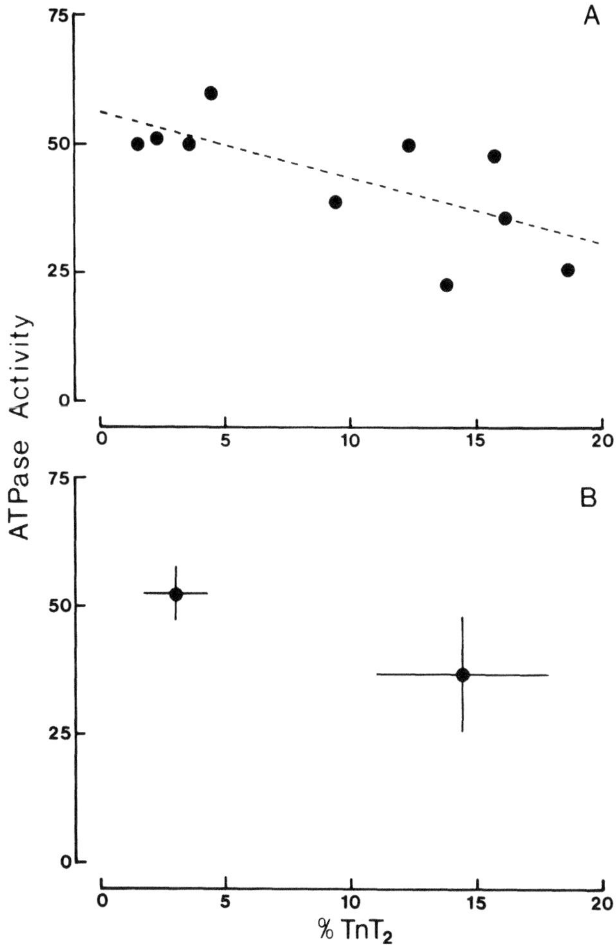

Fig. 6 Upper panel: The myofibrillar ATPase activities of individual normal and failing left ventricles are illustrated as a function of the percentage of total troponin T that is comprised of the isoform, TnT$_2$. A linear inverse relationship (dashed line) was found, $r = 0.7$, $p < 0.02$. ATPase activity is in nmole Pi/mg protein/min. Lower panel: The myofibrillar ATPase activities for the normal (left) and heart failure (right) patient groups as a function of the percentage of troponin T comprised of TnT$_2$ are illustrated as means $\pm$ one standard deviation; the means are significantly different. ATPase activity is the same as in A.

experiments support this interpretation: the large acidic spots of TnT$_1$ and TnT$_2$ were markedly decreased while the basic spots of TnT$_1$ and TnT$_2$ were increased. Whether the minor troponin T isoforms that are unaffected by the dephosphorylation treatment represent products of alternative splicing or post-translational modifications remains to be resolved.

Proteolysis is not the basis of normal and failing left ventricle having different relative amounts of TnT$_2$. The spots on the two-dimensional immunoblots are discrete and have a narrow range of molecular weights, suggesting that proteolysis is not the cause. Our demonstration that TnT_2 does not increase in amount in myocardium that is undergoing proteolysis further supports our conclusion that proteolysis is not the origin of TnT_2.

In that skeletal muscle contractile protein genes are expressed transiently in the stressed adult heart (18), the possibility exists that TnT_2 is encoded by a skeletal muscle gene. MAb 13-11, which was raised against a rabbit cardiac troponin T isoform, does not recognize its determinant in skeletal muscle, but recognizes both TnT_1 and TnT_2. These findings suggest that neither TnT_1 nor TnT_2 are skeletal troponin T isoforms.

We propose that the changes in human myocardial troponin T isoform expression are an adaptive responsive to heart failure. The correlation that we found between the troponin T isoforms and myofibrillar ATPase activity is consistent with the findings of Tobacman and Lee (19). Future studies, e.g., biochemical or biophysical assays using myofilaments in which the expression of individual troponin T isoforms is enhanced or suppressed while the expression of other contractile proteins remains unchanged are needed to test critically whether the fall in myofibrillar ATPase activity with heart failure follows from the change in troponin T isoform expression. Although this lowered ATPase activity may decrease the energy requirements for performing work, the biochemical and biophysical advantages that TnT_2 conveys to the function of the failing heart remain to be elucidated.

Acknowledgments. The authors thank Dr. Frederick Schoen, Division of Cardiac Pathology, Dr. Verdi Disesa and Dr. Lawrence Cohen, Division of Cardiothoracic Surgery, and Dr. Rosemarie Maddi, Dr. John Fox, and Mrs. Santawana Muckherjee, Department of Anesthesia, Brigham and Women's Hospital, and the New England Organ Bank, Boston Massachusetts for their help in procuring the tissue used in this study.

References

1. Alpert NR, Gordon MS (1962) Myofibrillar adenosine triphosphatase activity in congestive heart failure. Am J Physiol 202:940–946
2. Anderson PAW, Moore GE, Nassar R (1988) Developmental changes in the expression of rabbit left ventricular troponin T. Circ Res 63:742–747
3. Anderson PAW, Oakeley AE (1989) Immunological identification of five troponin T isoforms reveals an elaborate maturational troponin T profile in rabbit myocardium. Circ Res 65:1087–1093
4. Breitbart RE, Nguyen HT, Medford RM, Destree AT, Mahdavi V, Nadal-Ginard B (1985) Intricate combinatorial patterns of exon splicing generate multiple regulated troponin T isoforms from a single gene. Cell 41:67–82
5. Briggs MM, Lin J J-C, Schachat FH (1987) The extent of amino-terminal heterogeneity in rabbit fast skeletal muscle troponin T. J Muscle Res Cell Motil 8:1–12
6. Bronson DD, Schachat FH (1982) Heterogeneity of contractile proteins: Differences in tropomyosin in fast, mixed, and slow skeletal muscles of the rabbit. J Biol Chem 257(7):3937–3944
7. Cooper TA, Ordahl CP (1985) A single cardiac troponin T gene generates embryonic and adult isoforms via developmentally regulated alternate splicing. J Biol Chem 260:11140–11148
8. Holroyde MJ, Howe E, Solaro RJ (1979) Modification of calcium requirements for activation of cardiac myofibrillar ATPase by cyclic AMP dependent phosphorylation. Biochim Biophys Acta 586:63–69

9. Laemmli UK (1970) Cleavage of structural proteins during the assembly of the head of bacteriophage T_4. Nature 227:680–685
10. Malouf NN, Taylor S, Gillespie GY, Bynum JM, Wilson PE, Meissner G (1986) Monoclonal antibody specific for the T-tubule of skeletal muscle. J Histochem Cytochem 34:347–355
11. Mercadier J-J, Bouveret P, Gorza L, Schiaffino S, Clark WA, Zak R, Swynghedauw B, Schwartz K (1983) Myosin isoenzymes in normal and hypertrophied human ventricular myocardium. Circ Res 53:52–62
12. O'Farrell PH (1975) High resolution two-dimensional electrophoresis of proteins. J Biol Chem 250:4007–4021
13. Pagani ED, Alousi AA, Grant AM, Older TM, Dziuban Jr SW, Allen PD (1988) Changes in myofibrillar content and Mg-ATPase activity in ventricular tissues from patients with heart failure caused by coronary artery disease, cardiomyopathy, or mitral valve insufficiency. Circ Res 63:380–385
14. Schier JJ, Adelstein RS (1982) Structural and enzymatic comparison of human cardiac muscle myosins isolated from infants, adults, and patients with hypertrophic cardiomyopathy. J Clin Invest 69:816–825
15. Schachat FH, Bronson DD, McDonald OB (1985) Heterogeneity of contractile proteins: A continuum of troponin-tropomyosin expression in mammalian skeletal muscle. J Biol Chem 260:1108–1113
16. Schachat FH, Diamond MS, Brandt PW (1987) Effect of different troponin T-tropomyosin combinations on thin filament activation. J Mol Biol 198:551–554
17. Spach MS, Dolber PC, Anderson PAW (1989) Multiple regional differences in cellular properties that regulate repolarization and contraction in the right atrium of adult and newborn dogs. Circ Res 65:1594–1611
18. Swynghedauw B (1986) Developmental and functional adaptation of contractile proteins in cardiac and skeletal muscles. Physiol Rev 66:710–771
19. Tobacman LS, Lee R (1987) Isolation and functional comparison of bovine cardiac troponin T isoforms. J Biol Chem 262:4059–4064

Author's address:
Page A. W. Anderson, M.D.
P.O. Box 3218
Duke University Medical Center
Durham NC, 27710, USA

Effects of different expression and posttranslational modifications of myosin light chains on contractility of skinned human cardiac fibers

I. Morano

II. Physiologisches Institut der Universität Heidelberg, FRG

Summary: In the human ventricle two isoforms of the phosphorylatable myosin light chain (MPLC) are expressed. These two forms are designated with increasing acidity as LC-2 and LC-2*. In the normal human heart the relation between LC-2/LC-2*-expression is 70/30, suggesting the existence of three different myosin isoenzymes (MPLC-polymorphism) in the normal human ventricle. Both ventricular MPLC-iso forms are monophosphorylated, the LC-2 being higher phosphorylated than the LC-2*. In some patients with heart failure both MPLC isoforms were found to be completely dephosphorylated. In the human atrium a MPLC isoform is expressed which is different from the ventricular MPLC isoforms. The atrial MPLC isoform is mono- and diphosphorylated. Mono-phosphorylation of both the ventricular MPLC isoforms and the atrial MPLC isoform increased responsiveness as well as sensitivity of isometric tension generation of skinned fibers to Ca^{2+}. Part of this effect could be explained by changing the cross-bridge-cycling rate: MPLC increased f_{app}, the rate-constant for the transition of cross-bridges from the non-force into the force-generating state, thus increasing the amount of force-generating cross-bridge states at a given $[Ca^{2+}]$. Mono-phosphorylation of the MPLC isoforms did not change maximal shortening velocity.

Key words: Human ventricle; phosphorylatable myosin light chain; human atrium; skinned fibers; mono-phosphorylation

Myosin isoenzymes in the human ventricle: polymorphism and phosphorylation of the myosin P-light chain (MPLC) isoforms

Cardiac myosin consists of two heavy chains (MHC) and 2 pairs of light chains (MLC) (5, 29, 32). In the human atrium and ventricle mainly the α-MHC and β-MHC, respectively, are expressed (10, 2, 8, 16). The MLC were designated as alkali LC's (LC-1,) and regulatory LC's (LC-2, DTNB-LC) (33). The MLC-2 of striated (MW 18,000–19,000) (9) and the 20 kDa myosin LC of smooth (13) muscle can be reversibly phosphorylated and have therefore been designated as the MPLC. Two monophosphorylated LC-2 forms having different isoelectric points but the same molecular weight exist in the ventricle of rabbit, cow, pig, chicken, and humans (15, 30, 20). These two isoforms will be designated herein with increasing acidity as LC-2 and LC-2*. In the bovine (29, 30) and chicken (15) heart both LC-2 forms have distinct primary sequences both containing only one recognition site for MLCK in the N-terminal region (11). The same holds for the sole ventricular rat LC-2 (14). Expression of MPLC isoforms in the rabbit ventricle is developmentally controlled: in the fetal and newborn rabbit the LC-2* isoform is below 10% of the total MPLC and increases with age up to approximately 30% (29).

In addition to the two ventricular-specific MPLC-isoforms the expression of an atrial-specific LC-2 isoform which is distinct from the two ventricular LC-2 isoforms has been described in the human heart (20, 24). This atrial-specific MPLC may be mono- and diphosphorylated (20). Evidence for this assumption came from the observations that: 1) only one MPLC form exists in the complete dephosphorylated state (atrial tissue incubated for 3 h in cold cardioplegic solution); 2) in the partially phosphorylated state (quickly frozen tissues, skinned fibers incubated with MLCK) two acidic derivatives (in addition to the one LC-2 form which exists in the dephosphorylated state) appear; 3) these two additional forms are phosphorylated derivatives as revealed by incorporation of ^{32}P; 4) the change in the charge induced by diphosphorylation is exactly twice as high as that induced by monophosphorylation (Figs. 1, 2; cf. 20).

The existence of two different MPLC isoforms in the human ventricle suggest the existence of three different myosin isoenzymes in the human ventricle: a LC-2/LC-2-homodimer, LC-2/LC-2*-hetero-dimer, and LC-2*/LC-2*-homodimer ("MPLC-polymorphism; Fig. 3). This is in contrast to the rat ventricle containing three isoenzymes on the basis of MHC-polymorphism ($\alpha\alpha$-homodimer, $\alpha\beta$-heterodimer, $\beta\beta$-homodimer).

The physiological function of different MPLC isoenzymes in the human heart is still not understood. It is puzzling that only slow (soleus, ventricle) but not fast contracting (vastus, atrium) muscle types showed a MPLC polymorphism (20, 31). There may be a functional role of the LC-2 in the regulation of cross-bridge cycling kinetics. Extraction of LC-2 from skinned rabbit psoas fibers decreased maximal (unloaded) shortening velocity (V_{max}) while subsequent reconstitution with LC-2 normalized V_{max} (22). A Drosophila mutant lacking the LC-2 revealed a reduced rate of stretch activation (34).

In the normal human heart mainly the LC-2 isoform is expressed: we found a LC-2/LC-2*-ratio of 70/30. In most patients with limited cardiac functions the LC-2/LC-2* ratio remained at its normal level. One patient (No. 14) revealed a distinct MPLC isoenzyme pattern (transition to the VLC-2* form: 55/45; Table 1). The cause of this change is unclear since a patient with similar symptoms (No. 2) maintained a normal

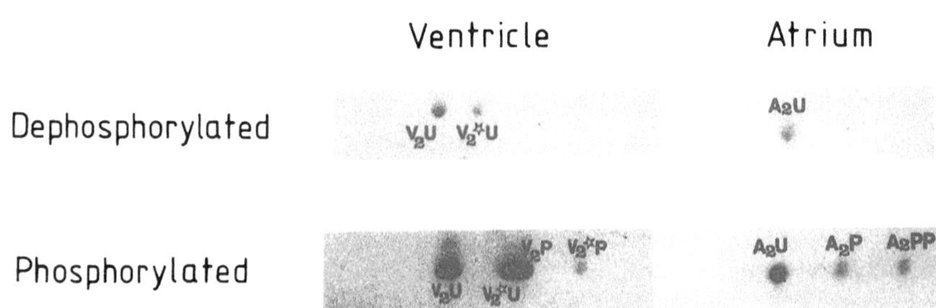

Fig. 1. Analysis of atrial and ventricular LC-2 isoforms by 2D-PAGE in the dephosphorylated state (after incubation of tissue in cardioplegic solution) and partially phosphorylated (tissue quickly frozen in liquid nitrogen) state. A2U, A2P, and A2PP refer to the unphosphorylated, monophosphorylated, and diphosphorylated atrial MPLC form. V2U, V2*U, V2P, and V2*P refer to the two unphosphorylated and their respective monophosphorylated derivatives (Coomassie-stained protein spots).

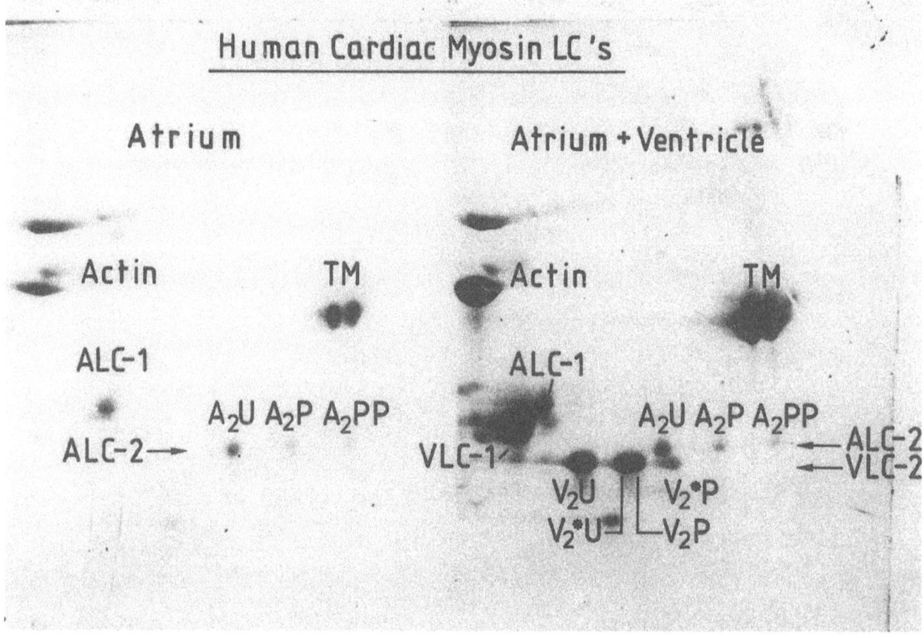

Fig. 2. Analysis of myosin light chains of human atrium and ventricle. Left: human atrial TCA-extract; right: a comigration of the same atrial extract and a ventricular TCA-extract. Human atrial and ventricular tissues were obtained from Prof. Saggau, Department of Surgery, University Hospital Heidelberg, FRG. Tissues were immediately frozen to obtain LC-2 in its partially phosphorylated state, powdered in liquid nitrogen, and homogenized in 15% TCA. TCA was extracted (ether/ethanole 8/2) after centrifugation. MLCs were separated by high-resolution two-dimensional polyacrylamide gel electrophoresis (2D-PAGE), as reported previously (13). Isoelectric focusing was performed overnight at pI 4.5–5.4 (Pharmacia, Sweden) at 400 V constant. Gels were stained with Coomassie-blue and scanned densitometrically. ALC-1 and VLC-1 refer to the atrial and ventricular alkali LC. A2U, A2P, and A2PP refer to the unphosphorylated, monophosphorylated, and diphosphorylated atrial P-LC (= ALC-2) form, respectively. V2U, V2*U, V2P, and V2*P refer to the two unphosphorylated ventricular P-LC (= VLC-2) forms and their corresponding phosphorylated derivatives.

isoenzyme pattern. This observation, however, demonstrates that, under some conditions still not understood, the expression of MPLC isoenzymes changes in the human ventricle. The importance of the LC-2 for the cross-bridge cycling kinetics suggests an important functional role of the existence of MPLC isoenzymes in the human heart.

Phosphorylation of the MPLC increases force of cardiac contraction, but not shortening velocity

The MPLC of striated muscle is phosphorylated by a Ca^{2+}-calmodulin dependent myosin light chain kinase (MLCK) and dephosphorylated by a LC-phosphatase (3, 1). The physiological role of MPLC phosphorylation in the striated muscle was

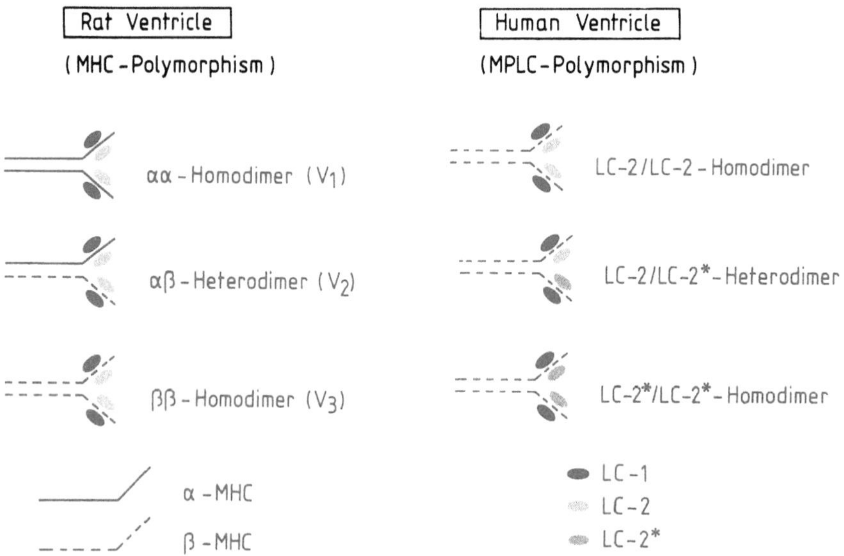

Fig. 3. Schematic representation of myosin isoenzymes based on a MHC-polymorphism (rat ventricle) and MPLC-polymorphism (human ventricle).

Table 1. Comparison of the relative distribution (in % of total) of the ventricular LC-2 forms (VLC-2 and VLC-2*) of patients with different cardiac diseases

Patient	Age (years)	Sex	Diagnosis	% VLC-2	% VLC-2*
No. 1	56	f	VHD	72	28
No. 34	49	f	VHD	73	27
No. 14	58	m	VHD	55	45
No. 2	56	m	VHD	78	22
No. 26	67	m	ICM	67	33
No. 27	30	m	ICM	75	25
No. 36	30	m	Normal	70	30

Valvular heart disease = VHD, ischemic cardiomyopathy = ICM. f = female, m = male. Tissues were provided by Prof. E. Erdmann, Klinikum Großhadern, Munich, FRG.

poorly understood up to the mid-1980s mainly due to the fact that MPLC phosphorylation in the striated muscle is no prerequisite for contraction. This is in contrast to the smooth muscle where MPLC phosphorylation initiates contraction (13). Furthermore, the in vivo MPLC phosphorylation level cannot be changed by alpha- (17) or beta-adrenergic (18, 12) stimulation, and remains unchanged during systolic and diastolic rhythm of the mammalian heart (26). It is, therefore, of particular interest to study the role of MPLC phosphorylation on the regulation of cardiac contractility.

We studied the effect of MPLC phosphorylation on shortening velocity and isometric tension development of skinned human atrium and pig ventricular myocardium. Using high glycerol concentrations and the detergent Triton X-100 all membraneous structures were dissolved, leaving the contractile apparatus functionally intact. The contractile structures were thus accessible to enzymes and ions.

Incubation of skinned atrial fibers with MLCK increased both the level of MPLC phosphorylation and the Ca^{2+} sensitivity of the tension-pCa relation. Results of an experiment with a skinned fiber of the non-hypertrophied atrium of a patient with coronary heart disease is shown in Figure 4. Ca^{2+} sensitivity increased from pCa 5.1 to pCa 5.21 upon addition of MLCK (100 nM) and calmodulin (1 µM) to the activation solutions. MPLC phosphorylation rose from an initial complete dephosphorylated state to a monophosphorylated state of 15% after incubation with MLCK (upper trace in Fig. 4). Since the diphosphorylated state could not be detected in these experiments, we have no idea about the mechanical effects of both mono- and diphosphorylation of the atrial MPLC isoform. The absence of the diphosphorylated state could be due to the fact that we used MLCK prepared from pig ventricle. MLCK prepared from pig atrium, however, seems to be different from ventricular MLCK, according to their kinetic characteristics (25). The increase of Ca^{2+} sensitivity upon MPLC (mono)phosphorylation was the same in both hypertrophied and

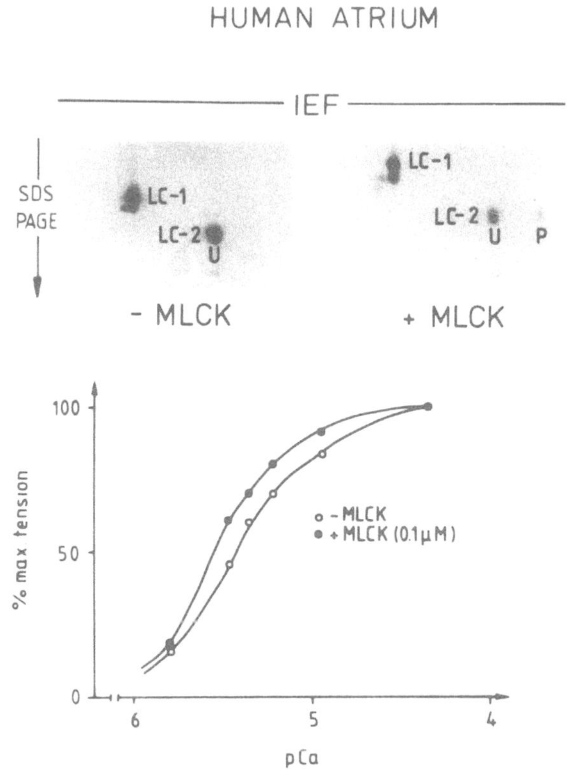

Fig. 4. Ca^{2+} sensitivity and MPLC phosphorylation of skinned human atria (lower part) and analysis of MPLC phosphorylation by 2D-PAGE (upper trace) before and after addition of MLCK in the activation solutions.

non-hypertrophied atria, although the change of the initial Ca^{2+} sensitivity of hypertrophied atria was higher than those from normal atria (19).

To study the effect of MPLC phosphorylation on the Ca^{2+} sensitivity of the ventricle, we used skinned fibers of the pig ventricle. Figure 5 shows the effects of incubation of skinned ventricular preparations on isometric tension generation and MPLC phosphorylation. In the lower part of Figure 5 an original registration of isometric tension generation before and after addition of MLCK into the contraction solutions is shown. MLCK increased isometric tension generation at submaximal Ca^{2+}; activation while tension achieved at maximal activation was hardly elevated. If normalized to the maximal response, addition of MLCK caused a leftward shift of the tension-pCa relation by 0.1 pCa-units (from pCa 5.72 to pCa 5.82; Table 2). Analysis of the phosphorylation state of the MPLC before and after addition of MLCK by two-dimensional (2D)-PAGE and densitometrical evaluation is shown as an inset in Figure 5. Similar to the human ventricle there are two MPLC isoforms in the pig ventricle (33, 20), leading to four MPLC forms in the partially phosphorylated state after analysis by 2D-PAGE. These forms have been designated with increasing

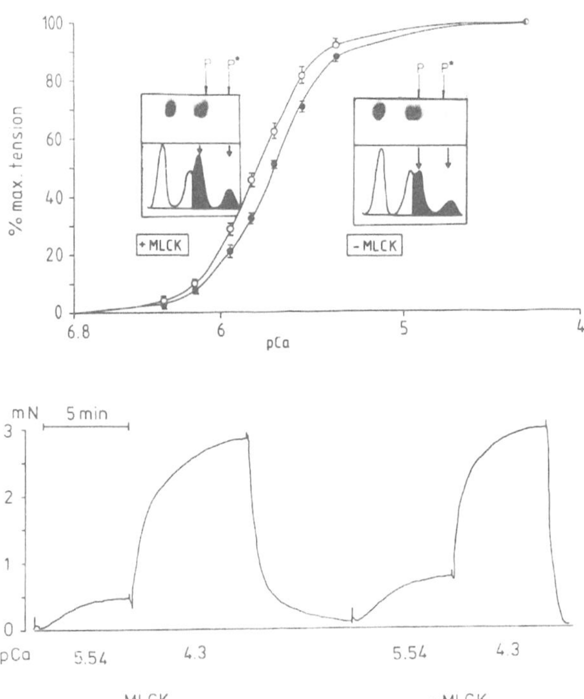

Fig. 5. Isometric tension generation (lower trace; original record), tension-pCa relation (middle trace), and MPLC phosphorylation levels (insets; protein staining after 2D-PAGE and densitometrical scans below) of skinned pig ventricular fibers before and after incubation with MLCK (100 nM) and calmodulin (1 μM). Ca^{2+} concentrations are given as pCa (negative decadic logarithm of free $[Ca^{2+}]$). The phosphorylated MPLC forms are labeled with a P and the corresponding peaks are shown in black.

Table 2. Ca^{2+} sensitivity (given as pCa50) and MPLC phosphorylation levels (mol Pi per mol MPLC) of chemically skinned ventricular fibers of the pig before and after incubation with MLCK or LC-phosphatase (LC-P)

	pCa50	MPLC phosphorylation (mol Pi/mol MPLC)	
		LC-2	LC-2*
Before MLCK	5.72	0.34	0.20
After MLCK	5.82*	0.445*	0.30*
Before LC-P	5.55	0.30	0.20
After LC-P	5.45*	0.16*	0.12*

Values are means of six to eight fibers each. SD was always within 10–15% of the mean value. * $p < 0.05$. Experiments with MLCK were performed at pH 7.0, those with LC-P at pH 6.7, which may explain the lower basal Ca^{2+} sensitivity of the skinned fibers used in the LC-P experiments.

acidity as LC-2, LC-2*, LC-2P, and LC-2*P. The level of MPLC phosphorylation before addition of MLCK was 0.34 mol Pi/mol LC-2 and 0.2 mol Pi/mol LC-2*. After addition of MLCK MPLC phosphorylation rose to 0.45 and 0.3 mol Pi/mol LC-2 and LC-2*, respectively (Table 2).

If an increased MPLC phosphorylation level sensitizes the contractile apparatus for Ca^{2+}, then a decrease of the MPLC phosphorylation should desensitize it. We tested this hypothesis using a LC-phosphatase prepared by DiSALVO (PCM-phosphatase (6)).

In the lower part of Figure 6 an original record of a contraction relaxation cycle of a skinned pig ventricular fiber before and after incubation with the LC-phosphatase is shown. Subsequent to a first submaximal and then maximal isometric tension generation the fiber was incubated with 160 U of PCM-Phosphatase in relaxing solution. An additional contraction/relaxation cycle under the same conditions as before led to a decrease of tension at submaximal, but not at maximal Ca^{2+} activation. The upper part of Figure 6 shows the tension normalized to maximal tension obtained at maximal Ca^{2+}. Incubation of cardiac fibers with LC-phosphatase caused a rightward shift of the tension-pCa relation by 0.1 pCa-units (Table 2). Analysis of the MPLC phosphorylation level by 2D-PAGE and the densitometrical evaluation before and after treatment of the fiber with LC-phosphatase is shown as inset. After addition of phosphatase the phosphorylated MPLC forms decreased from 0.3 and 0.2 to 0.2 and 0.12 mol Pi per mol LC-2 and LC-2*, respectively (Table 2).

In addition, we studied unloaded shortening velocity (Vmax) of skinned pig ventricular fibers before and after MPLC phosphorylation. We used the slack-test method according to EDMAN 1979 (7) to determine Vmax. Figure 7 shows the principle of the method (inset) and the results of the experiments. During isometric steady-state tension (temperature was 21 °C) quick releases (1 ms duration) of different amplitudes (given in % of initial muscle length; $\Delta L(\%)$) were applied to the preparations. Thus, tension dropped to zero and the fiber shortened without external load. The time required for tension development (slack-time; $\Delta t(ms)$) depends on the

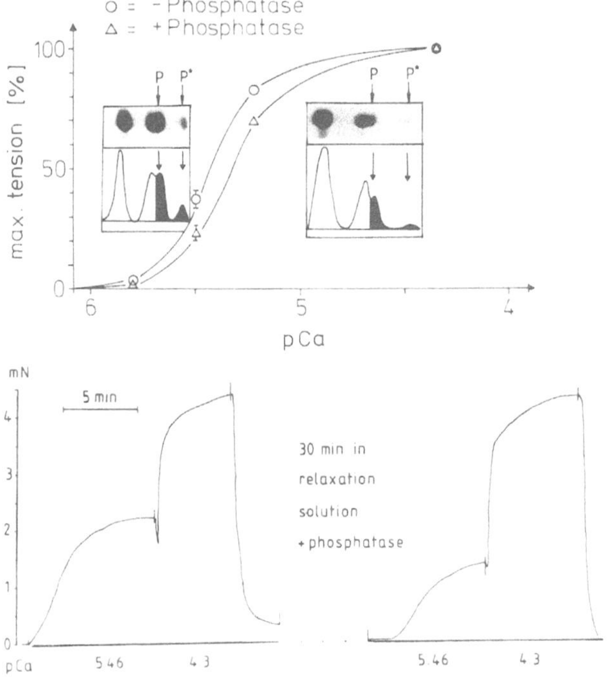

Fig. 6. Isometric tension generation (lower trace; original record), tension-pCa relation (middle trace), and MPLC phosphorylation levels (insets; protein staining after 2D-PAGE and densitometrical scans below) of skinned pig ventricular fibers before and after incubation with LC-Phosphatase (160 U). For details see Fig. 5.

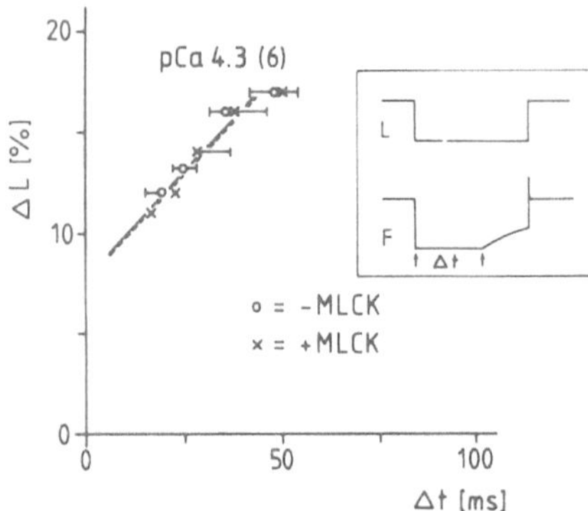

Fig. 7. Maximal (unloaded) shortening velocity of chemically skinned pig ventricular fibers before and after incubation with MLCK (pCa 4.5, 21 °C). Inset schematically shows the principle of the method.

release amplitude in that the higher the release amplitude the longer the slack-time. The slope of the relation between release amplitude and slack time gives a measure of Vmax. Figure 7 clearly demonstrates that MPLC phosphorylation has no influence on Vmax: before and after MLCK incubation Vmax was 2 muscle length/s (means of 6 fibers; SEM < 10%). This result suggest that at least the rate-limiting step which determines Vmax (g; see below) is not modified by MPLC phosphorylation.

MPLC phosphorylation modulates cross-bridge cycling kinetics

Figure 8 shows an over-simplified scheme of the cross-bridge cycle. Myosin cross-bridges attach to the thin filament (state 2), produce tension (state 3) by a conformational change, and detach (state 1) in a cyclic manner. Therefore, during isometric steady state a fraction of active cross-bridges is in the force and a fraction in the non-force-generating state. Huxley (11) designated the rate-constant for the transition from the non-force into the force-generating state as "f" and the reverse reaction as "g" (f_{app} and g_{app}, respectively, according to Brenner (3)). The fraction of force-generating cross-bridges during steady-state tension is given by the relation $f_{app}/(f_{app} + g_{app})$. Both an increase of f_{app} or a decrease of g_{app} will, therefore, elevate the fraction of cross-bridges, producing force and, thus, tension generation of a muscle fiber.

To study whether MPLC phosphorylation increases f_{app}, we used a previously developed method (3). The principle of this method is schematically shown as an inset in Fig. 9. During isometric steady-state tension the fiber is allowed to shorten without load and then restretched to the initial sarcomere length. The tension first drops to near zero and then, upon restretch, redevelops isometric steady-state tension. The time course of tension redevelopment can be described by an exponential function and the rate-constant (k_{redev}) of this function equals ($f_{app} + g_{app}$). These experiments have been performed with glycerinated rabbit psoas fibers. k_{redev} was measured at two Ca^{2+}-concentrations, a submaximal (15–20% of maximal) and a near-maximal (90–95% of maximal) activation, first in the absence of and then in the presence of MLCK. Figure 8 shows that at low Ca^{2+}-activation k_{redev} nearly doubled upon addition of MLCK from $1.6\,s^{-1}$ to $2.8\,s^{-1}$ (p < 0.01). At high Ca^{2+}-activation k_{redev} increased from $4.9\,s^{-1}$ to $6.2\,s^{-1}$ (p < 0.01) (means of six to eight experiments; SD < 10%). Since g_{app} is not influenced by MPLC phosphorylation (no change of AT Pase/force relation (27)), the increase of k_{redev} upon MPLC phosphorylation is mainly due to a selective influence on f_{app}.

Isometric steady-state tension is shown in the right hand side of Figure 9. Tension doubled after addition of MLCK at low Ca^{2+}. This effect was much lower at high activation levels. If normalized, it will explain perfectly the sensitization effect of

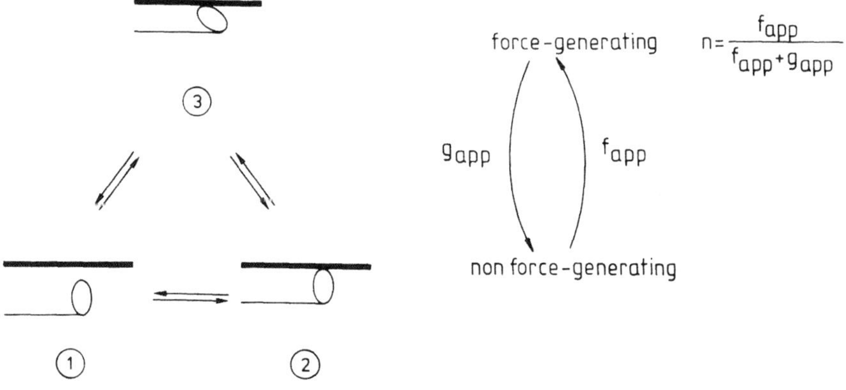

Fig. 8. Simplified scheme of the cross-bridge cycle (see text for explanations).

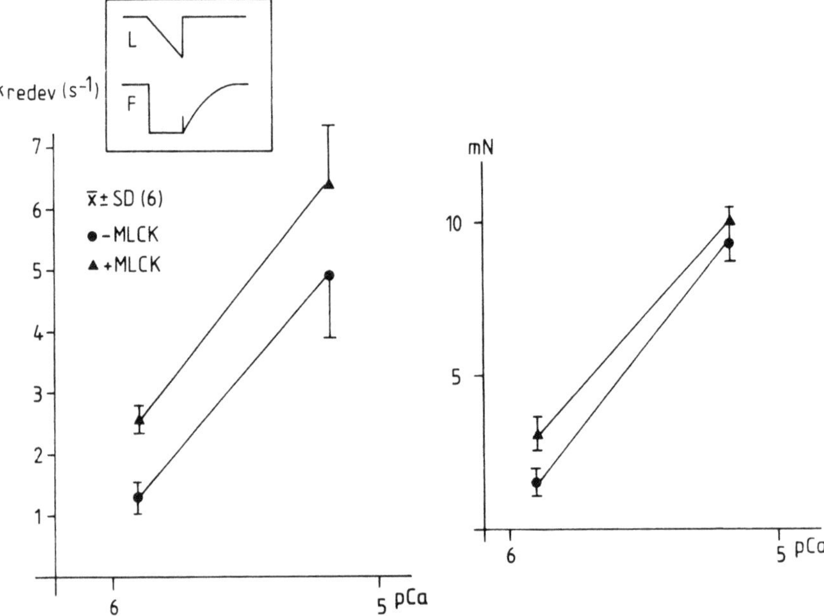

Fig. 9. Changes of k_{redev} (s^{-1}) and isometric tension (mN) before and after incubation of glycerinated rabbit psoas fiber with MLCK. Inset shows the principle of the method for determination of k_{redev} schematically. Experiments were performed by **B. Brenner**, Ulm (33).

Table 3. In vivo phosphorylation levels of the human ventricular MPLC isoenzymes

Patient	NYHA class	Phosphorylation level (%)	
		VLC-2	VLC-2*
5	I	36	25
6	II	39	26
7	II	37	33
8	I	36	15
12	III	47	27
13	III	0	0
20	II	0	0
22	II	38	29
23	II	0	0
32	II	0	0

Values arc given in % of total LC-2 or LC-2*. Biopsies of the human ventricle were quickly frozen in liquid nitrogen (Dr. Katus, University of Heidelberg, Dept. of Cardiology, FRG). MPLC isoenzymes were analyzed by 2D-PAGE and silver staining.

MPLC phosphorylation on the tension-Ca^{2+} curve. These experiments show that MPLC phosphorylation increases the ability of the heart to create force and that at least part of the effect can be explained by a modulation of the cross-bridge cycling kinetics, namely, by an increase of f_{app}.

In vivo MPLC phosphorylation levels in the human heart

The MPLC isoforms in the human ventricle were always partially phosphorylated in vivo. The phosphorylation levels of both MPLC isoforms were different, the LC-2

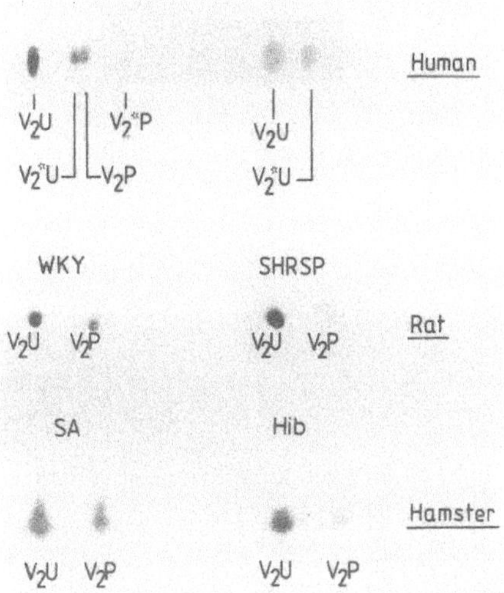

Fig. 10. In vivo MPLC phosphorylation levels in the left ventricle of humans (upper trace), rats (middle trace), and European hamster (lower trace). Tissues were quickly frozen in liquid nitrogen (biopsies) or freeze-clamped with a Wollenberger clamp precooled in liquid N_2 (beating hearts; rats and hamster) and stored at $-80\,°C$. For MPLC analysis the tissue was powdered in liquid nitrogen, denatured with 15% TCA, and subjected to 2D-PAGE as described (13).
Upper trace: Human MPLC forms were silver stained upon 2D-PAGE. V2U and V2*U correspond to the unphosphorylated, V2P and V2*P to the respective phosphorylated MPLC forms. Please note that in some patients (right part) the two MPLC isoforms were completely dephosphorylated, whereas the phosphorylation pattern of the MPLC of the patient shown on the left side appears to be normal. Biopsies were made by Dr. H. Katus, Department of Cardiology, University of Heidelberg, FRG.
Middle trace: MPLC of 44-week-old normotensive (WKY) and spontaneously hypertensive (SHRSP) male rats. In contrast to the human ventricle, only one MPLC isoform exists in the rat ventricle, leading to two MPLC forms after 2D-PAGE (the unphosphorylated (V2U) and its phosphorylated (V2P) derivative). In older SHRSP the relative amount of the V2P form decreased from 40% (WKY) to 18%. Animals were obtained from Prof. Ganten, German Institute for High Blood-Pressure Research, Department of Pharmacology, University of Heidelberg.
Lower trace: MPLC of European hamster during hibernation (Hib) and summer activity (SA). MPLC phosphorylation level decreased from 45% during summer to 23% during hibernation. Animals were obtained from Dr. Agostini, Max-Planck-Institute for Medical Research, Department of Physiology, Heidelberg.

isoform being phosphorylated to a greater degree than the LC-2* isoform (Table 3, Fig. 10). In the normal rabbit ventricle, we observed phosphorylation levels of 36 $\pm$ 0.05% and 15 $\pm$ 0.03% for the LC-2 and the LC-2*, respectively (means $\pm$ SD of six animals) (17). In some patients, however, the two MPLCs were completely dephosphorylated (Fig. 10, Table 3). Although this failure of a post-translational process in the heart with limited functions is interesting enough, we cannot yet correlate this phenomenon with any hemodynamic parameter. A statistically significant decrease of the in vivo MPLC phosphorylation level has also been described in the hypertrophied, chronic hypertensive rat heart (18). The level of MPLC phosphorylation in the European hamster, measured during summer, declined from 45% to 25% during hibernation (Fig. 10; c.f. 21).

Acknowledgements. I thank Prof. J.C. Rüegg and Prof. B. Brenner for helpful discussions, and Bettina Sogl for excellent technical assistance. This work is part of the author's *habilitation* supported by DFG Mo 362/5-1.

References

1. Adelstein RS, Eisenberg E (1980) Regulation and kinetics of the actin-mysoin ATP interaction. Ann Rev Biochem 49:921–956
2. Bouvagnet P, Leger J, Pons F, Dechesne C, Leger JJ (1984) Fiber types and myosin types in human atrial and ventricular myocardium. Circ Res 55:794–808.
3. Brenner B (1988) Effect of Ca^{2+} on cross-bridge turnover kinetics in skinned single rabbit psoas fibers: implications for regulation of muscle contraction. Proc Natl Acad Sci 85:3265–3269
4. Brenner B, Morano I (1990) Effects of myosin light chain phosphorylation on isometric force and cross-bridge turnover kinetics. Pflügers Arch 415(1):R7339
5. Delcayre C, Swynghedauw B (1975) A comparative study of heart myosin ATPase and light subunits from different species. Pflügers Arch 355:39–47
6. DiSalvo J, Gifford D, Kokkinakis A (1984) Modulation of aortic protein phosphatase activity by polylysine. Proc Soc Exp Biol Med 177:24–32
7. Edman KAP (1979) The velocity of unloaded shortening and its relation to sarcomere length and isometric force in vertebrate muscle fibers. J Physiol 291:143–159
8. Everett, AW (1986) Isomyosin expression in human heart in early pre- and post-natal life. J Mol Cell Cardiol 18:607–615
9. Frearson N, Perry VS (1975) Phosphorylation of the light chain components of myosin from cardiac and skeletal muscles. Biochem J 151:99–107
10. Gorza L, Sartore S, Schiaffino S (1982) Myosin types and fiber types in cardiac muscle. II. Atrial myocardium. J Cell Biol 95:838–835
11. Huxley AF (1957) Muscle structure and theories of contraction. Prog Biophys Chem 7:255–318
12. Jeacocke SA, England PJ (1980) Phosphorylation of myosin light chains in perfused rat heart. Biochem J 188:763–768
13. Kamm KE, Stull JT (1985) The function of myosin and myosin light chain kinase phosphorylation in smooth muscle. Ann Rev Pharmcol Toxicol 25:593–620
14. Kumar CC, Cribbs L, Delaney P, Chien KR, Siddiqui MAQ (1986) Heart myosin light chain 2 gene. Nucleotide sequence of full length cDNA and expression in normal and hypertensive rat. J Biol Chem 261:2866–2872
15. Matsuda G, Maita T, Kato Y, Chen JJ, Umegane T (1981) Amino acid sequence of the cardiac L-2A, L-2B, and gizzard 17000 Mr light chains of chicken muscle myosin. FEBS Lett 136:232–236
16. Mercadier JJ, De La Bastie D, Menasche P, N'Guyen A, Caa V, Bouveret P, Lorente P, Piwinca A, Slama R, Schwartz K (1987) Alphamyosin heavy chain and atrial size in

patients with various types of mitral valve dysfunction: a quantitative study. J Am Coll Cardiol 9:1024–1030
17. Morano I, Wiesner R, Rösen P, Rüegg JC (1987) Alpha-adrenergic stimulation of the rat heart does not influence the level of myosin P light chain phosphorylation. J Muscle Res Cell Mot 8:90
18. Morano I, Lengsfeld M, Ganten U, Ganten D, Rüegg JC (1988) Chronic hypertension changes myosin isoenzyme pattern and decreases myosin phosphorylation in the rat heart. J Mol Cell Cardio 20:875–886
19. Morano I, Bächle-Stolz C, Katus A, Rüegg JC (1988) Increased calcium sensitivity of chemically skinned human atria by myosin light chain kinase. Basic Res Cardiol 83:350–359
20. Marano I, Wankerl M, Böhm M, Erdmann E, Rüegg JC (1989) Myosin P-light chain isoenzymes in the human heart: evidence for diphosphorylation of the atrial P-light chain isoform. Basic Res Cardiol 84:298–305
21. Morano I, Agostini B, Katus H, Ganten D, Rüegg JC (1991) Changes of the in vivo level of myosin P-light chain phosphorylation in the human and animal heart. Biophys J 59:228a
22. Moss RL, Julian GG, Greaser ML (1982) Physiological effects accompanying the removal of LC2 from skinned skeletal muscle fibers. J Biol Chem 257:8588–8591
23. Perry WT, Smilie LB, Perry SV (1973) A phosphorylated light chain component of myosin from skeletal muscle. Cold Spring Harbor Symp. Quant Biol 37:17–19
24. Price KM, Littler WA, Cummins P (1980) Human atrial and ventricular myosin light chain subunits in the adult and during development. Biochem. J. 191:571–580
25. Rinne B, Morano I (1991) Different activities of myosin light chain phosphatase and distinct isoenzymes of myosin light chain kinase in the heart. J Muscle Res Cell Mot 12:108
26. Stull JT, Sanford CF, Manning DR, Blumenthal DK, High (1981) CW Phosphorylation of myofibrillar proteins in striated muscles. Cold Spring Harbor Conf Cell Prolif 8:823–839
27. Sweeney HL, Stull JT (1990) Alteration of cross-bridge kinetics by myosin light chain phosphorylation in rabbit skeletal muscle: implications for regulation of actin-myosin interaction. Proc Natal Acad Sci 87:414–418
28. Westwood SA, Perry SV (1981) The effect of adrenaline on the phosphorylation of the P-light chain of myosin and troponin. Biochem J 197:185–193
29. Westwood SA, Perry SV (1982) Two forms of the P light chain of myosin in rabbit and bovine hearts. FEBS Lett 142:31–34
30. Westwood SA, Perry SV (1982) The effect of adrenaline on the phosphorylation of the P light chain in rabbit and bovine heart. FEBS Lett 142:31–34
31. Westwood SA, Hudlicka O, Perry SV (1984) Phosphorylation in vivo of the P-light chain of myosin in rabbit fast and slow skeletal muscles. Biochem J 218:841–847
32. Weeds AG, Lowey S (1971) Substructure of the myosin molecule. J Mol Biol 61:701–725
33. Weeds AG, Pope B (1971) Chemical study on light chains from cardiac and skeletal muscle myosins. Nature 234:85–88
34. Yamakawa M, Warnke J, Falkenthal S, Maughan D (1990) Properties of skinned muscle fibers from myosin light chain 2 deficient flightless mutants of Drosophila melanogaster. Biophys J 57:411a

Author's address:
I. Morano
II. Physiologisches Institut
Universität Heidelberg
Im Neuenheimer Feld 326
6900 Heidelberg, FRG

Responsiveness of the myofilaments to Ca^{2+} in human heart failure: implications for Ca^{2+} and force regulation

R.J. Hajjar, W. Grossman, J.K. Gwathmey

Charles A. Dana Research Institute and the Harvard-Thorndike Laboratory of Beth Israel Hospital, Cardiovascular Division, Department of Medicine, Beth Israel Hospital (WG, JKG), Department of Cellular and Molecular Physiology (JKG), Medical Services, Massachusetts General Hospital (RJH), and Harvard Medical School, Boston, Massachusetts, USA

Summary: Myofilament calcium sensitivity and maximal calcium-activated force are fundamental properties of the contractile proteins in the heart. We examined these properties in normal human right-ventricular trabeculae carneae obtained from hearts of brain-dead patients with no known cardiac disease, and from patients with end-stage heart failure undergoing cardiac transplantation. There were no differences in calcium-activation of the control and myopathic muscles from chemically-skinned trabeculae or from intact tetanized preparations. We then tested the effect of DPI 201-106 (4-[3-(4-diphenylmethyl-1-piperazinyl)-2-hydroxypropoxy]-1H-indole-carbonitrile), a new inotropic agent, in both preparations. In myopathic muscles, 1 μM DPI sensitized the myofilaments to Ca^{2+}, as evidenced by a significant shift of the $[Ca^{2+}]$-force relationship towards lower $[Ca^{2+}]$, in both skinned and intact preparations. On the other hand, the same concentration of DPI did not affect the calcium-activation in control muscles in both preparations. We also found that the twitch $[Ca^{2+}]$-force relationship, which has been used as an indication of myofilament sensitivity, was dissociated from the steady-state $[Ca^{2+}]$-force relationship, and was shifted along the $[Ca^{2+}]$ axis by modulation in the time-course of the twitch and $[Ca^{2+}]_i$, and not by the sensitivity of the myofilaments to Ca^{2+}. Protein kinase C stimulation differentially altered the responsiveness of the myofilaments to Ca^{2+} in normal and myopathic muscle fibers. We propose that even though calcium activation and maximal calcium-activated force are unaltered in myopathic hearts there are changes in thin filament regulation in myopathic hearts that result in altered responses to agents that directly act on the thin filaments, and that the potential for force development is similar in normal and myopathic human hearts.

Key words: Myofilaments; calcium; human; protein kinase C; cardiomyopathy

Introduction

The human heart contracts and relaxes rhythmically while constantly adapting its output to the hemodynamic constraints of the body. In working myocardium, the contractile mechanism is activated and deactivated by the rise and fall of intracellular calcium. This activating calcium released from the sarcoplasmic reticulum (SR) into

JKG is an Established Investigator of the American Heart Association. This work is supported by Glaxo Inc., Upjohn Company, HL 39091 and HL 36797 to JKG, a Beth Israel Biomedical Research Grant to RJH and JKG, and HL 38189 to WG.

the myoplasm does not maximally stimulate the contractile machinery, thus allowing cardiac muscle to have a large contractile reserve so that its performance can be increased by a variety of maneuvers. These include an increase in the frequency of stimulation (Treppe effect), an increase in intracellular sodium concentration, an increase in muscle length (Frank-Starling mechanism), and the application of compounds with a "positive inotropic" action. In most of these cases, the increase in force is due to an increase in intracellular free Ca^{2+} concentration reached during contraction (26). In considering the modulation of force by calcium in mammalian myocardium it is important to note that the decisive parameter determining contractility is not the calcium concentration per se, but the calcium occupancy of troponin. Thus contractility may also be altered by changing the responsiveness of the myofilaments to calcium ions (Ca^{2+}) or the duration of troponin C-calcium interaction. The responsiveness of the myofilaments to Ca^{2+} encompasses: the concentration of Ca^{2+} associated with 50% activation of the myofilaments (i.e., force) normalized to the maximum at saturating levels of free Ca^{2+}, the shape and slope of the calcium-force relationship, and maximal activity at saturating Ca^{2+} levels. Recent thinking about design of inotropic agents centers around mechanisms involving altered responsiveness of the myofilaments to Ca^{2+} (35). New cardiotonic compounds have been developed that target the myofilaments for their inotropic action. These include sulmazole, pimobendane, isomazole, MCI-154, and DPI 201-106, which increase the sensitivity of the myofilaments to Ca^{2+} as part of their mechanism of action (7, 8, 28, 35, 40). These agents either increase the affinity of troponin C for Ca^{2+}, or act directly on the contractile proteins (36). It thus became important to investigate whether there are differences in the responsiveness of the myofilaments to Ca^{2+} in control and myopathic hearts; we first studied the calcium-activation of both tissues. We then examined the effects of agents that act directly on the thin filaments in both preparations.

Methods

Muscle Preparation

Muscles were obtained from 12 patients undergoing transplantation due to end-stage heart failure as described earlier (18, 21), and were labeled myopathic. Seven of the patients from this group were diagnosed as idiopathic dilated cardiomyopathy, and five were patients with ischemic cardiomyopathy. Muscles were also obtained from seven brain-dead organ donors without known cardiac disease and were labeled control. All experimental protocols were reviewed and approved by an Internal Review Board.

Fibers with diameters less than 200 μm were selected for skinned fiber preparations and fibers with diameters less than 800 μm were selected for intact preparations. The muscles were clamped at one end of a muscle holder and were attached at the other end to a servocontrolled electromagnetic lever system. The muscle fibers were superfused with a bicarbonate-buffered oxygenated solution with the following composition (mM): NaCl 120, KCl 5.9, $NaHCO_3$ 25, NaH_2PO_4 1.2, $MgCl_2$ 1.2, $CaCl_2$ 2.5, Dextrose 11.5, bubbled with 95% O_2 and 5% CO_2, pH of 7.4, at 30°C. Muscles were stimulated with a square wave pulse of 5 ms duration at threshold voltage at a frequency of 0.33 Hz.

Measurements of Intracellular Calcium Concentration

Aequorin was introduced intracellularly by a chemical-loading technique described previously (17). The light emitted by aequorin was detected with a photomultiplier tube (9635QA, Thorn EMI Inc, Fairfield, NJ) attached to a collecting apparatus similar to the one described by Blinks et al. (4).

An in-vitro calibration curve was used to convert normalized light signals obtained from the preparations into calcium concentrations, as shown by the following equation:

$$\frac{L}{L_{max}} = \left(\frac{1 + KR[Ca^{2+}]_i}{1 + K_{TR} + K_R[Ca^{2+}]_i} \right)^3$$

where L is the luminescence signal, L_{max} is the maximal light emitted after exposing the muscle preparation to a solution containing 2% Triton X-100 and a saturating extracellular $[Ca^{2+}]$. K_R and K_{TR} which were measured to be 3.07×10^{-6} M^{-1} and 119.8 in an in-vitro medium containing 150 mM KCl, 2 mM $MgCl_2$, and 5 mM 1,4 piperazine-diethanesulfonic acid (PIPES), at pH of 7.1 and at 30°C. The rate constant for aequorin consumption measured in this in-vitro medium was 1.98 s^{-1}.

Tetanizations of Muscles

Muscle tetani were induced by stimulating intact muscles at 15–20 Hz for 4–6 s using stimuli pulses of 50 ms in duration (29, 43). Steady-state measurements were made during the first 4–6 s of tetani. Tetani were elicited at varying $[Ca^{2+}]_0$ from 0.5 mM to 16 mM.

Skinning Procedure

Trabeculae carneae were selected from human hearts for skinning as previously described (17, 21). They were chemically skinned by exposure to a solution containing saponin 250 µg/ml (10), K_2ATP 5 mM, $MgCl_2$ 7 mM, EGTA (ethyleneglycol-bis N,N,N,N'-tetra-acetic acid) 5 mM, KCl 60 mM, imidazole 60 mM, phosphocreatine 12 mM, creatine phosphokinase 15 units/ml, pH = 7.1 at 22°C. The muscles were then exposed to this solution for 30 min.

The total salt concentrations necessary for obtaining the desired pCa ($-\log_{10}[Ca^{2+}]$), pMg ($-\log_{10}[Mg^{2+}]$), pMgATP ($-\log_{10}[MgATP]$), and pH at a constant ionic strength were calculated using the program described by Fabiato and Fabiato (11, 12). The solutions were prepared at a temperature of 22°C with a pMg of 2.5, pMgATP of 2.5, an EGTA concentration of 10 mM, an ionic strength of 0.16 M, and a pH of 7.1 adjusted using 30 mM TES. The solutions also contained 12 mM phosphocreatine and 15 units/ml creatine phosphokinase. All stock solutions were prepared in plasticware: 5 mM $K_2CaEGTA$ was prepared with $CaCO_3$, EGTA, and KOH at 80°C; 5 mM K_2EGTA was prepared with EGTA and KOH. Creatine phosphate and creatine phosphokinase were added immediately before the experiment. The relaxation solution had a pCa > 8.0 and EGTA was replaced with HDTA

(K_2-2,6-diaminohexane-N,N,N',N'-tetraacetic acid) in the solution, whereas the activation solution had a pCa of 4.0. In the relaxed state the muscle was adjusted to a length L_0, at which an increase in resting tension was first observed.

Force-$[Ca^{2+}]$ Analysis

The force versus $[Ca^{2+}]$ curves were fit to a modified Hill equation:

$$\frac{F}{F_{max}} = \frac{[Ca^{2+}]^n}{[Ca^{2+}]_{50\%}{}^n[Ca^{2+}]^{n'}}$$

where F is developed force, F_{max} is the maximal force developed, n is the Hill coefficient, and $[Ca^{2+}]_{50\%}$ is the $[Ca^{2+}]$ required for 50% activation.

Statistical Analysis

Results are presented as mean $\pm$ SEM. Statistical significance was determined by Student's *t*-test or one-way analysis of variance (ANOVA). The level of statistical significance was set at a probability of 0.05.

Chemicals

DPI 201-106 (4-[3-(4-diphenylmethyl-1-piperazinyl)-2-hydroxypropoxy]-1H-indole-carbonitrile) was generously supplied by Sandoz Pharmaceutical Corp., East Hanover, New Jersey, USA. 12-deoxyphorbol 13 isobutyrate 20 acetate (DPBA) was purchased from LC Services, Woburn, Massachusetts, USA. All other chemicals were purchased from Sigma Co., St. Louis, Missouri, USA. The aequorin used in these experiments was purchased from the laboratory of Dr. J. R. Blinks in Rochester, Minnesota, USA.

Results

Skinned Fiber Preparations

The activation range in both control and myopathic bundles was 10^{-7} M to 10^{-4} M free calcium ion. The isometric force developed at saturating free $[Ca^{2+}]$ in skinned preparations for both groups (corrected for the cross-sectional area of each muscle) was not significantly different (p > 0.1). As illustrated in Figure 1, there were no differences in the $[Ca^{2+}]$-force relationships between control and myopathic muscles. The $[Ca^{2+}]$ required for half-maximal activation was 1.53 ± 0.32 µM (n = 11) and 1.33 ± 0.19 µM (n = 12) for control and myopathic muscles, respectively. The Hill coefficients were 2.05 ± 0.19 and 2.36 ± 0.21 in the control and myopathic groups, respectively. The slope of the pCa-force curve in the myopathic muscles was not significantly (p > 0.1) steeper than the control's slope.

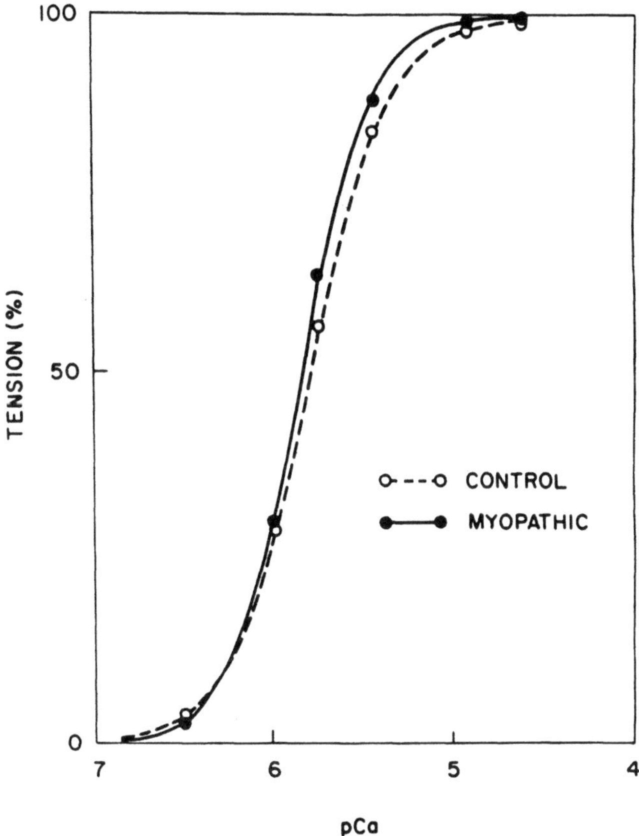

Fig. 1. $[Ca^{2+}]$-force relationship for control and myopathic muscles. Mean values for 11 control and 12 myopathic fibers. The solid curves are approximated with the Hill relationship. (Reproduced with permission from Gwathmey JK, Hajjar RJ (1990) Circulation 82:1266–1278 by copyright permission of the American Heart Association)

Steady Levels of $[Ca^{2+}]_i$ *and Force*

Trabeculae carneae from both control and myopathic human hearts were tetanized at 30°C, without chemical interventions, producing steady-state levels of force and $[Ca^{2+}]_i$, as shown in Figure 2. The tetanus force reached a higher level than twitch, and both force and aequorin luminescence reached a plateau. At a given $[Ca^{2+}]_0$, the tetani were superimposable when separated by periods of > 3 min, during which the muscles were stimulated at 0.33 Hz. The level of force and light signals during the plateau phase of the tetanus was determined from a single tetanus. Increasing $[Ca^{2+}]_0$ from 1 mM to 16 mM increased the amplitude of the twitch and tetanus forces. Figure 2 illustrates the effect of increasing $[Ca^{2+}]_0$ on force and $[Ca^{2+}]_i$ in a preparation from a myopathic heart. Increasing $[Ca^{2+}]_0$ was associated with a progressive increase in the magnitude of steady-state $[Ca^{2+}]_i$ levels along with steady-state force levels which increased from 1 mM to 8 mM $[Ca^{2+}]_0$, but reached a plateau at 10 mM $[Ca^{2+}]_0$ in both control and myopathic muscles. We verified that

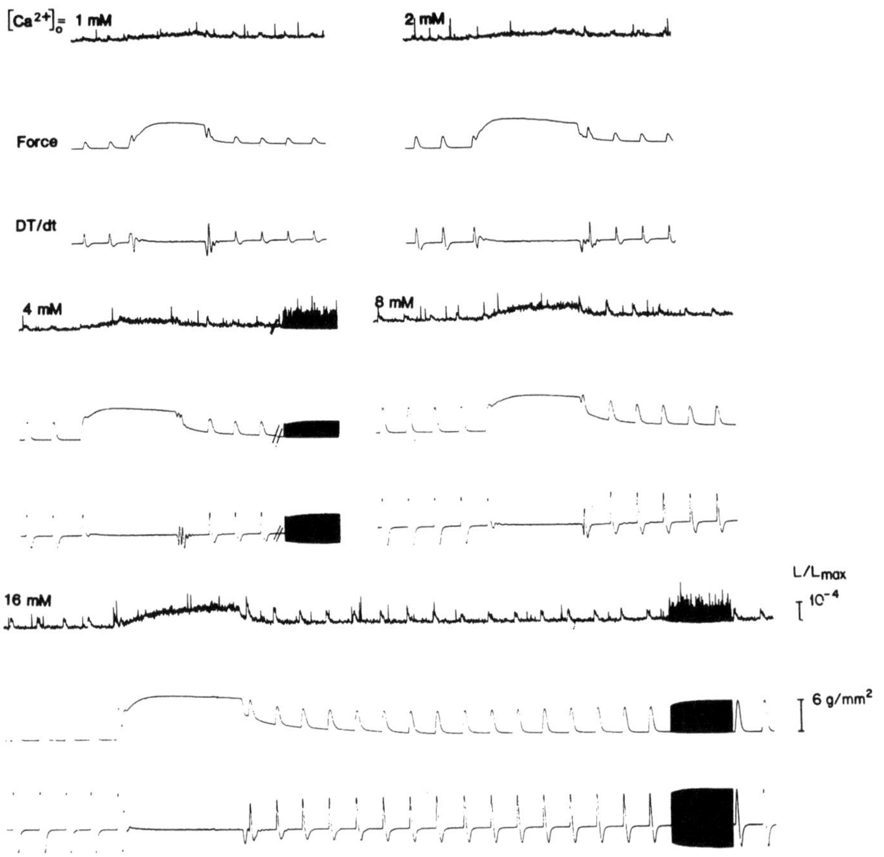

Fig. 2. Different steady-state levels of developed force at various $[Ca^{2+}]_0$ from a myopathic muscle tetanized by a 15-Hz stimulation frequency, 50-ms square pulse duration, at 30°C. DT/dt demonstrates the abbreviation of the time-course of contraction in post-tetanic twitches. (Reproduced with permission from Gwathmey JK, Hajjar RJ (1990) Circulation 82:1266–1278 by copyright permission of the American Heart Association)

maximal force was attained at 16 mM $[Ca^{2+}]_0$ by adding 0.3 µM BAY K 8644. No further increase in force was observed as compared to 16 mM $[Ca^{2+}]_0$, but $[Ca^{2+}]_i$ increased.

We related the mean levels of force and $[Ca^{2+}]_i$ at the time when both the force and $[Ca^{2+}]_i$ reached a plateau, allowing us to plot steady-state force vs. $[Ca^{2+}]_i$ relationships, as shown in Figure 3. Hill functions were fitted to the data from each experiment using a non-linear estimation method, and values for the Hill coefficient and the $[Ca^{2+}]_i$ at half-maximal force were obtained for each individual experiment. Mean values for the various parameters of the Hill functions were derived from curves fitted to the data of individual experiments. In Figure 3, we plotted the curves for control and myopathic muscles which were essentially superimposable, with no significant differences in Hill parameters (p > 0.2). Peak force and peak $[Ca^{2+}]_i$ were measured at varying $[Ca^{2+}]_0$. Figure 4 illustrates the peak twitch force-peak $[Ca^{2+}]_i$

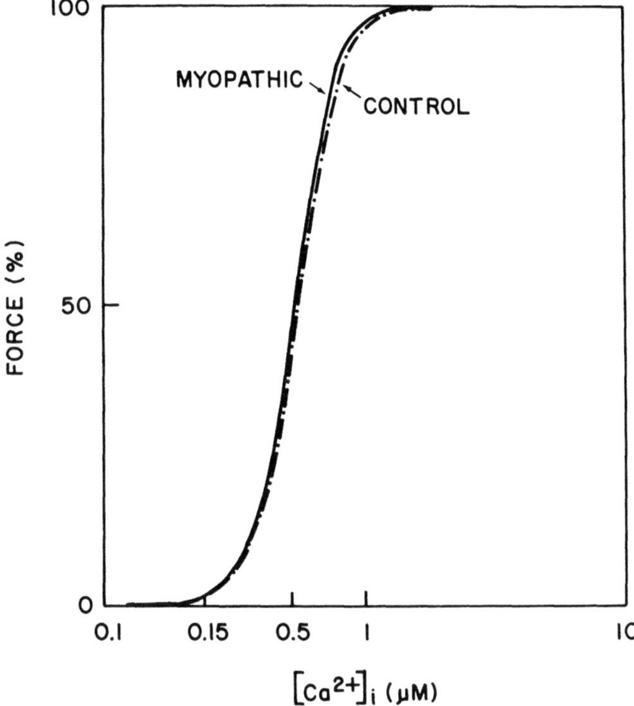

Fig. 3. Steady-state relationship between force and $[Ca^{2+}]_i$ in control (n = 6) and myopathic (n = 10) muscles derived from curves fitted individually to the data of each experiment. The mean values for various parameters of the Hill functions were derived from curves fit to the data of individual experiments. $[Ca^{2+}]_{50\%}$, Hill coefficient, and F_{max} were 0.56 ± 0.05 µM, 5.21 ± 0.20, and 1.23 ± 0.49 g/mm² in control muscles, and 0.54 ± 0.09 µM, 5.61 ± 0.60, and 1.84 ± 0.74 g/mm² in myopathic muscles. (Reproduced with permission from Gwathmey JK, Hajjar RJ (1990) Circulation 82:1266–1278 by copyright permission of the American Heart Association)

relationship in control and myopathic muscles. The peak $[Ca^{2+}]_i$-peak twitch force curve of the myopathic heart is shifted to the left of the $[Ca^{2+}]_i$-force curve of control myocardium. These data reflect the longer $[Ca^{2+}]_i$ transient and higher resting $[Ca^{2+}]_i$ in myopathic muscles.

$[Ca^{2+}]_i$ was measured during tetani at varying levels of $[Ca^{2+}]_0$ from control and myopathic muscles and plotted vs. $[Ca^{2+}]_0$. Separate linear fits were derived for control and myopathic relationships (Fig. 5). The lines fitted for control and myopathic data had the following equations, respectively: y = 0.260 + 0.036 × (R = 0.76) and y = 0.385 + 0.069 × (R = 0.86). If we consider this intercept to be an index of resting $[Ca^{2+}]_i$, this result would be in accordance with the findings of Gwathmey et al. (15, 18), who demonstrated an abnormal intracellular calcium handling in myopathic hearts. It is important to note that at any $[Ca^{2+}]_0$, the mean levels of $[Ca^{2+}]_i$ depend upon sarcolemma Ca-permeability, the duration of stimulation, and SR function.

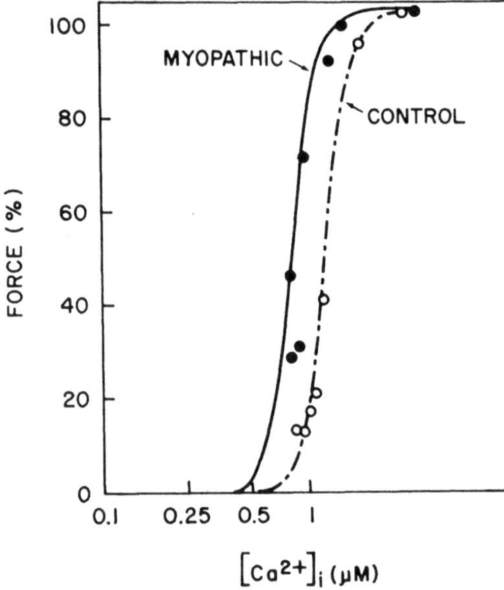

Fig. 4. Relationship between peak twitch force and peak $[Ca^{2+}]_i$ in control and myopathic muscle. (Reproduced with permission from Gwathmey JK, Hajjar RJ (1990) Circ Res 67:744–752 by copyright permission of the American Heart Association)

We also measured resting intracellular calcium concentration at $[Ca^{2+}]_o$ = 2.5 mM. In control muscles, resting $[Ca^{2+}]_i$ was 0.225 ± 0.052 μM, whereas in myopathic muscles, resting $[Ca^{2+}]_i$ was 0.361 ± 0.068 μM (p < 0.01).

Effect of the Inotropic Agent DPI 201-106 on Ca^{2+}-Activation

DPI 201-106, a piperazinyl-indole, has been reported to act as a cardioactive agent with a novel mechanism of positive inotropic action. We examined the effect of DPI first on skinned muscles preparations.

The relationship between force and free calcium was not significantly altered by the addition of 1 μM of DPI in control muscles (Fig. 6), but in myopathic muscles the $[Ca^{2+}]_{50\%}$ was shifted to the left by 0.29 pCa units (p < 0.01). We then used steady-state $[Ca^{2+}]_i$-force relationship in tetanized muscle fibers to evaluate the effects of DPI 201-106 on normal and myopathic hearts. Figure 7 shows $[Ca^{2+}]_i$-force relationships obtained from tetani in myopathic and control human muscles in the presence of 1 μM DPI. In control muscle, DPI does not affect the $[Ca^{2+}]_i$-force relationship, whereas in myopathic muscle, DPI shifts the $[Ca^{2+}]_i$-force relationship to the left, indicating an increase in the sensitivity of myofilaments to Ca^{2+}. These data are in accordance with our previous findings in skinned fiber preparations (22).

Effect of Protein Kinase C Activation

Protein kinase C regulates the activity of a diverse group of cellular proteins. It has been shown to phosphorylate both the inhibitory subunit of troponin I (TnI) and the

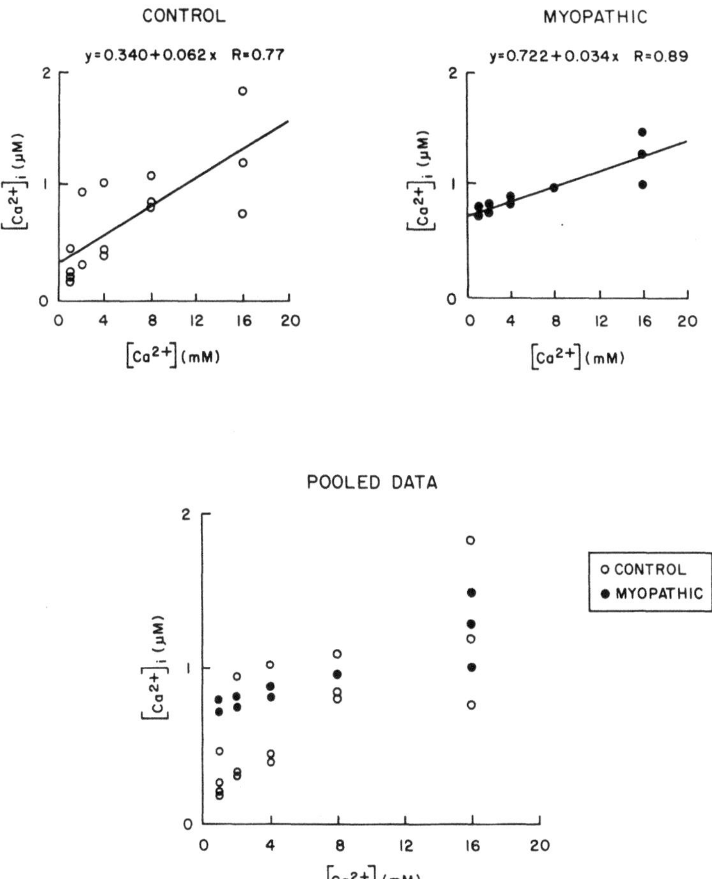

Fig. 5. Dependence of $[Ca^{2+}]_i$ on $[Ca^{2+}]_o$ during tetani pooled from both control (n = 6) and myopathic (n = 10) muscles. The control and myopathic data were segregated and fitted separately using the least-squares method to yield simple regressions. The intercepts were 0.260 µM and 0.385 µM in the control and myopathic muscles, respectively, whereas the slopes were 0.036 µM/mM and 0.069 µM/mM. (Note that some points are not visible on the plot due to overlap).

tropomyosin-binding subunit troponin T (TnT). We examined the effect of protein kinase C by stimulating its activity with 12-deoxyphorbol 13 isobutyrate 20 acetate (DPBA). Figure 8 shows the effect of DPBA on the steady-state $[Ca^{2+}]_i$-force relationships in tetanized preparations of control and myopathic hearts. In the presence of 1 µM DPBA, the steady-state $[Ca^{2+}]_i$-force relationship in both control and myopathic muscles was shifted to the right, indicating a decrease in the sensitivity of the myofilaments to Ca^{2+}. DPBA also decreased F_{max} in both control and myopathic muscles. However, 1 µM DPBA increased the steepness of the $[Ca^{2+}]_i$-force relationship in myopathic muscle fibers, whereas it significantly decreased the steepness of the relationship in normal myocardium.

Discussion

Myopathic hearts have different metabolic and structural properties and function under load conditions that are different from those of normal hearts (27). When compared with normal heart preparations, myopathic preparations show differences in rates of tension development (18), creatine kinase activity (24), and sensitivity to inotropic interventions (13, 34). These differences have been shown to be due to alterations in excitation-contraction coupling (18). It was therefore important to also investigate Ca^{2+}-activation in these myopathic hearts. In our preparations, we found no differences in the sensitivity of the myofilaments to Ca^{2+} and in the maximal Ca^{2+}-activated force between the control and myopathic groups. This result reveals that the dysfunction underlying diseased myocardium does not reside at the level of the myofilaments. This is comparable to pressure-overload hypertrophied hearts which display mechanical, biochemical, and calcium-handling changes, but do not differ in their response to Ca^{2+} (19).

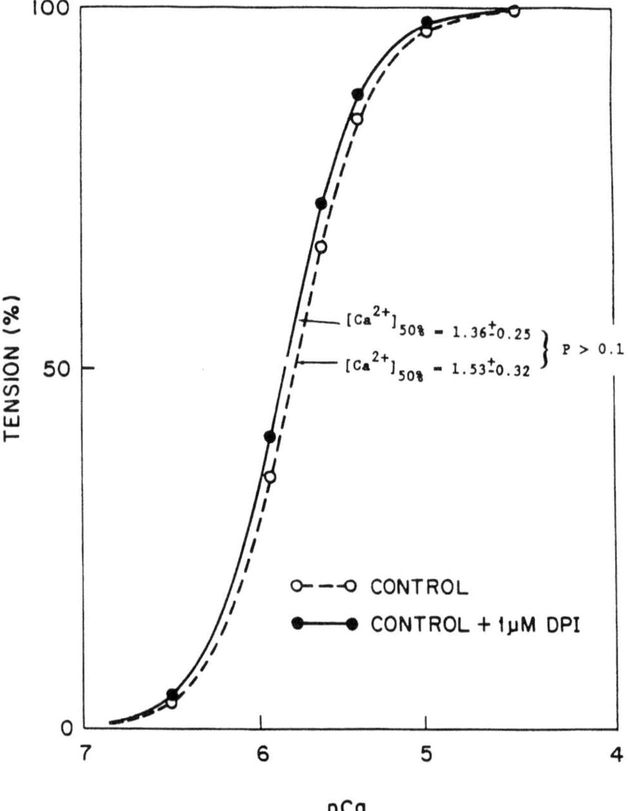

Fig. 6a. Effect of 1 µM DPI on the relation between pCa and percent maximal tension in control myofibers. A DPI concentration of 1 µM was added to each of the calcium buffer solutions. The solid curves are approximated with the Hill relationship.

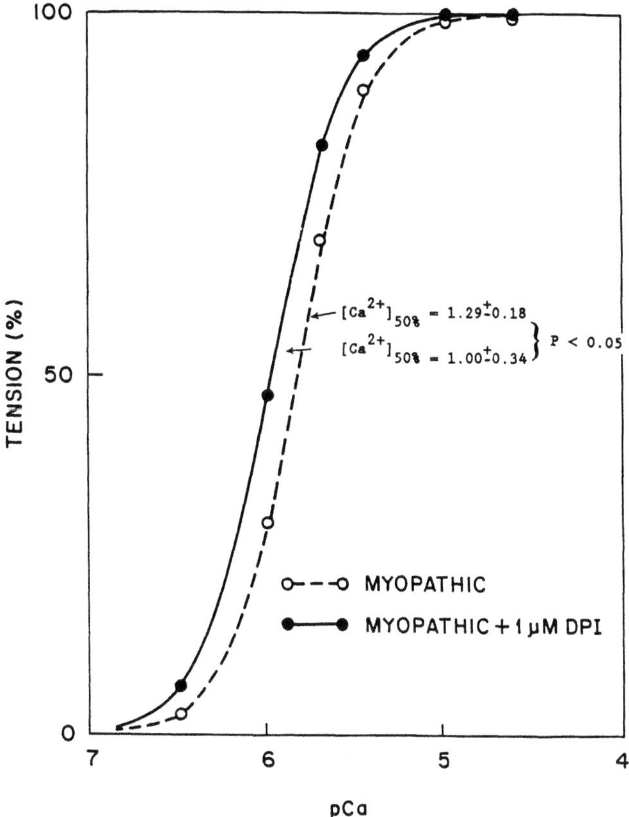

Fig. 6b. Effect of 1 µM DPI on the relation between pCa and percent maximal tension in myopathic myofibers. A DPI concentration of 1 µM was added to each of the calcium buffer solutions. The solid curves are approximated with the Hill relationship.

Since there are no differences in myosin ATPase isoenzymes between normal and myopathic hearts, we investigated the effects of agents that directly act on the myofilaments to further elucidate the $[Ca^{2+}]_i$-force relationships in normal and myopathic hearts. We first studied the effects of a new inotropic agent DPI 201-106 on myofilament calcium responsiveness. The results of our experiments described here indicate that DPI directly acts on human cardiac myofilaments and, further, it differentially alters the sensitivity of the myofilaments to Ca^{2+}. Protein kinase C stimulation by DPBA results in a similar decrease in the sensitivity of the myofilaments to Ca^{2+} in both control and myopathic myocardium, but in the presence of DPBA the Hill coefficient was increased in myopathic hearts and significantly decreased in normal hearts.

The observed changes in Ca^{2+} sensitivity and cooperativity between thick and thin filaments can be achieved through different mechanisms. The molecular mechanism by which calcium controls contraction is explained by the steric hindrance model. In this model, Ca^{2+} allows cross-bridge interaction with the thin filament by binding to troponin C (38), allowing tropomyosin to move on the thin filament away

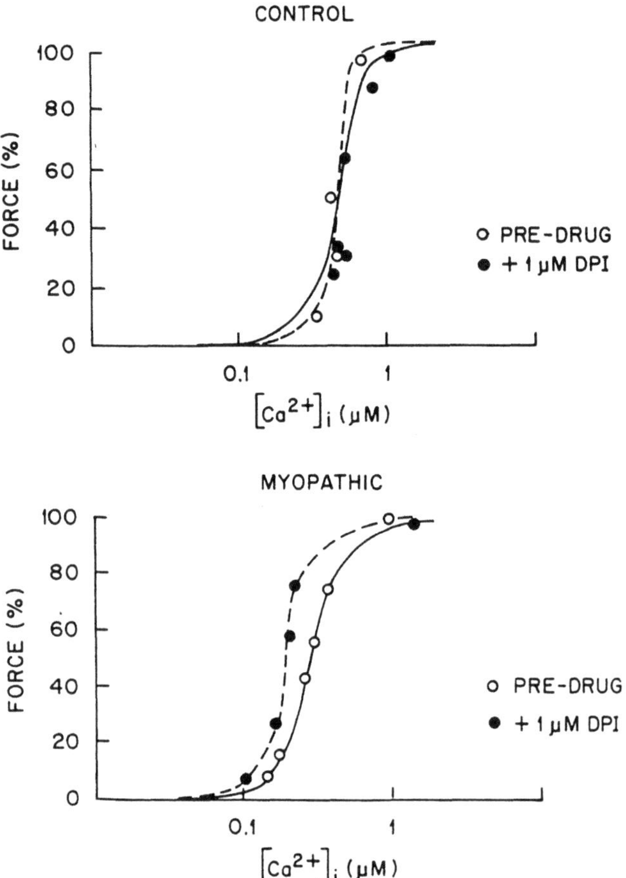

Fig. 7. Effect of 1 µM DPI 201-106 on the steady-state $[Ca^{2+}]_i$-force relationship obtained from tetani in a control muscle (upper panel), and a myopathic muscle (lower panel).

from its site. Phosphorylation of troponin-I, which in the resting state is bound to actin, inhibits actin and myosin from interacting and thus affects force production.

The $[Ca^{2+}]$-force relation in muscles from myopathic hearts exhibited a significant shift to the left, in the presence of DPI, as opposed to muscles from control hearts. This means that, at any concentration of Ca^{2+}, more force was developed by the addition of DPI to myocardium from myopathic hearts as compared to muscles from control hearts. A change in the cAMP-dependent phosphorylation of troponin-I can alter the sensitivity of the myofilaments to Ca^{2+}. DPI does not inhibit cAMP-dependent protein kinase in the concentration ranges studied in these experiments (37); therefore, the influence of DPI cannot be due to a decrease in cAMP-dependent phosphorylation of troponin-I.

Since the effects of DPI on cAMP-dependent phosphorylation have not been studied in myopathic cardiac tissue, we cannot rule out the possibility that troponin-I

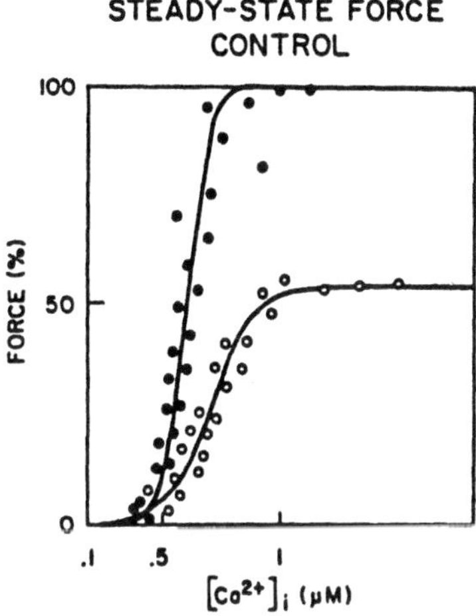

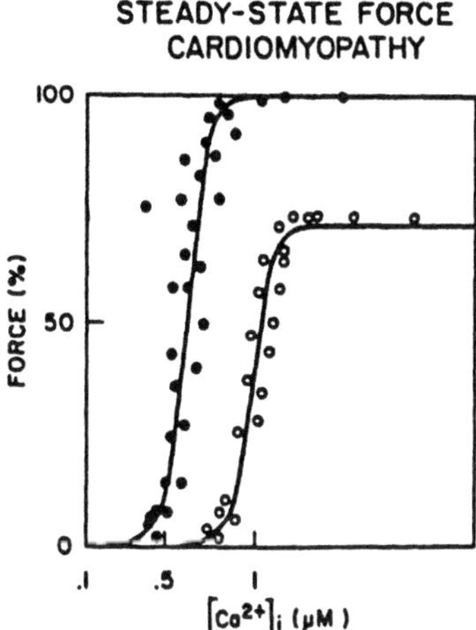

Fig. 8. The effect of 1 µM DPBA on the steady-state $[Ca^{2+}]_i$-force relationship in control (n = 5) and myopathic (n = 8) muscles. Force was normalized to the maximal force before the addition of DPBA. ● = pre-DPBA; ○ = post-DPBA. (with permission)

may exhibit different isozymes in myopathic hearts and be responsible for the sensitization (9). Further, protein kinase C phosphorylates both cardiac TnI and TnT. The results of TnI phosphorylation by protein kinase C would be expected to be a rightward shift of the $[Ca^{2+}]$-force relationship, which occurred in both control and myopathic hearts. Protein kinase C activation phosphorylates at least two sites of TnI (25), which may explain the relatively large shift to the right of the $[Ca^{2+}]$-force relationship.

Another site of action for the differential effect of agents acting on the myofilaments to consider is troponin-C, the cardiac myofibrillar receptor activating actin myosin interaction. Troponin-C might be another myofibrillar protein that is altered in myopathic hearts, however, only one type of troponin C has been found in cardiac muscle (38).

Finally, the troponin T-tropomyosin complex should be considered as another site where functional changes in the myocardium can be correlated with changes in isozyme. The role of TnT in Ca^{2+} activation of the myofilaments has been poorly defined. Recently, investigators have proposed that shifts in TnT isoforms can affect the myofilament Ca^{2+} responsiveness (21, 22, 41). In our study, we found that DPBA increased the Hill coefficient in myopathic hearts, whereas it decreased it in control hearts. In myopathic tissue there are two TnT isoforms, as opposed to one isoform in normal hearts (2, 3). Changes in troponin T isoforms found in developing rabbit hearts have been related to post-natal increase in Ca^{2+}-sensitivity and cooperativity (3). If different isoforms are phosphorylated by protein kinase C in control and myopathic hearts, then one would expect a different response from normal and myopathic myocardium.

Recently, we have shown that cross-bridges in skinned fiber preparations from myopathic hearts cycle at a slower rate than those from normal hearts (20). These findings are comparable with animal models where chronic pressure-overloading of the left ventricle significantly lowers the turnover rate of the cross-bridges (30, 42). Studies in skinned preparations from rat and rabbit models have shown that the shortening velocity is decreased in hypertrophied hearts. In these animal models the diminished cross-bridge cycling rate correlates with altered synthesis of myosin heavy chains which determine myosin ATPase activity and muscle shortening velocity. In response to overload, there is preferential synthesis of V_3 heavy chains, causing slow myosin to replace fast myosin V_1. Although the human ventricle primarily synthesizes a slow-myosin isoform (31, 39), overload has been found to alter the expression of myosin light chains. More recently, there has been evidence that troponin T isoforms may play a role in the regulation of force development and calcium-activation (2, 3). The slower cycling rates observed in human cardiomyopathy may be explained by lower myosin ATPase activity due to different isoforms of troponin T synthesized in the diseased state. Ca-ATPase and actin-activated Mg-ATPase of purified myosin isolated from hypertrophied ventricular muscle have been reported not to be different from values obtained in control human myocardium (31, 39). Further, Ca-ATPase of myosin isolated from human hearts with end-stage heart failure due to cardiomyopathy is also not different from that obtained in control human myocardium, whereas myofibrillar ATPase is reduced in myopathic hearts (33).

Tension generated during each systole facilitates ejection by the heart. A reduction in cross-bridge cycling rate is thought to reduce myocardial contractility. However, this may indeed not be the case as a decreased cross-bridge cycling rate allows longer interaction periods between actin and myosin, resulting in greater force development (5). Furthermore, even though the slowing of cross-bridge cycling may reduce contractility, most likely in the presence of normal intracellular calcium handling, it improves mechanical efficiency and so is sparing in energy (1). Brenner (6) has proposed that, with no change in troponin C calcium binding affinity, increases in the cross-bridge cycling rate would result in a rightward shift on the calcium axis of the pCa-force relationship with a resultant reduced apparent calcium sensitivity, whereas a slower cycling rate would result in a leftward shift with an increase in apparent calcium sensitivity. Therefore, in diseased human myocardium with a slower cross-bridge cycling rate there would be an increase in apparent calcium sensitivity. This would explain, in part, the finding of similar if not enhanced contractile performance reported in failing human myocardium when stimulated at relatively slow rates and under relatively hypothermic conditions (18).

Previous studies have shown that contractions and Ca^{2+} transients of muscles from failing hearts were markedly prolonged (14–16). These studies showed that abnormal Ca^{2+} handling plays a role in the diastolic and systolic dysfunction observed in heart failure (14). The time-courses of calcium transients are determined by the rate at which the SR uptakes released calcium and the rate at which troponin C releases calcium. The regulation of cross-bridge kinetics by Ca^{2+} does not directly correlate with Ca^{2+} binding to troponin C (6, 32). There is mounting evidence that the regulation of cross-bridge turnover kinetics depends on the balance between number of cross-bridges in the force-generating state (f_{app}) and the non-force-generating state (g_{app}). Slower cross-bridge cycling rates translate into a longer time spent in the force generating state. This will in turn increase the ratio of f_{app}/g_{app} which will lead to an increase in apparent Ca^{2+} sensitivity and cooperativity although Ca^{2+} binding to regulatory proteins remains unchanged. The increase in Ca^{2+}-sensitivity would then result in a slower release of Ca^{2+} during relaxation. So the result of a slower cross-bridge cycling rate is consistent with the slower time-course of calcium and force transients in heart failure.

Our findings show that there are no differences in maximal Ca^{2+}-activated force in normal and myopathic human skinned fiber preparations (22). A number of investigators have suggested that the systolic failure observed in cardiomyopathy may be due to a reduced number of cross-bridges (1, 23). Our studies, however, have not found any reduction in force development between control and myopathic muscles at low stimulation rates (14–16). Our results indicate that force development and the number of cross-bridges attached are not significantly different between control and myopathic tissue (20).

References

1. Alpert NR, Mulieri LA (1982) Increased myothermal economy of isometric force generation in compensated hypertrophy induced by pulmonary artery constriction in rabbit: A characterization of heat liberation in normal and hypertrophied right ventricular papillary muscles. Circ Res 50:491–500
2. Anderson PAW, Oakley A, Allen PD (1989) Human troponin T expression in normal and end-stage heart failure patients. Circulation 80 (Suppl II): 503 (Abs)
3. Anderson PAW, Moore GE, Nassar RN (1988) Developmental changes in rabbit left ventricular troponin T. Circ Res 63:742– 747
4. Blinks JR, Wier WG, Hess P, Prendergast FG (1982) Measurement of Ca^{2+} concentrations in living cells. Prog Biophys Mol Biol 40:1–114.
5. Brandt PW, Cox RN, Kawai M, Robinson T (1982) Regulation of tension in skinned muscle fibers. J Gen Physiol 997–1016
6. Brenner B (1988) Effect of Ca^{2+} on cross-bridge turnover kinetics in skinned single rabbit psoas fibers: Implications for regulation of muscle contraction. Proc Natl Acad Sci 85:3265–3269
7. Colucci WS, Wright RF, Braunwald E (1986) New positive inotropic agents in the treatment of congestive heart failure. Mechanisms of action and recent clinical developments. First of two parts. New Engl J Med 314:290–301
8. Colucci WS, Wright RF, Braunwald E (1986) New positive inotropic agents in the treatment of congestive heart failure. Mechanisms of action and recent clinical developments. Second of two parts. New Engl J Med 314:349–358
9. Dieckman LJ, Solaro RJ (1990) Effect of thyroid status on thin-filament Ca^{2+} regulation and expression of troponin I in perinatal and adult rat hearts. Circ Res 67:344–351
10. Endo M, Ilno M (1980) Specific perforation of muscle cell membranes with preserved SR functions by saponin treatment. J Muscle Res Cell Motility 1:89–100
11. Fabiato A, Fabiato F (1979) Calculator programs for computing the compositions of the solutions containing multiple metals and ligands used for experiments in skinned muscle cells. J Physiol (Paris) 1979;75:463–505
12. Fabiato A (1981) Myoplasmic free calcium concentration reached during the twitch of an intact isolated cardiac cell and during calcium induced release of calcium from the sarcoplasmic reticulum of a skinned cardiac cell from the adult rat or rabbit ventricle. J Gen Physiology 1981; 78:457–497
13. Feldman MD, Copelas L, Gwathmey JK, Phillips P, Warren SE, Schoen FJ, Grossman W, Morgan JP (1987) Deficient production of cyclic AMP: pharmocological evidence of an important cause of contractile dysfunction in patients with end-stage heart failure. Circulation 75(2):331–339
14. Gwathmey JK, Warren SE, Briggs GM, Copelas L, Feldman FD, Philips PJ, Callahan M, Schoen FJ, Grossman W, Morgan JP (1991) Diastolic dysfunction in hypertrophic

cardiomyopathy. Effect on active force generation during systole. J Clin Invest 1991;87:1023–1031

15. Gwathmey JK, Slawsky MT, Hajjar RJ, Briggs GM, Morgan JP (1990) Role of intracellular calcium handling in force-interval relationships of human ventricular myocardium. J Clin Invest 85:1599–1613

16. Gwathmey JK, Hajjar RJ (1990) Intracellular calcium related to force development in twitch contraction of mammalian myocardium. Cell Calcium 11:531–538

17. Gwathmey JK, Hajjar RJ (1990) Relation between steady-state force and intracellular [Ca^{2+}] in intact human myocardium: Index of myofibrillar response to Ca^{2+}. Circulation 82:1266–78

18. Gwathmey JK, Copelas L, MacKinnon, Schoen F, Feldman M, Grossman W, Morgan JP (1987) Abnormal intracellular calcium handling in myocardium from patients with end-stage heart failure. Circ Res 61:70–76

19. Gwathmey JK, Morgan JP (1985) Altered calcium handling in experimental pressure-overload hypertrophy in the ferret. Circ Res 57:836–843

20. Hajjar RJ, Gwathmey JK (1991) Contractile dysfunction in failing human hearts: Role of cross-bridge interactions. Circulation 84:447 (abstract)

21. Hajjar RJ, Gwathmey JK (1990) Modulation of calcium-activation in control and pressure-overload hypertrophied ferret hearts: Effect of DPI 201–106 on myofilament calcium responsiveness. J Mol Cell Cardiol 23:65–75

22. Hajjar RJ, Gwathmey JK, Briggs GM, Morgan JP (1988) Differential effect of DPI on the sensitivity of the myofilaments to Ca^{2+} in intact and skinned trabeculae from control and myopathic human hearts. J Clin Invest 82:1578–1584

23. Hasenfuss G, Mulieri LA, Blanchard EM, Holubarsch C, Leavitt BJ, Ittleman F, Alpert NR (1991) Energetics of isometric force development in control and volume-overload human myocardium. Circ Res 68:836–846

24. Ingwall JS, Kramer MF, Fifer MA, Lorell BH, Shemin R, Grossman W, Allen PD (1985) The creatine kinase system in normal and diseased human myocardium. N Engl J Med 313:1050–1054

25. Katoh N, Wise BC, Kuo JF (1983) Phosphorylation of cardiac troponin inhibitory subunit (TnI) and tropomyosin-binding subunit (troponin T) by cardiac phospholipid-sensitive Ca^{2+}-dependent protein kinase. Biochemistry 209:189–195

26. Katz AM (1983) *Physiology of the heart*. Raven Press, 1983, New York

27. Katz AM (1990) Cardiomyopathy of overload: A major determinant of prognosis in congestive heart failure. N Eng J Med 322:100–110

28. Kitada Y, Narimatsu A, Matsumura N, Endo M (1987) Contractile proteins: Possible targets for the cardiotonic action of MCI-154, a novel cardiotonic agent? Eur J Pharm 134:229–231

29. Marban E, Kusuoka H (1987) Maximal Ca^{2+}-activated force and myofilament sensitivity in intact mammalian hearts. J Gen Physiology 90:609–623

30. Maughan D. Use of functionally skinned tissue in studying altered contractility in hypertrophied myocardium (1983) In: Alpert NR (ed) Perspectives in Cardiovascular Research, Vol. 7 Myocardial Hypertrophy and Failure. Raven Press, New York, pp 337–343

31. Mecardier JJ, Bouveret P, Gorga L, Schiaffino S, Clark WA, Zak R, Swynghedauw B, Schwartz K (1983) Myosin isoenzymes in normal and hypertrophied human ventricular myocardium. Circ Res 53:52–62

32. Morano I, Bletz C, Wojciechowski R, Ruegg JC (1991) Modulation of cross-bridge kinetics by myosin isoenzymes in skinned human heart fibers. Circ Res 68:614–618

33. Pagani ED, Alousi AA, Grant AM, Older TM, Dziuban SW, Allen PD (1988) Changes in myofibrillar content and Mg-ATPase activity in ventricular tissues from patients with heart failure caused by coronary artery disease, cardiomyopathy, or mitral valve insufficiency. Circ Res 63:380–385

34. Perreault CL, Meuse AJ, Bentivegna LA, Morgan JP (1990) Abnormal intracellular

calcium handling in acute and chronic heart failure: role in systolic and diastolic dysfunction. Eur Heart J 11 (Suppl C):8–21
35. Rüegg JC (1986) Effects of new inotropic agents on Ca^{2+} sensitivity of contractile proteins. Circulation 73 (suppl III):78–84
36. Rüegg JC, Brewer S, Zeugner C, Trayer IP (1989) Peptides from the myosin heavy chain are calcium sensitizers of skinned skeletal muscle fibers. J Muscle Res Cell Motil 10:152–162
37. Salzmann R, Scholtysik G, Clark B, Berthod R (1986) Cardiovascular actions of DPI 201–106, a novel cardiotonic agent. J Cardiovasc Pharm 8:1035–1043
38. Scheuer J, Bhan AK (1979) Cardiac contractile proteins. Circulation Research 45:1–12
39. Schier JJ, Adelstein RS (1982) Structural and enzymatic comparison of human cardiac muscle myosin isolated from infants, adults, and patients with hypertrophic cardiomyopathy. J Clin Invest 69:816–825
40. Scholtysik G, Salzmann R, Berthold R, Herzig JW, Quast U, Markstein R (1985) DPI 201–106, a novel cardioactive agent. Combination of cAMP-independent positive inotropic, negative chronotropic, action potential prolonging and coronary dilatory effects. Arch Pharm 329:316–325
41. Tobacman LS, Lee R (1987) Isolation and functional comparison of bovine cardiac troponin T isoforms. J Biol Chem 262(9):4059–64
42. Ventura-Clapier R, Mekhfi H, Olivero P, Swynghedauw B (1988) Pressure-overload changes cardiac skinned-fiber mechanics in rats, not in guinea pigs. Am J Physiol 254:H517–H524
43. Yue DT, Marban E, Wier WG (1986) Relationship between force and intracellular [Ca^{2+}] in tetanized mammalian heart muscle. J Gen Physiol 17:223–242

Author's Address:
Judith K Gwathmey VMD, Phd
Beth Israel Hospital
330 Brookline Avenue
Boston, MA 02215
USA

The regulation of the human β myosin heavy-chain gene

H.-P. Vosberg, U. Horstmann-Herold, A. Wettstein

Max-Planck-Institute for Medical Research, Department of Cell Physiology, Heidelberg, FRG

Summary: The human myosin heavy-chain (MHC) genes for cardiac and skeletal muscle exist as a multigene family with eight or more non-allelic genes. Two of them code for the cardiac α and β myocin HCs. They are located on chromosome 14. The skeletal muscle myosin HC genes are on chromosome 17. The cardiac MHCs coexist in the heart, however, with a distinct distribution within cardiac tissue of the human adult. α-MHC is predominantly found in the atria and β-MHC is found in the ventricles. Both genes are also expressed in certain types of skeletal muscle fibers. We have sequenced the β-gene in its entire length and have further studied in detail its expression in muscle cells. Promoter activities were tested using DNA-mediated gene transfer in cultured chicken embryonic myoblasts. By deletion mapping of the 5′ flanking region of the β-gene a candidate signal sequence was identified in a region which stimulates the promoter in a tissue specific and differentiation dependent mode. The presumed signal was located about 210 bp 5′ to the basic promoter which, by itself, is almost inactive, even in muscle cells. The sequence of the signal (CAGCTG) has homology to known E-box sequences. E-boxes (consensus sequence CANNTG) constitute a family of transcription control sites frequently found upstream of muscle genes. In nuclear extracts of cardiac and skeletal muscle (of rabbit) a protein was identified which binds to the region containing the E-box like motif of the β-gene. Since this protein was present in both types of muscle, overlapping expression control patterns are assumed to operate in these tissues.

Key words: Cardiac muscle; myosin heavy chain genes; regulation

Introduction

The study of gene control in the myocardium contributes to an improved understanding of the normal biology of the heart and also of events which induce or accompany cardiac disease. Many functions of the myocardium are characterized by a high degree of phenotypic variability. Developmental stimuli, hormones, physical training, and other physiological or pathological conditions affect cardiac performance and structural composition. Functions which respond to changing conditions are, to name a few important ones, signal transduction, calcium and ion transport, ATP metabolism and contractility (for a comprehensive discussion, see (23)).

At present, it is largely unknown by which mechanisms functional variations are implemented in cardiac muscle. It can be assumed that altered gene expression patterns contribute significantly to these changes. Other events like posttranslational modifications of proteins or membrane components of the cardiocyte are probably also involved.

Major molecular tools for the control of gene expression are changing rates of transcription of a given gene, differential splicing of pre-mRNA and switching between indvidual genes within multigene families containing the information for a limited number of distinct but related proteins.

Contractility is one of the cardiac functions which is genetically represented by multigene families. Myosin light and heavy chains as well as actins are encoded by more than one gene, respectively. We have been analyzing the genetic control of contractile proteins by focusing on the genes and cDNAs coding for the cardiac myosin heavy chains. Two heavy and four light chains are the components of native myosin molecules which assemble within the myofiber to generate thick filaments. These form, together with the thin actin filaments, the core of the contractile apparatus. ATP-dependent sliding of thick along thin filaments is the basic process of muscle contraction.

Among the multigene families coding for myosin heavy chains, those of mammalian organisms have been studied most thoroughly. Two subfamilies have been identified in mouse, rat, and man. In the human genome one is located on chromosome 14, which consists of two genes coding for the cardiac heavy chains of type α and β (22). A second subfamily is on chromosome 17 with presumably six genes coding for the heavy chains of skeletal muscle cells (6). The α and β chains are not exclusively cardiac. The α gene is also transcribed in the masseter muscle, and the β chain is a component of slow skeletal muscle fibers (e.g., in m. soleus). Whether skeletal muscle genes are in turn expressed in cardiac cells is not known to date.

The two cardiac genes are clustered within a close distance of 5 kilobasepairs (kb) apart from each other (15). The exons of the two human genes and the proteins are, as far as they can be compared, rather similar (about 90%), suggesting that this two-gene-cluster resulted from a gene duplication during evolution of the mammalian genome (approximately 500 million years ago). Both human genes have been isolated. The sequence of the β gene is known completely, that of the α gene is partially known (13, 14, 27).

The α and β chains differ with respect to intracardial distribution and in their functional properties. In the adult human heart the β chains are most abundant in the ventricles. The α chain is predominantly found in the atria. Functional differences between the two types exist with respect to speed of contraction and rate of ATP cleavage. The α chain is fast and has a high rate of hydrolysis, the β chain is slow and has a low rate. Two α chains assemble to form native V1 (or fast) myosin, whereas two β chains result in V3 (or slow) myosin (12).

Knowledge about the molecular details of the controls of myosin gene activities is still incomplete. Quite well understood is the influence of the thyroid hormone trijodothyronin (T3) which affects the two cardiac myosin heavy chain genes in an antithetic fashion. The α gene is stimulated, by and the β gene is inhibited by T3 (8). The T3-receptor-binding signals responsible for these controls have been determined in the 5' flanking regions of the α- and β-genes of rat and man (18). Retinoic acid has also been shown to affect the activity of these genes (18). In addition, control functions have been assigned (in the rat) to testosterone and estrogen, which act phenotypically similar to T3: these hormones activate the α-gene and suppress the β-gene (16, 21). The mode of action of sexual hormones on myosin HC genes is not known. It cannot be ruled out that they act indirectly by controlling the level of unidentified regulatory factors, which in turn stimulate or inhibit the myosin genes.

The goal of our present study was the identification of cis-elements and trans-acting factors which are involved in transcriptional control of the β-gene in skeletal and cardiac muscle. As far as gene expression is concerned, our results describe activities in muscle cells. The respective activities in cardiocytes require additional investigation. A protein binding to a DNA sequence apparently involved in promoter regulation was detected in both skeletal and cardiac muscle nuclei.

We note that the β-gene has been implicated in a different field of molecular analysis, i.e., the etiology of inherited cardiomyopathies. Two mutations in the α/β-gene cluster on chromosome 14 have been identified in two families transmitting the dominant familial hypertrophic cardiomyopathy (FHC) (10, 24). Thus, a complete study of the cardiac MHC genes is not only of biological, but also of clinical relevance.

Methods and Materials

The experiments were based on the known sequence of the promoter of the β myosin heavy chain gene and additional 5' flanking regions. The complete sequence of the cloned human β gene has been reported (13, 14). For functional studies promoter and upstream sequences were subcloned into the plasmid vectors pCAT-O and pGCAT-C which contained the gene for chloramphenicol acetyl transferase (CAT) as a bacterial "reporter" gene, but no promoter (11, 9). A promoter containing control plasmid was pSV2-CAT. For transient expression the promoter-CAT gene constructions were transfected into myoblasts isolated from embryonic chicken muscle (m. pectoralis major). The isolation and growth conditions for myoblasts were according to established procedures (2). The measurement of CAT activities in extracts of transfected cells was based on acetylation of (^{14}C)-labeled chloramphenicol analyzed by thin-layer chromatography. For mapping of promoter and regulatory upstream sequences, we applied either random partial digestion with exonuclease III or targeted PCR amplification (19) of selected subregions of the DNA flanking the 5' end of the β-gene. Amplification primers were synthesized by the phosphoamidite procedure (4). For DNA-cloning standard protocols were used (20). Nuclear proteins from muscle and non-muscle cells were extracted essentially as reported (3), with modifications as required for the handling of muscle tissues. Mobility shift assays of (^{32}P)-labeled DNA complexed with nuclear protein were investigated on 5% non-denaturing polyacrylamide gels (28). Radioactive end-labeling of DNA with (^{32}P) was achieved using polynucleotide kinase and gamma-labeled ATP as phosphate donor.

Results

The human β-myosin heavy-chain gene has been cloned and sequenced in its entire length. The arrangement of exons and introns is schematically shown in Figure 1. The gene consists of 40 exons. Two exons at the 5' end are untranslated. The mRNA has a length of 6008 bases and codes for 1945 amino acids. The β-protein sequence shows a generally high degree of homology with other human and non-human vertebrate myosin heavy chains of cardiac or skeletal muscle. As far as differences to other myosins have been identified, they appear to be clustered in regions which are presumably important for the isotype specific properties of the β-chain (13).

Expression control activities of the β-gene were analyzed with 5' flanking DNA sequences. The transcription start point (which is identical with the 5' end of exon 1) was assigned – with a remaining ambiguity of a few basepairs – using β-specific

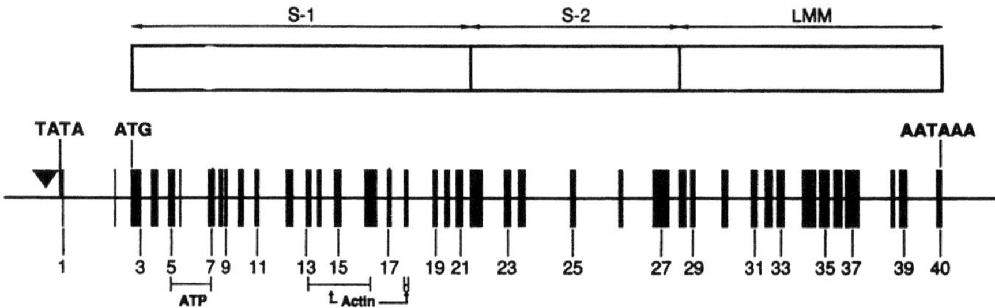

Fig. 1. Structure of the human cardiac β-myosin heavy-chain gene. The TATA box (TATA), the translation start codon (ATG) and the poly-adenylation signal (AATAAA) are indicated. Exons are drawn as black boxes. The triangle points to the promoter region. The major protein domains S-1, S-2 and LMM of the heavy chain are shown on top. Exons coding for ATP and actin binding domains are indicated.

mRNA from human cardiac tissue and enzymatic amplification of its 5' end (13). This result was corroborated by in vitro reverse runoff transcription (not shown). Two basic promoter signals TATA and CTAAAT have been identified upstream of the transcription start point in positions -51 and -106, respectively. (It is not clear whether a CCAATT sequence in position -199, which we previously believed to be part of the basic promoter (26), is involved in the control of the β-gene.)

The 5' flanking region of the β-gene is depicted in Figure 2. The sequence is documented from position -330 to -91 relative to the transcriptional start site (position $+1$).

Transient expression of transfected genes

Transcription control experiments were initiated by fusing a 1.3 kb *Hind*III fragment extending from position -1283 to $+103$ of the β-gene to the plasmid pCAT-O containing the CAT reporter gene. The fusion construct was transiently transfected into freshly prepared chicken myoblasts 18 h after plating. Cells started to fuse after about 50 h, and myotube formation was essentially complete after 80 h. Control transfections included the vector pGCAT-C without inserted DNA and plasmid pSV2-CAT which contains the CAT gene under the control of the SV40 promoter and enhancer.

The vector pCAT-O with the 1.3 kb upstream sequence was able to induce CAT activity if this sequence was inserted in transcriptional orientation. No activity was obtained with the insert in the opposite orientation. pCAT-O alone failed to induce CAT. The level of induction by the β-gene promoter was about 20-fold higher than with the negative control and represented about 40–50% of the strong positive control pSV2-CAT. No CAT expression was observed after transfection of the β-promoter construct into non-muscle cells (HeLa) (results not shown).

It was concluded from these data that the 1.3 kb upstream region of the β-gene contained a promoter and, presumably, at least one cis-acting sequence (an enhancer-like element) able to stimulate the promoter in a muscle-specific manner.

The muscle specificity of the promoter was confirmed in two additional experiments. The first one was based on measuring CAT activities before and after the

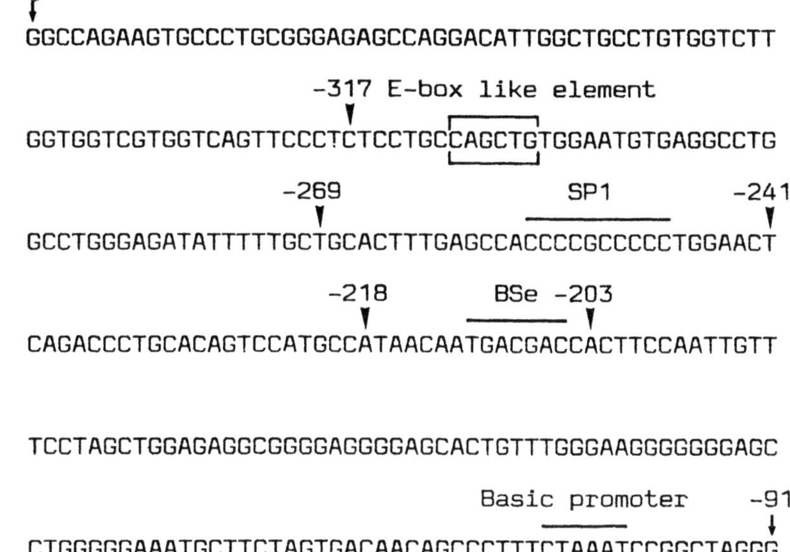

Fig. 2. A segment of the 5' flanking region of the β-gene. The sequence between position -390 and -91 is shown. The CCAAAT box of the basic promoter and the E-box-like CAGCTG motif presumably involved in regulation of the β-promoter are indicated. In addition, two sequences homologous to reported transcription control elements, SP1 and BSe, are marked. The arrows label the 5' cutoffs points (with their position numbers) of deleted promoter upstream regions used in this study.

onset of myotube formation. In this experiment CAT expressing plasmids were added to myoblasts about 30 h after plating at the beginning of fusion. One set of cells was kept under growth conditions, and a second set was induced to form myotubes by lowering the serum concentration. The results indicated quite clearly that the β-promoter was selectively stimulated five- to 10-fold in cells allowed to differentiate terminally as compared to the undifferentiated controls (Fig. 3).

In the next assay the known inhibition of muscle differentiation by the nucleotide analogue 5'-bromodeoxyuridine (BrdU) was tested. Proliferating myogenic cells cultured in the presence of this thymidine analogue fail to fuse and do not initiate the synthesis of contractile proteins (1, 17). If myoblasts were, after exposure of the cells to BrdU (12 h, 20 μM), transfected with CAT plasmids containing the 1.3 kb flanking region or with truncated fragments thereof (see below) reporter gene activity was clearly lower than in the untreated control cells (Fig. 4). We note that no difference in the response to BrdU treatment was seen with inserts harboring the 1.3 kb region or partial sequences thereof with 5' cutoff points at -419 or -317. An insert with a 5' terminus at -203 did not show promoter stimulation, even in the absence of BrdU (see below).

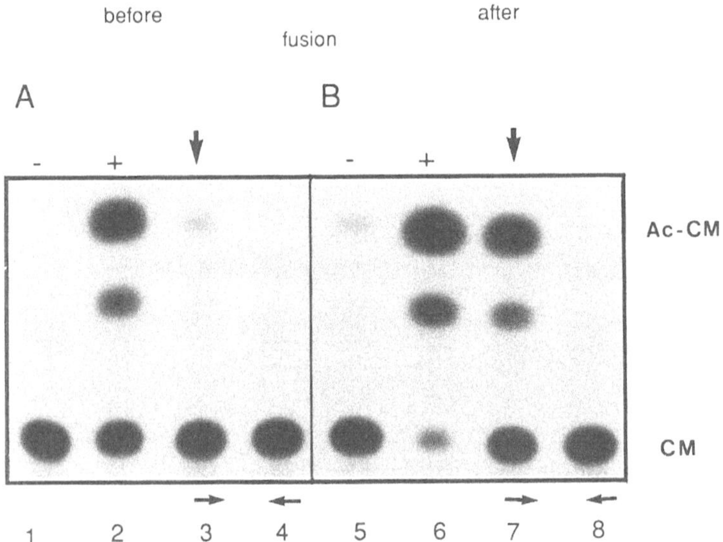

Fig. 3. Stimulation of CAT activity in transfected myoblasts before and after fusion. A 1.3 kb fragment of the 5′ flanking region was inserted into the expression plasmid pCAT-O in front of the CAT gene. CAT activity was monitored by thin-layer chromatography of acetylated (^{14}C)-labeled chloramphenicol (Ac-CM). CAT activity before fusion (panel A) and after fusion (panel B). The lanes show: 1 and 5, negative control (vector without promoter); 2 and 6, positive control (pSV2-CAT); 3, activity in unfused myoblasts with insert in transcriptional orientation; 4, the same as in lane 3, but with insert against transcriptional orientation; 7, activity in fused myoblasts with insert as in lane 3; 8, activity in fused myoblasts with insert as in lane 4. The arrows point to the products obtained in unfused and fused myoblasts.

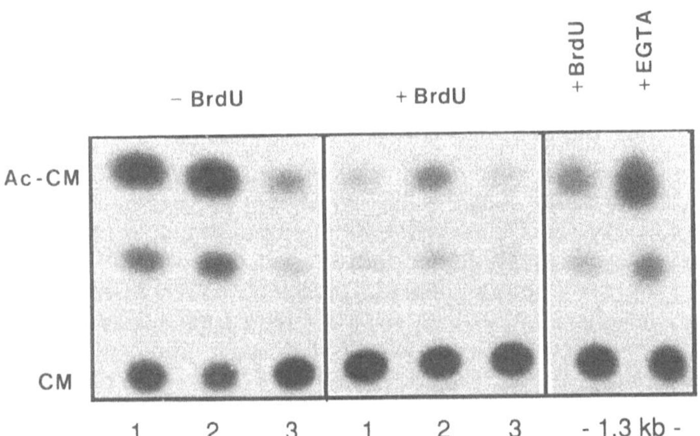

Fig. 4. Suppression of β-promoter controlled CAT activity by BrdU. Three flanking fragments with 5′ cutoff points at -419, -317, and -203 cloned in vector pGCAT-C were tested in lanes 1, 2, and 3, respectively. Extracts were from untreated ($-$ BrdU) and treated myotubes ($+$ BrdU). The effect of EGTA was measured against the inhibition by BrdU with the 1.3 kb insert in pCAT-O (right panel).

An inhibition of myoblast fusion by EGTA (1, 8 mM) did not repress the promoter-induced CAT activity after transfection. EGTA acts by depleting the medium of calcium which is required for the process of fusion (7) (see Fig. 4).

For narrowing the region responsible for CAT induction in muscle cells, upstream fragments with different cutoff points at their 5′ ends were made in two different ways. The first one involved orientation-specific limited digestion of cloned 1, 3 kb DNA with exonuclease III leading to fragments with 5′ ends at the positions − 1145, − 884, − 610 and − 515, respectively. All of these truncated fragments were able to trigger CAT induction if inserted in pGCAT-C plasmids in transcriptional orientation. They were all fusion-dependent and were inhibited by BrdU. These results indicate that critical regulatory signals were located 3′ to the shortest fragment (cutoff at − 515). (Results with exonuclease III trimmed fragments are not shown).

The second method for narrowing the control region was based on enzymatic amplification (PCR) of targeted sequences 3′ to the − 515 cutoff point using synthetic oligonucleotides. The PCR products were cloned into the transcription vector pGCAT-C. All fragments obtained by this procedure had the same 3′ end (position − 33) which included the TATA box, but not the β-specific transcriptional start. The 5′ cutoff points were positioned between − 419 and − 203, as indicated in Figure 5.

The CAT expression obtained with these constructions in fused myoblasts is shown in Fig. 6. The data indicate that removal of the region 5′ to position − 317 had no effect on the induction of CAT. A deletion beyond this point (extending to position − 269 or to − 241, respectively) showed a marked decrease in the CAT activity. Finally, if the flanking sequence was cut down to position − 218 or even further, to − 203, almost no CAT activity was observed.

These results imply that a sequence critical for regulation is located between position − 317 and − 269. They also suggest that a second, albeit less effective site stimulating CAT activity may be located between − 269 and − 241. Finally, the basic promoter is by itself rather weak (almost inactive) and presumably requires for its activity one or more cis-acting sequences in upstream positions. The region between − 317 and − 269 was tentatively called upstream regulatory element (URE).

Identification of nuclear binding proteins

The regulation of a promoter by a cis-acting DNA signal sequence depends on the interaction with a specific trans-acting binding factor. In the search for proteins recognizing sequences within the CAT stimulating URE region, nuclear extracts were isolated from cardiac muscle and analyzed for protein binding to appropriate DNA fragments. Since the β-gene is expressed not only in the heart, but also in slow skeletal muscle, we additionally prepared nuclear extracts from soleus muscle. We used rabbit heart and skeletal muscle as a source for nuclei. The choice of heterologous non-human mammalian tissues was justified because the sequences upstream of the β-genes of man and rabbit are about 80% homologous up to position − 390 (numbering according to the human sequence) with long stretches of perfect match in between (5). Therefore, we expected that rabbit β-gene binding proteins were able to recognize the corresponding human DNA sequences correctly.

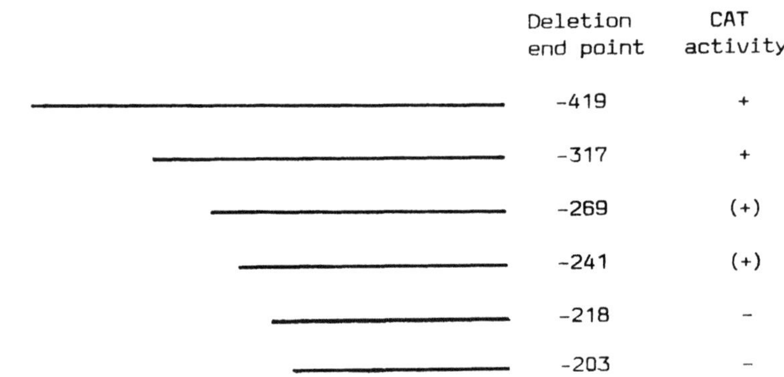

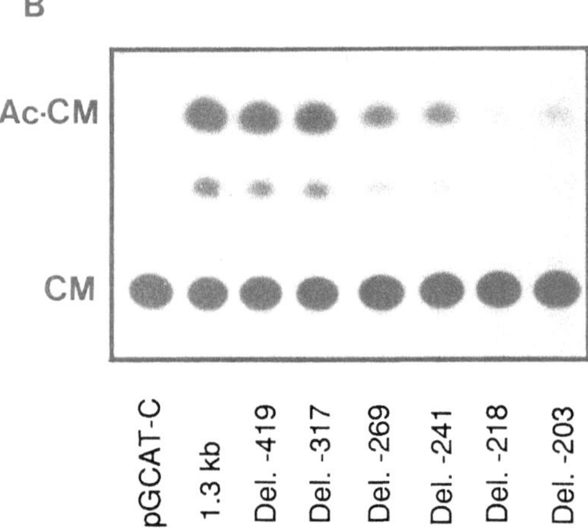

Fig. 5. Deletion mapping of the 5' upstream region. Panel A depicts schematically the deletion clones in pGCAT-C obtained by amplification of selected sequences with 5' cutoff points between − 419 and − 203. The 3' cutoff point for all clones was − 33. Panel B shows CAT activities obtained with the individual clones. The original 1.3 kb insert was included in this series as a positive control for promoter stimulation. The negative control was pGCAT-C. In all experiments β-galactosidase expressing plasmids were co-transfected to measure transfection efficiencies.

Protein binding was monitored by electrophoretic mobility shift assays with radioactively labeled DNA fragments. DNA complexed with protein migrates slower in non-denaturing gels than did free DNA. The probe DNA was a 114 bp fragment (synthesized by PCR) extending from position − 363 to − 250, a region which included the tentatively assigned URE.

The mobility shifts obtained with proteins from nuclear extracts of cardiac and skeletal muscle are depicted in Figure 6A and B. The two lanes in panel A show that,

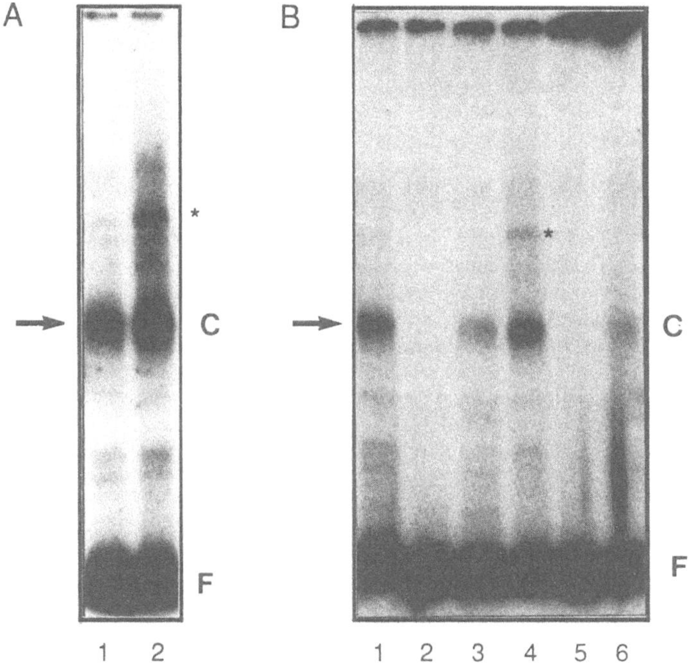

Fig. 6. Mobility shift assays with a DNA fragment containing the upstream regulatory element (URE) of the β gene. The probe DNA extending from − 317 to − 250 was radiolabeled at the 5′ end of the coding strand. The gel position of the DNA-protein complex is indicated by the arrow. A) Lane 1, nuclei from skeletal muscle; lane 2, nuclei from cardiac muscle. A minor band (asterisk) was formed with cardiac extracts only. B) Competition experiments: lane 1 and 4, uncompeted mobility shift with skeletal muscle and cardiac extracts, respectively; lane 2 and 5, complex formation in the presence of cold probe DNA in excess (100 ×); lanes 3 and 6, complex formation with competing DNA (76 bp) in excess (100 ×), extending from − 269 to − 194. Complex formation with labeled probe DNA was partially affected, but not completely suppressed by the competitor DNA. F, unbound DNA; C, DNA-protein complexes.

in both types of muscle nuclei, at least one protein exists which, upon binding, alters the mobility of the DNA to the same respective position (see the arrow). In cardiac nuclear extracts an additional minor DNA band appeared regularly in a position higher up in the gel (marked by an asterisk). In a competition-binding experiment with a mixture of labeled and unlabeled target DNA, complex formation with the labeled fragment was completely suppressed in both extracts (Fig. 6B, lanes 2 and 5). In a control experiment with a 120 bp DNA sequence entirely unrelated to the β-promoter region (a 123 bp long PCR fragment from intron 24 of the β-MHC gene) no competition was observed (not shown). These results suggested a specific interaction between a nuclear protein and the region containing the upstream regulatory sequence.

For defining more closely the DNA region required for binding, we made two competing DNA fragments which partially overlapped with the target region (− 363

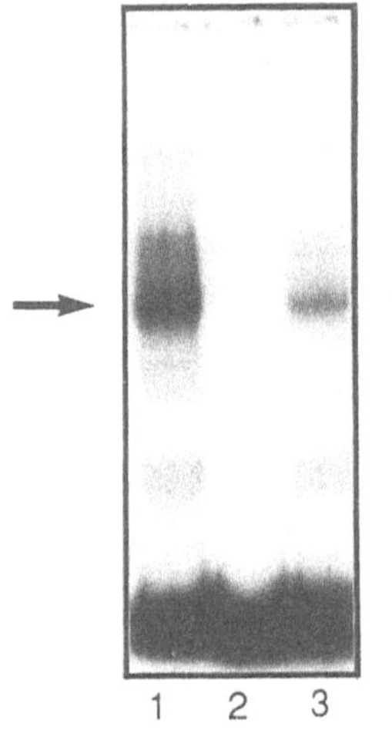

Fig. 7. Mobility shift assays with nuclear extracts of muscle. The radiolabeled probe DNA was as in Figure 6. Lane 1, shift without competing DNA; lane 2, competition by unlabeled probe DNA in excess (100 ×); lane 3, competition by a competing 106 bp fragment extending from position − 299 to − 194 in excess (100 ×). This fragment overlapped for 50 bp with the probe DNA. It decreased the amount of complex formed, but it did not completely suppress it. F, unbound DNA; C, DNA-protein complexes.

to − 250). The first competing fragment (− 269 to − 194) overlapped by 20 bp and the second one (− 299 to − 194) by 50 bp. Both competing sequences exhibited some, but definitely incomplete interference with protein binding of the target sequence (see Fig. 6B: 20 bp overlap in competing DNA, extracts from cardiac and skeletal muscle in lanes 3 and 6, and Figure 7: 50 bp overlap, extract from skeletal muscle).

These results suggest that protein binding occurs predominantly in a region which in functional assays is required for full induction of CAT in transfection tests. Removal of DNA 3′ to − 317 leads to a drastic decrease of CAT activities. Hence, a stimulatory enhancer-like sequence recognized by a muscle-specific regulatory protein (or proteins) may be located close to this position within the promoter upstream region of the β-gene.

Discussion

Studies on a growing number of eukaryotic genes have shown that they are regulated by promoters under the control of trans-acting factors interacting with defined binding sites in the DNA (cis-elements). These sites are frequently, although not exlcusively located upstream and adjacent to their respective promoters. The cardiac myosin HC genes are under various controls, but molecular details of the mechanisms are – with the exception of the T3 receptors – not understood.

The basic promoter of the β-gene is by itself rather weak, and almost inactive, even in muscle cells. We have identified in the 5′ flanking region of the human β-myosin heavy chain gene at least one cis-acting element which was able to stimulate the β-promoter in a muscle specific manner.

The region investigated extended from position − 1283 to the transcription start point. Removal of DNA down to position − 317 did not affect the stimulation of the CAT reporter gene in chicken myoblasts. This result suggests the presence of a regulatory signal, presumably an enhancer-like element, within about 200 bp upstream of the basic promoter and 3′ to − 317.

Removal of DNA between − 317 and − 269 induced a decrease in promoter strength. But the CAT gene was still active, indicating that a second site involved in transcription control might exist in this region. However, this result could also suggest that a trans-factor which binds strongly to a sequence in the vicinity of − 317, still binds to the region further downstream, albeit with a reduced strength.

A protein was detected in extracts of heart and skeletal muscle nuclei which binds in vitro to a sequence near the 5′ end of that region (3′ to − 317) which, upon transfection, appeared to be necessary for stimulation of the β-promoter in muscle cells.

The binding protein isolated from nuclear extracts was able to form complexes with a 114 bp DNA fragment which included the region around position − 317. The fact that a sequence extending from − 194 to − 299 showed only weak competition, even if applied in large excess, suggested that a signal required for binding is located 5′ to − 299.

Thus, the two critical observations are the decrease in promoter activity if sequences 3′ to − 317 were removed and the demonstration of a protein binding site 5′ to − 299. These results taken together could indicate that the 18 bp stretch between − 299 and − 317 is required for both promoter stimulation and protein binding. By searching for a candidate signal sequence in this region, we have located a hexameric motif CAGCTG in position − 312 to − 309. This motif matches a family of signals of the consensus composition CANNTG (also called E-boxes). They play a major role in the differentiation of muscle and in the control of muscle genes by proteins of the myoD family (25). It is conceivable that the β-gene is stimulated by an E-box like element about 200 bp upstream of the basic promoter.

A role of the CAGCTG sequence in complex formation with nuclear proteins was corroborated by DNAse "footprinting" which allows to identify in detail DNA regions protected by protein. This analysis showed that a binding protein has contact with the DNA in the region containing the presumed enhancer-like site (Wettstein, unpublished). Thus, evidence is accumulating that this part of the flanking DNA is involved in β-gene control.

It is not known whether the similarity of the CAGCTG motif with the E-box consensus sequence is fortuitous or not. A transcriptional control sequence in the corresponding upstream region of the rabbit β-myosin heavy chain gene (5) shows only loose homology to the human β-specific motif (CANGCNNTG versus CAGCTG). Thus, this element, although apparently involved in regulation, may in a strict sense not belong to the family of E-boxes.

By screening the upstream region of the β-gene for other possible regulatory signals two motifs were identified which closely resemble known eukaryotic transcription control sequences, SP1 and BSe, at positions − 57 and − 211, respectively (see Fig. 2). The orientation of the human BSe is inverse to that detected in the upstream region of the embryonic myosin IIC gene of rat (28). Whether these signals contribute to the expression of the β-gene is not known.

The β-promoter activities were assayed in this study in chicken skeletal myoblasts and not in human cardiocytes, which would have been the appropriate target cells. Human cells were not available for testing. To use skeletal muscle cells instead seemed justified because of the known appearance of β-myosin in cardiac and in slow skeletal muscle cells. Hence, both types of cells should have trans-factors able to recognize β-control signals. A comparsion of extracts of cardiac and skeletal muscle cells indicate that the two types of nuclei have, indeed, at least one factor in common which recognizes a β-specific signal, suggesting overlapping modes of transcriptional control in these tissues.

References

1. Bischoff R, Holtzer H (1970) Inhibition of myoblast fusion after one round of DNA synthesis in 5-bromodeoxyuridine. J Cell Biol 44:134–150
2. Blau HM, Webster C (1981) Isolation and characterization of human muscle cells. Proc Natl Acad Sci USA 78:5623–5627
3. Braun T, Tannich E, Buschhausen-Denker G, Arnold HH (1989) Promoter upstream elements of the chicken cardiac myosin light chain 2-A gene interact with *trans*-acting regulatory factors for muscle-specific transcription. Mol Cell Biol 9:2513–2525
4. Caruthers MH, Barone AD, Beaucage SL, Dodds DR, Fisher EF, McBride LJ, Matteuci M, Stabinsky Z, Tang JY (1987) Chemical synthesis of deoxynucleotides by the phosphoamidite method. In: Wu R, Grossman L (eds) Methods in Enzymology, Vol 154. Academic Press, San Diego, pp 287–326
5. Cribbs LL, Shimizu N, Yockey CE, Levin JE, Jakovcic S, Zak R, Umeda PK (1989) Muscle-specific regulation of a transfected rabbit myosin heavy chain β gene promoter. J Biol Chem 264:10672–10678
6. Edwards YH, Parker M, Povey S, West LF, Parrington MJ, Solomon E (1985) Human myosin heavy chain genes assigned to chromosome 17 using a human cDNA clone as a probe. Ann Hum Genet 49:101–109
7. Endo T, Nadal-Ginard B (1987) Three types of muscle-specific gene expression in fusion blocked rat skeletal muscle cells: Translational control in EGTA-treated cells. Cell 49:515–526
8. Everett AW, Sinha AM, Umeda PK, Jakovcic S, Rabinowitz M, Zak R (1984) Regulation of myosin synthesis by thyroid hormone: Relative change in the α- and β-myosin heavy chain mRNA levels in rabbit heart. Biochemistry 23:1596–1599
9. Frebourg T, Brison O (1988) Plasmid vectors with multiple cloning sites and CAT-reporter gene for promoter cloning and analysis in animal cells. Gene 65:315–318
10. Geisterfer-Lowrance AAT, Kass S, Tanigawa G, Vosberg HP, McKenna W, Seidman CE, Seidman JG (1990) A molecular basis for familial hypertrophic cardiomyopathy: A β cardiac myosin heavy chain gene missense mutation. Cell 62:999–1006
11. Gorman CM, Moffat GT, Howard BH (1982) Recombinant genomes which express chloramphenicol acetyltransferase in mammalian cells. Mol Cell Biol 2:1044–1051
12. Hoh JHY, Yeoh GPS, Thomas MAW, Higginbottom L (1979) Structural differences in the heavy chain of rat ventricular myosin isoenzymes. FEBS Letters 97:330–334
13. Jaenicke T, Diederich KW, Haas W, Schleich J, Lichter P, Pfordt M, Bach A, Vosberg HP (1990) The complete sequence of the human β-myosin heavy chain gene and a comparative analysis of its product. Genomics 8:194–206
14. Liew CC, Sole MJ, Yamauchi-Takihara K, Kellam B, Anderson DH, Lin L, Liew JC (1990) Complete sequence and organization of the human cardiac β-myosin heavy chain gene. Nucleic Acids Res 18:3647–3651
15. Mahdavi V, Chambers AP, Nadal-Ginard B (1984) Cardiac α- and β-myosin heavy chain genes are organized in tandem. Proc Natl Acad Sci USA 81:2626–2630
16. Morano I, Gerstner J, Rüegg JC, Ganten U, Ganten D, Vosberg HP (1990) Regulation of myosin heavy chain expression in the hearts of hypertensive rats by testosterone. Circ Res 66:1585–1590
17. O'Neill MC, Stockdale FE (1974) 5-Bromodeoxyuridine inhibition of differentiation. Kinetics of inhibition and reversal in myoblasts. Dev Biol 37:117–132
18. Rottmann JN, Thompson WR, Nadal-Ginard B, Mahdavi V (1990) Myosin heavy chain gene expression: Interplay of *cis* and *trans* factors determines hormonal and tissue

specificity. In: Pette D (Ed) The dynamic State of Muscle Fibers. de Gruyter, Berlin New York, pp 3–16

19. Saiki, RK, Gelfand DH, Stoffel S, Scharf SJ, Higuchi R, Horn GT, Mullis KB, Erlich HA (1988) Primer-directed enzymatic amplification of DNA with a thermostable DNA polymerase. Science 239:487–494

20. Sambrook J, Fritsch EF, Maniatis T (1989) Molecular cloning: A Laboratory Manual. 2nd edition. Cold Spring Harbor Laboratory Press, Cold Spring Harbor, New York

21. Scheuer J, Malhotra A, Schaible TF, Capasso J (1987) Effects of gonadectomy and hormonal replacement on rat hearts. Circ Res 61:12–19

22. Solomon SD, Geisterfer-Lowrance A, Vosberg HP, Hiller G, Jarcho JA, Morton CC, McBride WO, Mitchell AL, Bale AE, McKenna W, Seidman JG, Seidman CE (1990) A locus for familial hypertrophic cardiomyopathy is closely linked to the cardiac myosin heavy chain genes, CRI-L436 and CRI-L329 on chromosome 14 at q11-q12. Am J Hum Gen 47:389–394

23. Swynghedauw B (Ed) (1990) Cardiac Hypertrophy and Failure. John Libbey Eurotext Ltd/INSERM, London Paris

24. Tanigawa G, Jarcho JA, Kass S, Solomon SD, Vosberg HP, Seidman JG, Seidman CE (1990) A molecular basis for familial hypertrophic cardiomyopathy: An α/β cardiac myosin heavy chain hybrid gene. Cell 62:991–998

25. Tapscott SJ, Weintraub H (1991) MyoD and the regulation of myogenesis by helix-loop-helix proteins. J Clin Invest 87:1133–1138

26. Vosberg HP, Horstmann-Herold U, Jaenicke T, Morano I, Rüegg JC, Eldin P, Leger JJ (1990) Regulation of the human β-myosin heavy chain gene and an approach to a functional analysis of recombinant protein subregions of the β-chain. In: Pette D (Ed), The Dynamic State of Muscle Fibers. de Gruyter, Berlin New York, pp 45–60

27. Yamauchi-Takihara K, Sole MJ, Liew J, Ing D, Liew CC (1989) Characterization of human cardiac myosin heavy chain genes. Proc Natl Acad Sci USA 86:3504–3508

28. Yu, YT, Nadal-Ginard B (1989) Interaction of nuclear proteins with a positive cis-acting element of rat embryonic myosin heavy-chain promoter: Identification of a new transcription factor. Mol Cell Biol 9:1839–1849

Author's address:
Dr. H.-P. Vosberg,
Max-Planck-Institute for Medical Research,
Department of Cell Physiology,
Jahnstr. 29
D-6900 Heidelberg
FRG

Mutations in cardiac myosin heavy-chain genes cause familial hypertrophic cardiomyopathy

C.E. Seidman[1], J.G. Seidman[2]

[1]Cardiovascular Division, Brigham and Women's Hospital, Boston, USA
[2]Department of Genetics and Howard Hughes Medical Institute Harvard Medical School, Boston, USA

Summary: Familial hypertrophic cardiomyopathy (FHC) is a genetically inherited disorder of heart muscle. Over the past 40 years many studies have detailed the clinical presentation of this disease and its associated pathophysiologic consequences. The primary focus of this review is to discuss more recent studies involving the genetic mapping of one locus on chromosome 14 which causes FHC and then to summarize studies demonstrating that this locus contains mutations in the cardiac myosin heavy-chain genes. The chromosomal location of other putative FHC loci will also be considered. Finally, the implications of results which demonstrate that cardiac myosin heavy-chain defects produce the pathophysiology of FHC will be considered from both a clinical and basic research perspective.

Key words: Hypertrophic cardiomyopathy; myosin heavy chains genetic heterogeneity. Familial hypertrophic cardiomyopathy (FHC); chromosome 14; cardiac myosin heavy-chain genes

Introduction

Pathophysiology

Familial hypertrophic cardiomyopathy (FHC) is a primary and inherited disorder of heart muscle that is characterized by increased ventricular mass, hyperkinetic systolic function and impaired diastolic relaxation (14). The pathologic features of this disorder are well established (21). In addition to the classical finding of asymmetrical thickening of the interventricular septum, hypertrophy of the adjacent left-ventricular anterior free wall, apex, or right ventricle can also occur. Hence the anatomical distribution and severity of hypertrophy can vary considerably (23). Fibrosis occurs within the hypertrophied ventricle, and a fibrotic plaque is frequently demonstrable over the septal region that apposes the anterior mitral valve leaflet during systole. Other gross pathologic findings include atrial dilation and thickening of the mitral valve leaflets (32). The most characteristic histologic abnormalities seen in FHC are myocyte and myofibrillar disarray (7). Myocytes can be hypertrophied to 10–20 times the diameter of a normal cardiac cell and may contain hyperchromatic, bizzare nuclei (2). Cells are arranged in a disorganized fashion with abnormal bridging of adjacent muscle fibers and intercellular contacts, producing whorls. Ultrastructural organization is also distorted; myofibrils and myofilaments are disoriented with irregular z bands (11). While the histopathologic features overlap with those seen in hypertrophy that is secondary to other diseases, the extent of ventricular involvement and severity of myocyte and myofibrillar disarray are considerably greater in FHC.

The pathology of FHC results in physiologic consequences of both systolic and diastolic dysfunction (24). Systolic abnormalities include rapid ventricular emptying, a high ejection fraction, and the development of a dynamic pressure gradient. Reduced left-ventricular compliance results from an increase in the stiffness of the hypertrophied left ventricle and an increase left ventricular mass. Impaired relaxation produces elevated diastolic pressures in the left ventricle as well as in the left atrium and pulmonary vasculature.

Clinical Features

The clinical symptoms in individuals with FHC are variable and may reflect differences in the pathophysiologic manifestations of this disease (12). Affected individuals frequently present with exertional dypsnea, reflecting the diastolic dysfunction that characterizes this disease. Angina pectoris is a common symptom, despite the absence of coronary artery disease. Ischemia may result from increased myocardial demand as well as inappropriately reduced coronary flow due to increased left ventricular diastolic pressures. Sudden, unexpected death is the most serious consequence of FHC, and occurs both in symptomatic and asymptomatic individuals. An annual mortality from sudden death in FHC families approximates 4% annually (26). The mechanism for sudden death is often unknown, and both primary arrhythmias and hemodynamic factors have been proposed. The hypertrophied myocardium is predisposed to ventricular tachycardia and this arrhythmia occurs more commonly in those with sudden death. While the precipitating event of sudden death may be hemodynamic in nature, the vulnerability of the hypertrophic myocardium to arrythmias appears to contribute significantly to the ultimate outcome.

Studies of the natural history of FHC have frequently focused on symptomatic individuals with progressive deterioration of cardiac function. Recent data on ambulatory patients suggests a somewhat more benign prognosis (35). In a series of over 250 affected individuals reported by McKenna et al. (27), average age of onset of symptoms was 29 years, with half of these patients limited in their exercise capacity by dypsnea and angina. Symptomatic deterioration was slow, and often occurred in association with the development of arrythmias. In addition to sudden death, mortality can be attributed to congestive heart failure, cerebral embolisms, endocarditis, and perioperative death following myotomy or myectomy.

The diagnosis of FHC relies on the presence of typical clinical symptoms and the demonstration of unexplained ventricular hypertrophy (21, 28). Two-dimensional echocardiography and Doppler ultrasonography are used to quantitate ventricular wall thickness, cavity dimensions, and demonstrate the presence or absence of systolic anterior motion of the mitral valve. Electrocardiographic findings include bundle-branch block, abnormal Q waves and left-ventricular hypertrophy with repolarization changes.

The precise prevalence of FHC is unknown, however a recent estimate of the prevalence of hypertrophic cardiomyopathy in the United States is 19.7/100 000 (6). The proportion of these cases which is familial is uncertain. Difficulties in assessing disease incidence and prevalence relate to diagnostic methods and criteria which have changed since Teare's description in 1958. The development of two-dimensional echocardiography has particularly improved the sensitivity of diagnostic techniques applicable to FHC. Despite this modality, diagnosis can be difficult, particularly in the young who may exhibit hypertrophy only after adolescent growth has been completed (24).

Genetics of familial hypertrophic cardiomyopathy

FHC has been recognized as a genetic disorder since the late 1950s (37). Family studies have clearly demonstrated an autosomal dominant mode of inheritance (30, 5). That is, 50% of the children of an affected individual would be expected to carry the defective gene in their genomes. Further, the disease gene is highly penetrant and almost all affected individuals exhibit signs and symptoms of the disorder past the age of puberty. These features suggested that the disorder was admirably suited for cosegregation or linkage analyses to identify the chromosome bearing a mutation which causes FHC.

A genetic approach has led to the identification of the defective genes responsible for a number of inherited disorders including neurofibromatosis type I (3) and (39), cystic fibrosis (17) and retinitis pigmentosa (9). The first step in the genetic analysis of FHC involved identifying large kindreds affected by the disease. DNA probes corresponding to loci throughout the entire genome were then screened to identify one which identified a restriction fragment length polymorphism that is co-inherited with the disease. When using this genetic approach to identify the chromosomal location of a disease-causing mutation, one must distinguish between a chance (or random) association between the disorder and a particular DNA polymorphism. The more times that a particular DNA polymorphism is found to be associated with a disease trait among family members the less likely the association is to be random. Geneticists use the LOD score as the index to determine if the association is due to genetic linkage or merely due to random association of two traits (LOD is an abbreviation for logarithm of the odds. This is the logarithm of a probability ratio: the probability of obtaining a particular result assuming genetic linkage between the disease gene and the polymorphic gene divided by the probability of obtaining that result assuming no association between the two genes). A LOD score greater than + 3.0 represents an association that did not occur by chance, whereas a LOD score less than − 2.0 reflects a result that occurred by random chance.

The chance that an autosomal dominant disorder such as FHC will be inherited by a child of an affected parent is 50%. The probability that an affected individual will inherit a given DNA polymorphism from a particular parent is 50%. Thus, each informative individual who co-inherits a particular polymorphism and the disease (or vice versa) increases the probability that these are linked by a factor of 2, which increases the LOD score by 0.30. A mimimum of 10 informative individuals, each of whom has inherited the corresponding pair of traits, is therefore required to achieve a LOD score of 3.0. This problem is compounded by the fact that many nuclear families are not informative with a particular DNA probe. Thus, a large family with about 40 individuals who are the descendants of an affected parent was required for the genetic analysis.

To identify a DNA probe that detected a polymorphism linked to FHC, a large Canadian kindred with over 100 members was clinically assessed, by physical examination, electrocardiogram, and two-dimensional echocardiogram (15). To ensure accuracy of diagnosis, only individuals past the age of 16 were studied. A sample of blood was obtained to develop transformed lymphoblast cell lines which provided a source of DNA for genetic analyses. Southern blot analysis employing a radiolabeled DNA probe was used to define the genotype of each family member. Fortunately, a large number of DNA probes that recognize DNA polymorphisms at loci throughout the genome are now available (8). DNA probes were randomly screened until one was identified that detected a polymorphic locus that co-segregated with the disease locus. Although this could require the use of more than 200 probes, the FHC locus on chromosome 14 was identified with the 43rd probe. In the large Canadian kindred, the LOD score achieved with probe CRI-L436 was + 10.7 (34). A second smaller family showed linkage to the same region of chromosome 14, with a LOD score of + 4.35 (34). Demonstration that the FHC locus was on chromosome 14 (Fig. 1A) gave a very powerful clue to the subsequent identification of the gene responsible for this disorder.

Review of experimental data: a molecular basis for FHC

Cardiac myosin heavy chain gene mutations cause FHC

About 100 of the approximately 10000 genes encoded on chromosome 14 have been identified. Of these 100 genes only two, α and β cardiac myosin heavy chain (MHC) genes were good candidates for the FHC gene (Fig. 1B). These genes encode a major component of cardiac muscle, the myosin heavy chain polypeptides which are abundant in cardiac tissue. Thus, they appeared to have the expression pattern appropriate for an FHC-gene. Each cardiac myosin heavy chain gene is approximately 30000 basepairs long and is divided into 40 exons (33; 25, Fig. 2A). β MHC is expressed at high levels in human adult ventricle, at low levels in atria, and even lower levels in other skeletal muscle tissues. α MHC is expressed at high levels in the human adult atria, low levels in cardiac ventricles, and is not expressed in skeletal muscle tissues.

There are two methods for establishing that defects in a candidate gene cause a disease. One method is to demonstrate that a mutationally altered gene product is present in the affected tissue and adversely effects a known function. This method is not suitable for studies of FHC because cardiac tissues from affected individuals are not obtainable. An alternative method of demonstrating that a putative gene is responsible for FHC would be to demonstrate that several families whose disease locus maps to chromosome 14, have mutations in their cardiac myosin heavy chain genes.

To date, cardiac myosin heavy chain mutations have been defined in affected individuals from two unrelated families with FHC. Figure 2 (B, C) demonstrates the two mutations that have been identified in individuals with FHC. One mutation involves a large rearrangement of the α and β MHC genes. This mutation was identified initially by Southern blot analyses (36). The other FHC-producing mutation involves an alteration of a single β MHC nucleotide. This mutation was identified by more laborious procedures (13).

A single missense mutation in exon 13 of the β cardiac MHC causes FHC in one family (Fig. 2B). The β cardiac MHC produced by this mutationally altered gene would be expected to have a glutamine residue at position 403 instead of the arginine residue normally found at this position. All characterized myosin heavy chain

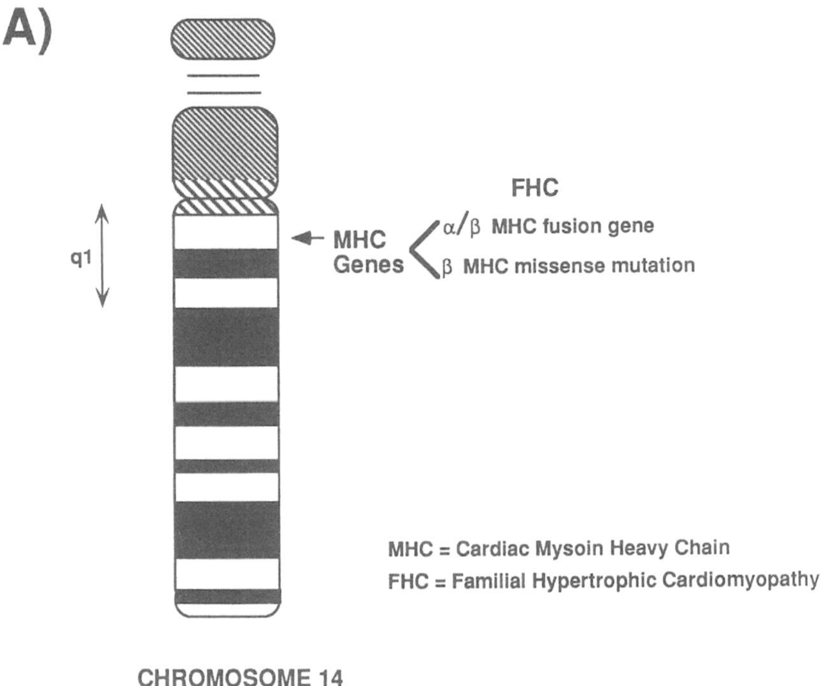

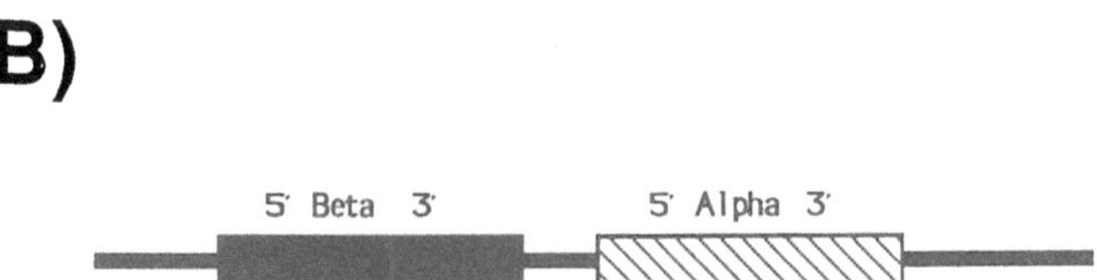

Fig. 1. A) The location of the FHC-1 locus on chromosome 14 band q1. B) The cardiac myosin heavy-chain genes on chromosome 14. Each gene is about 30 000 basepairs long and the genes are separated by about 4500 basepairs.

polypeptides (Fig. 2B) have an arginine residue at position 403. Only in individuals with FHC from one family is this highly conserved residue altered. Because glutamine residues differ in size and charge from arginine residues, we would expect this change to have a significant effect on the structure of β cardiac myosin heavy chain.

The other mutation demonstrated in individuals with FHC is a hybrid α/β cardiac myosin heavy chain gene (Fig. 2C). In all species examined to date, the α and β cardiac myosin heavy chain genes are highly homologous, and differ from one another by only 5–10 % (20, 19). These differences are scattered throughout the length of the molecules. Differences between the rat α and β MHC polypeptides occur in regions associated with ATPase activity, actin binding, myosin light-chain binding,

A)

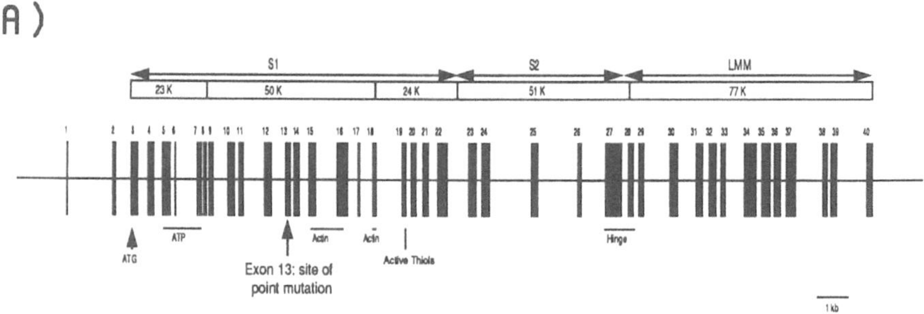

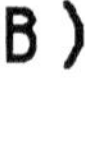

B)

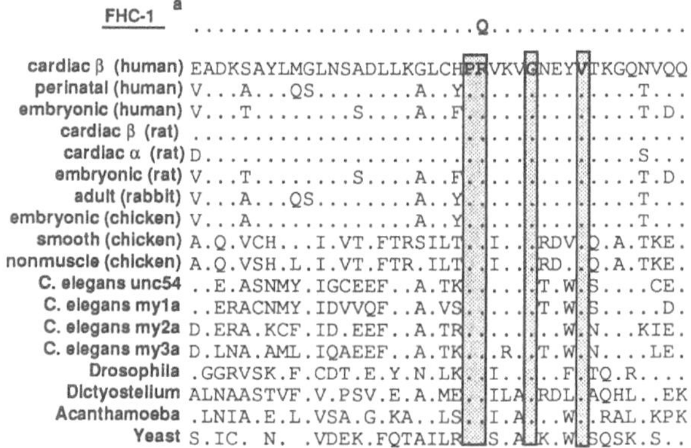

C)

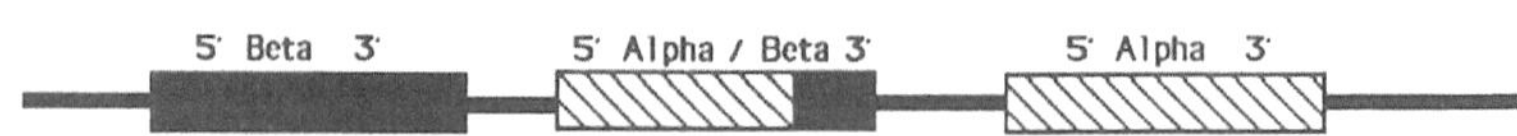

Fig. 2. Two cardiac myosin heavy-chain gene mutations that cause FHC. A) The location of a point mutation that causes FHC in patients from one family is indicated on a map of a cardiac myosin heavy-chain gene (13). Each cardiac myosin heavy-chain gene consists of 40 exons. B) The amino acid sequence encoded by exon 13 of myosin heavy-chain genes from a variety of species, and the amino acid sequence of exon 13 in patients with a point mutation in this exon. The location of the conserved residues is indicated (13). C) A second cardiac MHC mutation involves the formation of a hybrid α/β cardiac myosin heavy-chain gene and is found in FHC patients from a second family.

force production, and thick filament formation, as well as other portions of these molecules (29). The molecule produced by the hybrid gene is α-like MHC for the first 27 exons, and β-like for the remaining 13 exons and would be expected to have a novel structure. The effects of this unusual structure on protein function are at present unknown.

The mechanism by which these mutations cause FHC remains uncertain. Both of the mutationally altered MHC genes would be expected to produce cardiac MHC polypeptides. Two different mechanisms by which these mutationally altered polypeptides might cause FHC can be proposed. Models describing these mechanisms are outlined in Figure 3. The poison peptide model (Fig. 3A) suggests that the defective polypeptide is incorporated into the sarcomere and prevents appropriate assembly and/or organization of myofibrils. An example of this model is demonstrated by the missense mutations of the unc-54 MHC gene in Caenorhabditis, in which a defective protein binds to other components of the muscle fiber and prevents further fiber synthesis (16). Alternatively, another model (Fig. 3B) suggests that the FHC mutation effectively inactivates the MHC gene and causes the muscle cell to contain an abnormal ratio of α and β MHC polypeptides. For simplicity the model indicates that the ratio of α MHC to β MHC is altered in affected individuals. However, the relevant ratio might be between any structural muscle polypeptide (i.e., actin, myosin light chain, or troponin) and β MHC. (This model would explain how a hybrid gene might cause hypertrophic cardiomyopathy, because the existence of an extra gene in the genome of the affected individuals would be expected to produce larger than normal amounts of MHC). This model predicts that an aberrant ratio of α and β

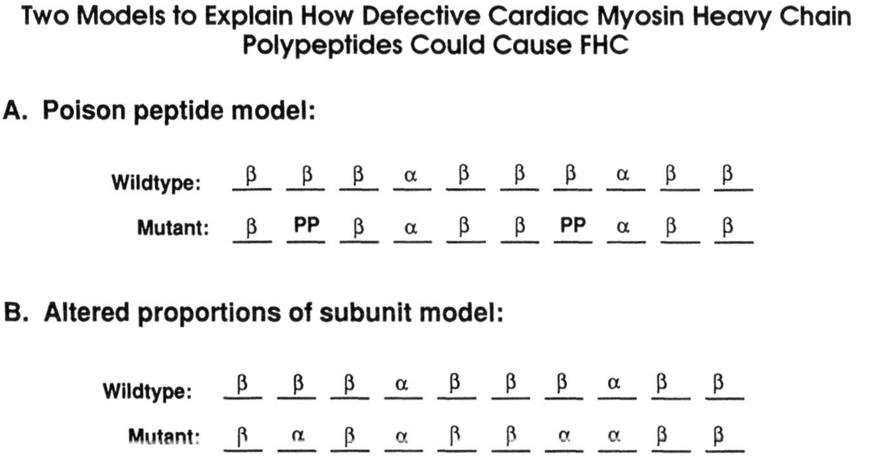

Two Models to Explain How Defective Cardiac Myosin Heavy Chain Polypeptides Could Cause FHC

A. Poison peptide model:

Wildtype:	β	β	β	α	β	β	β	α	β	β

Mutant:	β	PP	β	α	β	β	PP	α	β	β

B. Altered proportions of subunit model:

Wildtype:	β	β	β	α	β	β	β	α	β	β

Mutant:	β	α	β	α	β	β	α	α	β	β

Fig. 3. Two models describing mechanisms by which FHC mutations might alter cardiac muscle development. A) The poison peptide model suggests that a defective polypeptide (PP) is incorporated into growing muscle and prevents its proper development. B) This model suggests that individuals with a defective gene have aberrant amounts of functional polypeptide in the developing myocyte. The model suggests that the ratio of α cardiac MHC to β cardiac MHC is altered in affected individuals, but the physiologically important ratio might be between cardiac myosin heavy chains and troponin, myosin light chain, actin or other sarcomeric peptides.

MHC polypeptides to other muscle polypeptides would alter contractile function. The dominant effect of a heterozygous null mutation in a MHC genes of flightless Drosphila exemplify this model (1) in which gene dosage is critical for normal function.

While we cannot do so at present, we believe that distinguishing between these two models will provide important information about the disease. First, approaches to therapy for FHC patients will differ depending upon the mechanism by which the mutations cause the disease. Second, hypertrophic cardiomyopathy occurs secondary to a variety of conditions including hyperthyroidism and hypertension. If we understand the mechanism by which mutations in MHC genes cause FHC, we may gain insight into the molecular basis for secondary causes of cardiac hypertrophy.

FHC is a genetically heterogeneous disease

Prior to linkage analyses, physicians and geneticists could not predict whether FHC resulted from mutations in the same gene in unrelated families. Clinical studies of FHC patients have suggested that there is significant clinical variation among patients within a single family, all of whom must bear the same genetic defect (4). Other diseases that are inherited as autosomal dominant traits such as polycystic kidney disease (18) and neurofibromatosis (40) have been shown to be genetically heterogeneous; that is, mutations in different genes cause clinically similar diseases. Evidence now exists demonstrating that FHC is also genetically heterogeneous.

FHC has been shown to map to chromosome 14 in several families (34, 32). We have designated the disease due to mutations in the MHC genes as FHC-1. Studies of two other families have demonstrated that mutations in a gene not on chromosome 14 can cause FHC (34, 10). If these small numbers are representative of all FHC families, then one would expect that more than 50 % of all FHC is caused by defects in cardiac MHC genes. Future studies should define disease-causing loci that are not linked to FHC-1, and may facilitate identification of clinical traits that can distinguish between different genetic etiologies. While the identities of other genes which cause FHC are currently unknown, we expect that the unidentified genes will include genes which encode contractile or myosin-associated proteins.

Two mutations of cardiac MHC genes have been identified that cause FHC, and each of these occurs in only a single family. More than 30 different FHC families have been studied to date (our unpublished results), and none of these additional families have either of these two mutations. These data suggest that FHC mutations result from recent genetic events that have occurred within the past few generations. Thus, we expect that mutations that cause FHC have occurred on multiple occasions during human evolution rather than on a single or few occasions as occurred with the mutation that causes sickle cell anemia. FHC in multiple families, which we and others have studied, is caused by defects in their cardiac MHC genes (13, 36 and our unpublished results). However, preliminary data suggest that these defects are different from the mutations that have been identified previously and will probably be unique in each family. The finding that most FHC mutations are new mutations may explain the observation that approximately 6 % of all "idiopathic" hypertrophic cardiomyopathy is sporadic and occurs in individuals who do not have any affected first-degree relatives (5, 38). Perhaps these sporadic cases represent new mutations, which can be transmitted to subsequent generations.

Conclusions

Mutations within the cardiac myosin heavy-chain genes are responsible for some, but not all cases of FHC. There are several important clinical implications of these findings. First, individuals in whom the disease maps to FHC-1, can now obtain prenatal diagnoses and genetic counseling. Second, early and accurate diagnoses should be possible, even in young children who do not exhibit clinical signs of the disease. Third, since β cardiac myosin heavy chain is expressed at low levels in skeletal muscle, we expect that there may be skeletal muscle deficits in individuals who have mutations in the β cardiac MHC gene. Future study of patients with known β cardiac MHC defects may demonstrate a close correlation between these deficits and skeletal muscle alterations. Finally, as additional loci are identified that also cause FHC, an understanding of the clinical signs and prognostic significance of features of this genetically heterogeneous disorder should become clear.

Identification of MHC mutations that cause FHC should also improve our understanding of the structure and function of the myosin protein. Categorizing mutations within the MHC genes may suggest critical regions within the protein that are necessary for interactions with other sarcomeric proteins or are required for contractile functions. A more central question stemming from these results is how mutations in myosin molecules produce aberrant cardiac growth; what other gene products are altered by these mutations? and can the development of hypertrophy be attenuated or reversed? Application of transgenic technologies should facilitate the development of animal models for FHC that will provide the reagents necessary for answering these and other questions, and ultimately advance our ability to treat and prevent this disease.

Acknowledgements. This work was supported by grants from the National Institutes of Health (HL42467: CES), the American Heart Association (Clinician-Scientist Award: CES) and the Howard Hughes Medical Institutes (JGS). This contribution appears in the November 1991 issue of the journal *Molecular Biology and Medicine* and is reproduced here by the kind permission of the publisher.

References

1. Beall CJ, Sepanski MA, Fyrberg EA (1989) Genetic dissection of Drosophila myofibril formation: effects of actin and myosin heavy chain null alleles. Genes Devel 3: 131–40
2. Becker AE (1988) Pathology of Cardiomyopathies. In: Shaver JA Cardiomyopathies: Clinical Presentation, Differential Diagnosis, and Management. F.A. Davis Co., New York, pp 9–31
3. Cawthon RM, Weiss R, Gangfeng X, Viskochil D, Culver M, Stevens J, Robertson M, Dunn D, Gesteland R, O'Connell P, White R (1990) A major segment of the neurofibromatosis type 1 gene: cDNA sequence, genomic structure and point mutations. Cell 62: 193–201
4. Ciro E, Nichols, PF III, Maron, BJ (1983) Heterogeneous morphologic expression of genetically transmitted hypertrophic cardiomyopathy. Circulation 67: 1227–33
5. Clark CE, Henry WL, Epstein SE (1973) Familial prevalence and genetic transmission of idiopathic hypertrophic subaortic stenosis. N Engl J Med 289: 709–714
6. Codd MB, Sugrue DD, Gersh BJ, Melton LJ, III (1989) Epidemiology of idiopathic dilated and hypertrophic cardiomyopathy. A population-based study in Olmsted County Minnesota. Circulation 80: 564–72

7. Davies MJ (1984) The current status of myocardial disarray in hypertrophic cardiomyopathy. Br Heart J 51: 361–3
8. Donis-Keller H, Green P, Helms C, Cartinhour S, Weiffenbach B, Stephens K, Keith TP, Bowden DW, Smith DR, Lander ES, Botstein D, Akots G, Rediker KS, Gravius T, Brown VA, Rising MB, Parker C, Powers JA, Watt DE, Kauffman ER, Bricker A, Phipps P, Muller-Kahle H, Fulton TR, Ng S, Schumm JW, Braman JC, Knowlton RG, Barker DF, Crooks SM, Lincoln SE, Daly MJ, Abrahamson J (1987) A genetic linkage map of the human genome. Cell 51: 319–337
9. Dryja TP, Mcgee TL, Reichel E, Hahn LB, Cowley GS, Yandell DW, Sandberg MA and Berson EL (1990) A point mutation of the rhodopsin gene in one form of retinitis pigmentosa. Nature 343 (6256): 364–6
10. Epstein N, Fananapazir L, Lin H, Lalouel JM, Nienhuis A, Leppert M (1990) Genetic heterogeneity in hypertrophic cardiomyopathy: Evidence that HCM maps to chromosome 2p. Circulation 82: III-399 (abstract)
11. Ferrans VJ, Morrow AG, Roberts WC (1972) Myocardial ultrastructure in idiopathic hypertrophic subaortic stenosis: a study of operatively excised left ventricular outflow tract muscle in 14 patients. Circulation 45: 769–92
12. Frank S, Braunwald E (1968) Idiopathic hypertrophic subaortic stenosis: clinical analysis of 126 patients with emphasis on the natural history. Circulation 37: 759–88
13. Geisterfer-Lowrance AAT, Kass S, Tanigawa G, Vosberg H.-P., McKenna W, Seidman CE, Seidman JG (1990) A molecular basis for familial hypertrophic cardiomyopathy: A β cardiac myosin heavy chain gene missense mutation. Cell 62: 999–10006.
14. Goodwin JF, Gordon H, Hollman A, Bishop MB (1961) Clinical aspects of cardiomyopathy. Br Med J 21: 69–79
15. Jarcho JA, McKenna W, Pare JAP, Solomon SD, Holcombe RF, Dickie S, Levi T, Donis-Keller H, Seidman JG, Seidman CE (1989) Mapping a gene for familial hypertrophic cardiomyopathy to chromosome 14ql. N Eng J Med 321:1372–8
16. Karn J, Brenner S, Barnett L (1983) Protein structural domains in Caenorhabditis elegans unc-54 myosin heavy chain gene are not separated by introns. Proc Natl Acad Sci USA 80: 4253–57
17. Kerem B, Rommens JM, Buchanan JA, Markiewicz D, Cox TK, Chakravarti A, Buchwald M. Tsui L-C (1989) Identification of the cystic fibrosis gene: Genetic analysis. Science 245: 1073–80
18. Kimberling WJ, Fain PR, Kenyon JB, Goldgar D, Sujansky E, Gabow PA (1988) Linkage heterogeneity of autosomal dominant polycystic kidney disease. N Engl J Med 319: 913–18
19. Kurbayashi M, Tsuchimochi H, Komuro I, Takaku R, Tazaki Y (1988) Molecular cloning and characterization of human cardiac α- and β-form myosin heavy chain complementary DNA clones. J Clin Invest 82: 524–31
20. Mahdavi V, Periasamy M, Nadal-Ginard B (1982) Molecular characterization of two myosin heavy chain genes expressed in the adult heart. Nature 297: 659–64
21. Maron BJ, Epstein SE (1980) Hypertrophic cardiomyopathy: a discussion of the nomenclature. Am J Cardiol 43: 1242–4
22. Maron BJ, Epstein SE (1980) Hypertrophic cardiomyopathy. Recent observations regarding the specificity of three hallmarks of the disease: asymmetric septal hypertrophy, septal disorganization and systolic motion of the anterior mitral leaflet. Am J Cardiol 45: 141–154
23. Maron BJ, Gottdiener JS, Epstein SE (1981) Patterns and significance of distribution of left ventricular hypertrophy in hypertrophic cardiomyopathy: A wide angle, two dimensional echocardiographic study of 125 patients. Am J Cardiol 48: 418–28
24. Maron BJ, Bonow RO, Cannon RO, III Leon MB, Epstein SE (1987) Hypertrophic cardiomyopathy: Interrelations of clinical manifestations, pathophysiology and therapy. N Engl J Med 316: 780–9
25. Matsuoka R, Yoshida MC, Kanda N, Kimura M, Ozasa H, Takao A (1989) Human cardiac myosin heavy chain gene mapped within chromosome region 14q11.2-q13. Am J Hum Genet 32: 279–284

26. McKenna WJ, Goodwin JF (1981) The natural history of hypertrophic cardiomyopathy. Curr Probl Cardiol 6: 5–26
27. McKenna W, Deanfield J, Faruqui A (1981) Prognosis in hypertrophic cardiomyopathy: Role of age and clinical, electrocardiographic and hemodynamic features. Am J Cardiol 47: 532–8
28. McKenna WJ, Kleinebenne A, Nihoyannopoulos P, Foale R (1988) Echocardiographic measurement of right ventricular wall thickness in hypertrophic cardiomyopathy: Relation to clinical and prognostic features. J Am Coll Cardiol 11: 351–8
29. McNally EM, Kraft R, Bravo–Zehnder M, Taylor DA Leinwand LA (1989) Full-length rat alpha and beta cardiac myosin heavy chain sequences. J Mol Biol 210: 665–71
30. Pare JAP, Fraser RG, Pirozynski WJ, Shanks JA, Stubington D (1961) Hereditary cardiovascular dysplasia: A form of familial cardiomyopathy. Am J Med 31: 37–62
31. Riordan JR, Rommens JM, Kerem B, Alon N, Rozmahel R, Grzelczak Z, Zielenski J, Lok S, Plavsic N, Chou JL, Drumm ML, Iannuzzi MC, Collins FS, Tsui L.-C. (1989) Identification of the cystic fibrosis gene: Cloning and characterization of complementary DNA. Science 245: 1066–73
32. Roberts WC, Ferrans VJ (1975) Pathologic anatomy of the cardiomyopathies. Idiopathic dilated and hypertrophic types, infiltrative types and endomyocardial disease with and without eosinophilia. Hum Pathol 6: 287–342
33. Saez LJ, Gianola KM, McNally EM, Feghali R, Eddy R, Shows TB, Leinwand LA (1987) Human cardiac myosin heavy chain genes and their linkage in the genome. Nucl Acids Res 15: 5443–59
34. Solomon SD, Geisterfer-Lowrance AAT, Vosberg H.-P., Hiller G, Jarcho JA, Morton CC, McBride WO, Mitchell AL, Bale AE, McKenna WJ, Seidman JG, Seidman CE (1990) A locus for familial hypertrophic cardiomyopathy is closely linked to the cardiac myosin heavy chain genes, CRI-L436 and CRI-L329 on chromosome 14 at q11–12. Am J Hum Genet 47: 389–94.
35. Spirito P, Chiarella F, Carratino L, Berisso MZ, Bellotti P, Vecchio C (1989) Clinical course and prognosis of hypertrophic cardiomyopathy in an outpatient population. N Engl J Med 320:749– 55
36. Tanigawa G, Jarcho JA, Kass S, Solomon SD, Vosberg H.-P., Seidman JG, Seidman CE (1990) A molecular basis for familial hypertrophic cardiomyopathy: A α/β cardiac myosin heavy chain hybrid gene. Cell 62: 991–8
37. Teare RD (1958) Asymmetrical hypertrophy of the heart in young adults. Br Heart J 20: 1–18
38. van Dorp WG, ten Cate FJ, Vletter WB, Dohmen H, Roelandt J (1976) Familial prevalence of asymmetric septal hypertrophy. Eur J Cardiol 4/3: 349–57
39. Wallace MR, Marchuk DA, Andersen LB, Letcher R, Odeh HM, Saulino AM, Fountain JW, Brereton A, Nicholson J, Mitchell AL, Brownstein BH, Collins FS (1990) Type I neurofibromatosis gene: Identification of a large transcript disrupted in three NF1 patients. Science 249: 181–6
40. Wertelecki W, Rouleau GA, Superneau DW, Forehand LW, Williams JP, Haines JL, Gusella JF (1988) Neurofibromatosis 2: Clinical and DNA linkage studies of a large kindred. N Engl J Med 319: 278–83

Author's address:

C. E. Seidman
Department of Genetics
Harvard Medical School
Boston MA 02115, USA

The membrane proteins of the overloaded and senescent heart

B. Chevalier, D. Charlemagne, F. Callens-el Amrani, F. Carre, J.M. Moalic,
C. Delcayre, P. Mansier, B. Swynghedauw

INSERM U 127, Hopital Lariboisière Paris, France

Summary: Cardiac hypertrophy which occurs during chronic mechanical overload is one of
the numerous examples of biological adaptation to environmental requirements. As such, it is
obtained at random by trial and error, and adaptation represents the sum of various
modifications in gene expression, including the shift in isoform of myosin or in iso Na^+,
K^+ATPase, the decrease in β-adrenergic and muscarinic receptors, ryanodine channels or SR
Ca^{2+} ATPase densities and the unchanged density in Ca^{2+}current. Some of these changes
are beneficial at the cellular level, but are finally detrimental for the organism as a whole, as
is the slowing of V_{max}. It was suggested that the calcium homeostasis of the hypertrophied
cardiocyte was fragile and that this modified cell was less able to buffer the changes in the
intracellular calcium, thus providing a biological basis for the arrhythmogenicity of the
hypertrophied heart. These various modifications may provide a new key for future pharmaceutical research.

Key words: Hypertrophy; membrane proteins; aging

Introduction

Hypertensive cardiac hypertrophy has been referred to as accelerated aging and, at
least from a biological point of view, the senescent heart has strong homologies with
the pressure-overloaded myocardium. Systolic blood pressure, but above all vascular
stiffness are increased with advancing age, and both contribute to cardiac overload
by increasing the normal impedance characteristic of the aorta (19, 29). Because
several adaptation factors are called upon, chronic overload is tolerated by the heart
over a long period, but biological adaptation proceeds at random, by trial and error,
and does not result from rational learning (28) (Table 1). The ultimate requirement is
a functional equilibrium, but obviously the process is not perfect since the final
outcome is death and the main cause of mortality is senescence. Cardiac failure is an
index of both the limits and the imperfections of the biological process of adaptation.

Physiological characteristics of the senescent heart compared to
the pressure-overloaded heart

Elderly normotensive subjects exhibit a resting cardiac output similar to that of
young adults, but both the aortic impedance and the left-ventricular early diastolic
filling rate are altered (11). During exercise young people experience a greater
increase in heart rate and less end-diastolic dilatation than do elderly individuals,
mainly because of a more efficient β-adrenergic system.

Table 1. The main characteristics of biological adaptation.

1.	Trends to improve thermodynamic status.
2.	But proceeds by trials and errors.
3.	Has limitations, imperfections and neutral changes . . . by definition.
4.	Induces not one, but several changes in gene expression.

Examples: Polyglobulia and hypertension are adaptive changes induced by high altitude, but they may have detrimental consequences such as thrombosis.

Table 2. The senescent rat heart. Anatomical data are adapted from (7).

	Young adult (< 4 months, n = 16)	Aged ($\geqslant$ 24 months, n = 15)
Body weight (BW) in g	341 ± 8	$676 \pm 20^{***}$
Left-ventricular Weight (LVW) in mg	596 ± 12	$1055 \pm 23^{***}$
LVW/BW in mg/g	1.74 ± 0.04	1.57 ± 0.05

Between 60 and 90 years of age, the left-ventricular weight increases by approximately 1 g per year in males and 1.5 g per year in females (20), but the heart weight/body weight index remains unchanged, suggesting that there is, in fact, no real hypertrophy at the organ level. Similar data were obtained in rats (Table 2). At the cellular level the above interpretation is incorrect. Morphometric studies have demonstrated that in senescent rats the cardiac myocytes are hypertrophied. This cellular hypertrophy is likely to be compensatory because it is accompanied by a 30% loss in the total number of left-ventricular cardiocytes (1). At least in rats, the loss in myocytes is exactly compensated for by the increased size of the remaining cells, as occurs after myocardial infarction, and the resulting cardiac weight is unchanged. Whether the loss of myocytes is a true indicator of senescence or a simple consequence of ischemia is presently unknown.

From a physiological point of view the membrane action potential, cytosolic calcium transient, and the resulting contraction are of longer duration in senescent than in young rats (17). Another consistent finding is a diminished responsiveness to β-adrenergic modulation observed in both human and animals. This modification may explain why in a senescent population the maximal heart rate declines during exercise despite an elevated plasma catecholamine level (17). These alterations are also observed after aortic banding in young rats (6, 32, 33) (Table 3). Similarly, age-related diminution in chronotropic response to atropine or vagotomy have also been reported (10, 15) and, at least in humans, both the chronotropic response to atropine and the baroreflex slowing of heart rate in response to an acute increase in arterial pressure are depressed, indicating an additional modification at the level of the parasympathetic system. The existence and the importance of alterations in the autonomic nervous system have recently been demonstrated by spectral and non-spectral analysis of the sinus rate variability (9).

Table 3. Calcium channels in chronic cardiac hypertrophy (data from this laboratory; see (2, 23, 25, 27)) and during cardiac development (35). LV: left ventricle. DHP: dihyropyridine. d: days.

	Neonatal (1–3 d)	Adult (> 30d)	Adult 4–6 wk of aortic stenosis
Density per g of LV			
Ryanodine channels	1.0	2.10	1.00
DHP channels	1.0	1.35	1.35
Quantity per LV			
LV weight in g	0.1	1.00	2.00
Ryanodine channels	0.1	2.10	2.00
DHP channels	0.1	1.35	3.70

[Recalculated and normalized from Wibo and Godfraind 1991, Mayoux and Charlemagne 1988, Naudin and Charlemagne 1991]

The general process of adaptation

The general process of adaptation involves anatomical and biological changes at the level of the myocardium and periphery. Hypertrophy is adaptational since the number of contractile elements increases and it corrects the increase in wall stress, but it is limited by anatomical considerations and because myocytes can no longer divide in the adult.

A complex mechanical approach is not necessary to investigate the response of the cardiac papillary muscle subjected to an afterload greater than that to which it is accustomed. The immediate consequence is a decrease in the economy of the system, the muscle using more ATP (or oxygen) than normal per gram of developed tension. Biological changes will allow the heart to continue to contract in an economical way, resulting in a shift in the shortening velocity/after load curve. The slowing of the maximal velocity of shortening appears to be mostly responsible for cardiac survival (32, 33).

Biological analysis has demonstrated in both animals and men qualitative changes in genetic expression which account, to some degree, for the qualitative changes in the contractile properties of the hypertrophied tissues and enable the hypertrophied cardiac fibers to develop a normal active tension at the expense of speed. The same process is used by skeletal muscle which has an obvious advantage over myocardial muscle since mechanical overload is intermittent rather than sustained. Under the effects of mechanical overload, the synthesis of the V3 myosin isoform is stimulated in the rat ventricle and in the human atrium which express mostly the myosin isoform V1 (21). This adaptation is species specific and can also be associated with several changes in membrane proteins which may explain the physiological modifications occurring in human ventricle or in the ventricle of guinea pig, ferret, cat, or dog (32, 33).

The decrease of V_{max} is the most important mechanism by which the cardiac myocyte adapts to new environmental requirements. The myosin structure is a major determinant of V_{max} in rat and rabbit ventricles and in mammalian atria: even in these species and tissues V_{max} is influenced by intracellular calcium movements; in other mammalian species the intracellular calcium transient is likely to be the major determinant of V_{max}. Until recently changes in the components which determine intracellular calcium flux have been documented mainly in the rat, in which sarcomeric proteins are also modified. One of the best arguments in favor of the involvement of a calcium transient as an adaptation process comes from studies in the guinea pig. The normal guinea pig ventricles are quite different from those of rat and rabbit since they express only the isomyosin V3 and since calcium-induced calcium release is poorly developed. Yet, in the guinea pig, aortic-banding-induced cardiac hypertrophy produces pronounced slowing of the unloaded contractile velocity. The latter change is not accompanied by an isomyosin change or by a diminution in the turnover of the myosin cross-bridges, as assayed on skinned fibers (8) (Figs. 1 and 2).

One of the arguments supporting the concept that the physiological process of adaptation has limits beyond which heart failure occurs, is provided by simple experimental models in which cardiac insufficiency can arise without any ultrastructural myocyte change, necrosis or irreversible biochemical signs of ischemia (11). The imperfections of the system are numerous. For example, pressure-overload also activates protein synthesis in the vascular system, which augments both vascular impedance and coronary resistance (31). Quantitative and qualitative changes in

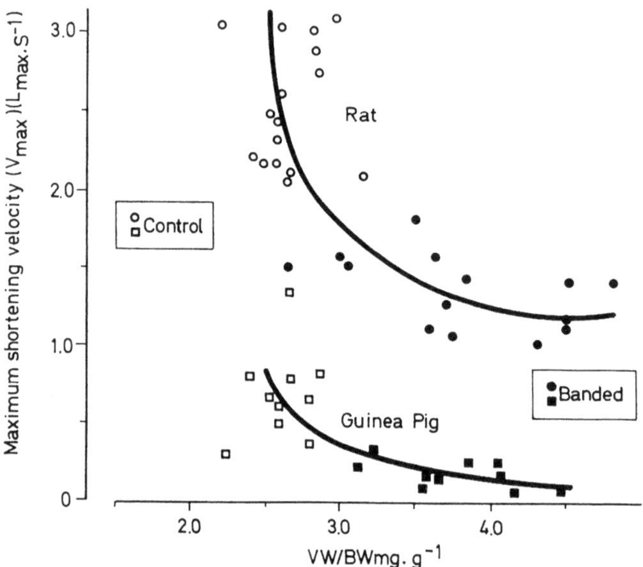

Fig. 1. Relationships between V_{max} and cardiac hypertrophy. Results are obtained with fresh papillary muscle from rat or guinea pig after abdominal aorta banding for 4–6 weeks. (These data are from this laboratory and that of Yves Lecarpentier; for the experimental work, see Lecarpentier in (32)). V_{max} represents the initial shortening velocity for an unloaded muscle. Cardiac hypertrophy is expressed as the left-ventricular weight/body weight ratio (LVW/BW).

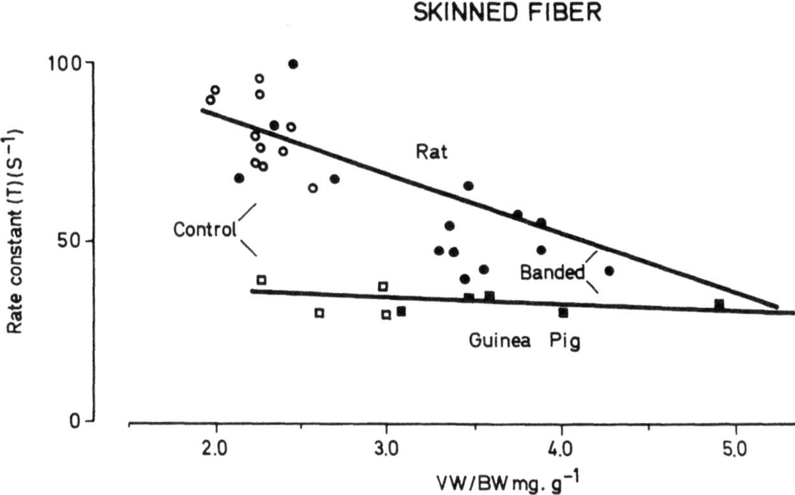

Fig. 2. Relationships between shortening velocity and cardiac hypertrophy in rat or guinea pig before and after abdominal aorta banding for 4–6 weeks. Results are obtained on skinned fibers (8). Note that the shortening velocity is not influenced by the degree of cardiac hypertrophy in this type of preparation in the guinea pig. LVW/BW: the left-ventricular weight/body weight ratio.

collagen content play a determining role in the genesis of compliance abnormalities and in diastolic dysfunction (34). Furthermore, although it is not perfectly clear, modifications of the action potential and the entering calcium current seem, to some degree, to account for the spontaneous arrhythmogenicity of hypertrophy myocytes.

Membrane proteins in the overloaded and senescent rat heart

Normal intracellular calcium homeostasis depends upon equilibrium with two pools of equilibrium: the extracellular calcium and the sarcoplasmic reticulum (SR) activity. Calcium homeostasis can also be modulated by stimulation of beta adrenergic or muscarinic receptors.

Calcium channels

There are two groups of calcium channels in the heart, one located on the external membrane specifically binds dihydropyridine, and the other belongs to the sarcoplasmic reticulum and binds ryanodine.

In cardiac hypertrophy of the rat the total number of *sarcolemnal calcium channels* per left ventricle has been calculated following specific labeling with radioactive dihydropyridine (23). The number of calcium channels increases proportionally with the degree of hypertrophy, however, the density of the calcium channels remains constant. The activity of these channels in cardiac overload has been measured by patch-clamp studies (30); the intensity of the calcium current per cell is increased, but the density normalized per unit of membrane surface remains unchanged, which also suggests that the total number of functional calcium channels per left ventricle increases proportionally with the degree of hypertrophy. The same results were obtained in another laboratory (16). These findings are also supported by other physiological data. The inotropic response of the isolated rat heart preparations to calcium, nifedipine, and Bay K 8644 was unchanged in the hypertrophied rat heart (3). Similar data have been reported in our laboratory with the guinea pig in which calcium metabolism is very different. In this species, as in the rat, pressure-overload does not alter the density of calcium channels, but as the total membrane surface is increased the total number of channels per left ventricle or cardiocyte increases proportionally (27). The adaptation of cardiac calcium channels to hypertrophy probably occurs via the same mechanism in different animal species and results in a maintained sarcolemnal density commensurate with the degree of hypertrophy.

Sarcoplasmic reticulum associated calcium channels specific for ryanodine have recently been quantitated in our laboratory in the rat myocardium after abdominal aorta banding (25). From a purely kinetic point of view, these receptors resemble the normal receptors with a Kd around 0.7 nM in both cases. In contrast to the sarcolemnal calcium channels, density is decreased by 25% in severe hypertrophy and, consequently, their total number per myocyte or left ventricle remains unchanged.

Informations on the L-type calcium current densities in aged rats are scarce, and those on the ryanodine receptors nonexistent. In senescent isolated myocytes only a small increase in the peak current density combined with reduced inactivation has been observed (17). Nevertheless, we know that the density of both types of channels increases during postnatal development (35). The density of ryanodine binding sites increases (from 44 to 92 pmol per g of wet weight) more than that of the dihydro-pyridine binding sites (from 10 to 13.5 pmol per g of wet weight). By comparison, during cardiac hypertrophy induced by pressure-overload in adults, the density of the ryanodine channels drops, while that of the sarcolemma calcium channels remains unchanged. In other words, the gene encoding the ryanodine receptors is probably activated during the neonatal period, but not activated by mechanically induced overload. These simple quantitations do not explain the localization of the channels. Developmental changes in calcium channel density are indeed accompanied by a redistribution of these channels. During the neonatal period the L-type calcium channels are concentrated in low-density membrane subfractions together with other sarcolemnal constituents, while in adults they are recovered predominantly in high-density membrane subfractions together with the ryanodine receptors (Table 3).

Calcium transporters

In sharp contrast, the density of the Ca^{2+}-ATPase of the sarcoplasmic reticulum is diminished in cardiac hypertrophy. The actual amount of protein has been calculated by measuring the ^{32}P-labeled Ca^{2+}-ATPase intermediate, which reflects the number of active Ca^{2+}-ATPase molecules (2). In normal hearts, the density of this protein in the sarcoplasmic reticulum is much higher than the sarcolemma receptors. During cardiac hypertrophy, the Ca^{2+}-ATPase density decreases, indicating that the number of molecules per myocyte is the same in normal and hypertrophied hearts. The decrease in density reflected by a diminution in intracellular calcium movement and a decrease in calcium uptake by isolated sarcoplasmic reticulum can partly explain the decrease in the speed of relaxation, which is a physiological finding in cardiac hypertrophy. The same results have also been obtained in the rabbit (24).

From a functional point of view, the Na^+-K^+, ATPase is coupled to the Na^+/Ca^{2+} exchanger. The two members of this system are modified during compensatory hypertrophy in rats. The number of Na^+-K^+, ATPase molecules present in the cardiac myocyte has been determined by binding of labeled oubain to membranes and to isolated myocytes. This technique also allows to differentiate ouabain high- and low-affinity sites, which correspond to two different proteins encoded by two different genes. The number of high-affinity sites increases by 61% during hypertrophy, whereas the low-affinity sites decreases proportionally. These results partly account for the particularities of the ouabain inotropic effects on hypertrophic rat hearts and are likely to reflect a switch from the adult $\alpha2$ subunit which diminishes to the fetal $\alpha3$ isoform (4, 5, 18).

The activity of the Na^+/Ca^{2+} exchanger has been measured in vitro on isolated membrane vesicles. Both activity and sensitivity to calcium are depressed in chronic cardiac hypertrophy in the rat heart (12).

During senescence the density of the SR calcium ATPase diminishes, as observed during pressure-overload (22), and the activity of the sodium/calcium exchanger is also altered (14). There are no data available on the sodium pump.

Receptors

In the rat myocyte, there are no $\beta2$ adrenergic receptors, less than 10 β-adrenergic receptors per μm^2 of surface membrane, and approximately 30 muscarinic receptors. The number of $\beta1$-adrenergic and muscarinic receptors per cell remains constant during cardiac compensatory hypertrophy; as a result, there is a large decrease in their density (30–50%) since the membrane surface of the myocyte increases. These results have a physiological counterpart: after exposure to isoproterenol, the inotropic response on the isolated heart and the increased amplitude of the calcium current measured on isolated myocytes by the "patch-clamp" technique, are reduced in the hypertrophic heart (6, 30). The levels of G protein subunits are not known in this model. By contrast, the human failing heart exhibits an increase in the $\beta2$ receptors while the density in $\beta1$-adrenergic receptors diminishes, but these changes are more pronounced and are not accompanied by a modification of the muscarinic receptor density. In fact the predominant finding in this case is an elevation of the Gi $\alpha2$ subunit which could likely explain the diminution of the inotropic effects of phosphodiesterase inhibitors.

During senescence, the inotropic response to isoproterenol is attenuated despite an elevation of the plasmatic catecholamine concentration (17). In the senescent rat heart the densities of both the β-adrenergic and muscarinic receptors are diminished, but the decrease in muscarinic receptors density is more pronounced than that in the β-adrenergic receptors, resulting in a drop of the β-adrenergic/muscarinic receptor ratio. In addition continuous infusion of propranolol or atropine upregulated both sympathetic and parasympathetic receptor densities in aged, but not in young adult rats (7) (Table 4).

The effects of mechanically induced cardiac hypertrophy on the density and the number of membrane proteins are equivocal. Two distinct groups of proteins can be

Table 4. The senescent rat heart. Data are adapted from (7). *, p < 0.05; **p < 0.01; ***p < 0.001.

	Young adult (< 4 months)	Aged (≥ 24 months)
β Adrenergic Receptors	(n = 8)	(n = 6)
Density in fmol/mg of LV	31 ± 2	23 ± 2*
Content in fmol per LV	2,515 ± 268	2,863 ± 479
Kd in pM	32 ± 5	34 ± 6
Muscarinic Receptors	(n = 8)	(n = 8)
Density in fmol/mg	104 ± 7	54 ± 3**
Content in fmol per LV	5,252 ± 524	5,349 ± 383
KD in PM	25 ± 3	27 ± 3

distinguished:
1) The first group includes the SR Ca^{2+}ATPase, the ryanodine channels and the β1-adrenergic and muscarinic receptors. The synthesis of this group of proteins are not activated by the hypertrophic process because their number per cell does not change and, as the cellular surface increases, their density decreases. The significance of these changes is not necessarily the same for each of these proteins. A decrease in the SR Ca^{2+} ATPase density reflects fairly well the adaptative slowing of relaxation,

The 2 components of intracellular Ca homeostasis in the hypertrophied cardiocyte

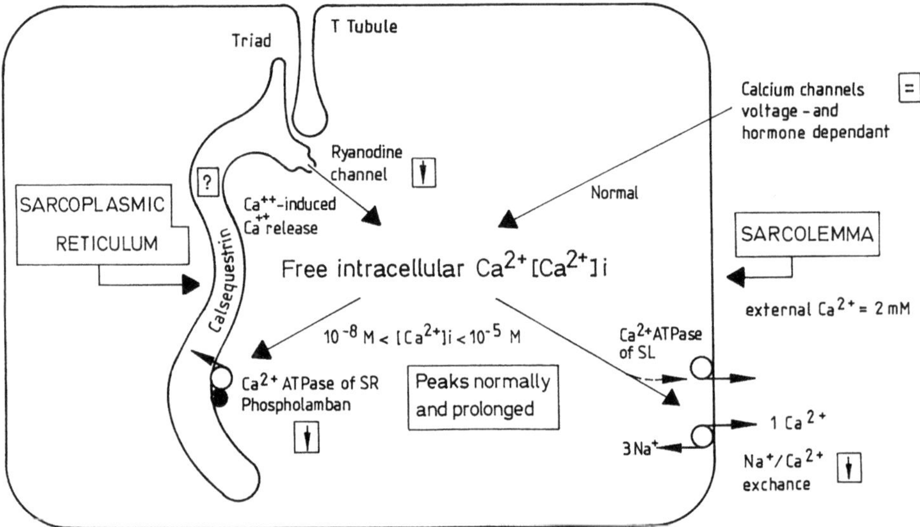

Fig. 3. The hypertrophied cardiocyte. This diagram shows the various components of calcium transient in the adult rat cardiocyte and how they are modified during compensatory cardiac hypertrophy (see text for the references).

whereas the "down regulation" of the β-receptors may have more of a protective role. The apparent insensitivity of each of these gene products to the adaptational process does not necessarily have the same origin: transcription of the SR Ca^{2+} ATPase does not seem to be activated, while the decrease in the β-receptor density may be the result of an increased breakdown.

2) The density of total and probably also functional calcium channels does not change, that is to say their number per cell increases and their increased synthesis is proportional to the hypertrophic process. In contrast to the preceding group of proteins, the genes encoding these proteins are probably activated during chronic mechanical overload.

3) In compensatory cardiac hypertrophy calcium transient peaks normally, and the resting tension is unchanged. Nevertheless, the above-described scheme suggests that calcium homeostasis is fragile since the sarcolemnal calcium channels responsible for the calcium inward current exhibit a normal density while the Na^+/Ca^{2+} exchanger in charge of the cellular extrusion of calcium is less active. In addition, at least in the rat, the role of the SR is attenuated since the densities of both the ryanodine channels and the calcium ATPase are lowered. The simplest hypothesis to explain the arrhythmogenicity of the hypertrophied cardiocyte is that a cell is less capable of buffering the changes in calcium concentration which normally occur during the cell life (Fig. 3).

References

1. Anversa P, Hilar B, Ricci R, Guideri G, Olivetti G (1986) Myocyte cell loss and myocyte hypertrophy in the aging rat heart. J Am Coll Cardiol 8:1441–1448
2. Bastie de la D, Levitsky D, Rappaport L, Mercadier JJ, Marotte F, Wisnewsky C, Brokovich V, Schwartz K, Lompre AM (1990) Function of the sarcoplasmic reticulum and expression of its Ca^{2+} ATPase gene in pressure overload-induced cardiac hypertrophy in the rat. Circ Res 66:554–564
3. Callens El Amrani F, Mayoux E, Mouas C, Clapier-Ventura R, Henzel D, Charlemagne D, Swynghedauw B (1990) Normal responsiveness to external Ca and to Ca-channel modifying agents in hypertrophied rat heart. Am J Physiol 258:H1727–1734
4. Charlemagne D, Orlowski J, Oliviero P, Lane L (1991) Expression of the α1, α2, α3, and β subunit in RNAs of the Na, K-ATPase in hypertrophied rat hearts. J Cell Biochem (abstract), suppl 15C, H305
5. Charlemagne D, Maixent JM, Preteseille M, Lelièvre L (1986) Ouabain binding sites and (Na^+, K^+)-ATPase activity in rat cardiac hypertrophy. Expression of the neonatal form. J Biol Chem 261:185–204
6. Chevalier B, Mansier P, Callens-El Amrani F, Swynghedauw B (1989) The β adrenergic system is modified in compensatory pressure overload in rats: Physiological and biochemical evidence. J Cardiovasc Pharmacol 13:412–420
7. Chevalier B, Mansier P, Teiger E, Callens-El Amrani F, Swynghedauw B (1991) Alterations in β adrenergic and muscarinic receptors in aged rat heart. Effects of chronic administration of propranolol and atropine. Mech Ageing Devel 60:215–224
8. Clapier-Ventura R, Mekhfi H, Oliviero P, Swynghedauw B (1988) Pressure overload changes cardiac skinned fibers mechanics in rats, not in guinea pigs. Am J Physiol 254:H517–H524
9. Coumel P (1990) Pathophysiology of ventricular arrhythmias during myocardial hypertrophy and cardiac insufficiency. (1990) In: B Swynghedauw (ed) Cardiac hypertrophy and failure. J Libbey/INSERM pub London/Paris. pp 647–664

10. Dauchot P, Gravenstein JS (1975). Effects of atropine on the electrocardiogram in different age groups. Clin Pharmacol Therap 12:274–280
11. Fleg JL (1986) Alterations in cardiovascular structure and function with advancing age; Am J Cardiol 57:33C–44C
12. Hanf R, Drubaix I, Lelièvre L (1988) Rat cardiac hypertrophy: altered sodium-calcium exchange activity in sarcolemmal vesicles. FEBS Lett 236:145–149
13. Hatt PY, Berjal G, Moravec J, Swynghedauw B (1970) Heart failure: an electron microscopic study of the left ventricular papillary muscle in aortic insufficiency in the rabbit. J Mol Cell Cardiol 1:235–247
14. Heyliger CE, Prakash AR, Mac Neill JH (1988) Alterations in membrane Na^+/Ca^{2+} exchange in the aging myocardium. Age 11:1–6
15. Kelliher GJ, Conahan ST (1980) Changes in vagal activity and response to muscarinic receptor agonists with age. J Gerontol 35:842–849
16. Kleiman RB, Houser SR (1988) Calcium current in normal and hypertrophied isolated feline ventricular myocytes. Am J Physiol 255:H1434–H1442
17. Lakatta EG (1987) Cardiac muscle changes in senescence. Ann Rev Physiol 49:1–22
18. Lelièvre L, Maixent JM, Lorente P, Mouas C, Charlemagne D, Swynghedauw B (1986) Prolonged responsiveness to ouabain in hypertrophied rat heart: physiological and biological evidence. Am J Physiol 250:H923–H931
19. Levy BI, Safar M (1990) Ventricular afterload and aortic impedance. In: B Swynghedauw (ed) Cardiac hypertrophy and failure. J Libbey/INSERM pub London/Paris, pp 521–530
20. Linzbach AJ, Akuamoa-Boateng E (1973) Die Altersveränderungen des menschlichen Herzens. I. Das Herzgewicht im Alter. Klin Wschr 51:156–163
21. Lompré AM, Schwartz K, Albis A, Lacombe G, Thiem NV, Swynghedauw B (1979) Myosin isozymes redistribution in chronic heart overloading. Nature 282:105–107
22. Maciel LMZ, Polikas R, Rohrer D, Popovich D, Dillmann WH (1990) Age-induced decreases in the messenger RNA coding for the sarcoplasmic reticulum. Ca^{2+} ATPase of the rat heart. Circ Res 67:230–234
23. Mayoux E, Callens F, Swynghedauw B, Charlemagne D (1988) Adaptational process of the cardiac Ca^{2+} channels to pressure overload: biochemical and physiological properties of the dihydropyridine receptors in normal and hypertrophied rat hearts. J Cardiovasc Pharmacol, 12:390–396
24. Nagai R, Zarain-Herzberg A, Brandl C, Fuji J, Tada M, Mac Lennan DH, Alpert N, Periasamy M (1989). Regulation of myocardial Ca^{2+} ATPase and phospholamban mRNA expression in response to pressure overload and thyroid hormone. Proc Natl Acad Sci USA 86:2966–2970
25. Naudin V. Oliviero P, Rannou F, Sainte-Beuve C, Primot I, Charlemagne D (1991) Ryanodine receptors in pressure-overload induced cardiac hypertrophy in the rat. FEBS Lett 285:135–138
26. Nordin C (1989) Abnormal Ca^{2+} handling and the generation of ventricular arrhythmias in congestive heart failure. Heart failure, 1:143–154
27. Primot I, Mayoux E, Oliviero P, Charlemagne D (1991) Effects of pressure overload on cardiac Ca^{2+} antagonist binding sites of guinea pig. Comparison with the adaptational response of the hypertrophied rat heart. Cardiovac Res 25:875–880
28. Ruffie J. De la Biologie à la Culture. Flammarion pub. Paris 1976
29. Safar M (1990) Ageing and its effects on the cardiovascular system. Drugs, 39 (suppl 1):1–8
30. Scamps F, Mayoux E, Charlemagne D, Vassort G (1990) Calcium current in single isolated cells from normal and hypertrophied rat heart. Effect of β adrenergic stimulation. Circ Res, 67:199–208
31. Strauer B. Hypertensive heart disease. In: B Swynghedauw (ed) Cardiac hypertrophy and failure. J Libbey/INSERM pub London/Paris, pp 444–446
32. Swynghedauw B (1990) Cardiac hypertrophy and failure. J Libbey/INSERM pub London/Paris
33. Swynghedauw B (1986) Developmental and functional adaptation of contractile proteins in cardiac and skeletal muscles. Physiol Rev, 66:710–771

34. Weber KT, Janicki JS, Shroff SG, Prick R, Chen RM, Bashey RI (1988) Collagen remodelling of the pressure-overloaded, hypertrophied non-human primate myocardium. Circ Res, 62:757–765
35. Wibo M, Bravo G, Godfraind T (1991). Post-natal maturation of excitation-contraction coupling in rat ventricle in relation to the subcellular localization and surface density of 1,4-dihydropyridine and ryanodine receptors. Circ Res 68:662–673

Authors' address:
Dr. B. Swynghedauw
INSERM-U 127
Hopital Lariboisière
41 Bd de la chapelle
F-75010 Paris,
France

Contraction frequency dependence of twitch and diastolic tension in human dilated cardiomyopathy (Tension-frequency relation in cardiomyopathy)

L.A. Mulieri, B.J. Leavitt,† G. Hasenfuss, P.D. Allen,[1], N.R. Alpert

Department of Physiology and Biophysics and †Department of Surgery University of Vermont College of Medicine Medical Center Hospital of Vermont, Burlington, Vermont, USA
[1]Department of Anesthesiology, Brigham and Women's Hospital Boston, Massachusetts, USA

Summary: We studied isometric twitch tension and diastolic tension at $37\,^{\circ}C$ as a function of stimulation frequency (12–240 $\min^{-1}$) in very thin (.07–.5mm^2), parallel fibered strips of left-ventricular myocardium. Non-failing control tissue (C) was obtained from epicardial biopsies taken during myocardial revascularization surgery on patients with normal ventricular function. End-stage failing tissue was obtained from endocardial and epicardial biopsies from explanted hearts with idiopathic dilated cardiomyopathy (DCM). The methods and apparatus for biopsy and dissection of myocardium are described. Maximal peak twitch tension at optimal stimulation frequency of $163 \pm 5\min^{-1}$ was $41.8 \pm 10mN/mm^2$ in non-failing myocardium and it was reduced by 70 % (p < .02) to $12.9 \pm 1.6mN/mm^2$ at an optimal frequency of $72 \pm 17\min^{-1}$ in DCM. The peaks of the tension-frequency curves occurred at frequencies between 12 and 60 $\min^{-1}$ in most DCM strips (5/9), while in C most of the peaks (8/9) fell between 156 and 180 $\min^{-1}$. The peaks from four DCM hearts fell in an intermediate range of frequencies (96–144 $\min^{-1}$) which also included one non-failing peak at 132 $\min^{-1}$. Diastolic tension declined in both groups as stimulation frequency increased above 12 $\min^{-1}$ and it began increasing when stimulation frequency rose above optimal frequency by 19 $\pm$ 5 % and 110 $\pm$ 50 % in C and DCM, respectively. Total duration of the isometric twitch diminished with tachycardia remaining shorter than stimulation intervals up to 140 $\pm$ 16 $\min^{-1}$ (3.1 $\pm$ 1 times optimal frequency) in DCM and up to 161 $\pm$ 14 $\min^{-1}$ (not different than optimal frequency) in C. Decline in peak twitch tension above optimal stimulation frequency was 4 to 6 times larger than the accompanying rise in diastolic tension in both groups. The premature decline in tension at lower than normal degrees of tachycardia in DCM does not arise from incomplete relaxation of the twitch response. The 70 % deficit in tension generating ability of DCM may be a major contributor to heart failure. Moderate shift in the peak of the tension-frequency curves to lower frequencies (130 $\min^{-1}$) in C does not appear to predispose end-stage failure, but it may make the ventricle more susceptible to dilation.

Key words: Idiopathic dilated cardiomyopathy; cardiac failure; twitch tension-frequency relation; diastolic tension-frequency relation; biopsy and dissection techniques

Introduction

In primary congestive heart failure, ventricular performance is depressed. In addition to low cardiac output at rest the necessary increase in cardiac output with exercise is

---

*Supported by National Institutes of Health Grant PHS 38001-05/P1.

greatly diminished or absent in cardiomyopathy, possibly as a result of absence of the normal decrease in end-systolic ventricular volume and a smaller than normal increase in stroke volume (14, 24). Measurements obtained by cardiac catheterization in cardiomyopathic as compared with normal patients show increased end-diastolic pressure, reduced rates of left-ventricular pressure generation and relaxation and a lowering of the positive slope of the pressure-rate vs. contraction frequency relation (5). The elevated end-diastolic pressures may also be attributed to reduced diastolic compliance of the ventricle, as well as to the reduced systolic emptying resulting from the diminished rates of pressure build-up and relaxation (7).

Evidence that these defects in cardiac performance may be related to alterations in myocardial contractility and its inotropic response to tachycardia has been accumulating (8, 9, 10, 19, 21). Measurements of twitch tension in isolated left-ventricular strips from explanted cardiomyopathic hearts compared with non-failing hearts show reductions in peak rates of generation and relaxation of twitch tension and a decrease in slope of the tension-rate vs contraction frequency relation which parallel the cath-lab measurements (19, 21).

Measurements of aequorin light emission during the twitch of isolated trabeculae from cardiomyopathic hearts are consistent with a decrease in rate of calcium uptake during relaxation and show an increase in end-diastolic calcium concentration is present in the failing myocardium (8, 9, 10). These changes may be related to a reduction in Ca^{2+} ATPase of the sarcoplasmic reticulum in the failing hearts (3).

We studied the stimulation frequency dependence of contraction and relaxation in isolated myocardial strips to further understand the role of the tension-frequency relation in heart failure and to determine if the onset of incomplete relaxation of the twitch above a critical contraction frequency might be the mechanism of reduction in twitch tension in myocardium from failing hearts. In order to compare the performance of endstage failing myocardium from heart transplant recipients with myocardium from non-failing hearts, a method was developed to safely obtain viable left-ventricular biopsies from patients with normal left-ventricular function who were undergoing myocardial revascularization surgery.

The results show reduced contractility and reduced inotropic response to tachycardia in failing myocardium and do not support the hypothesis that incomplete relaxation is the basis for the reduced contractility in DCM.

Methods

Severely failing myocardium (NYHA IV). Left-ventricular tissue was obtained from the explanted hearts of five patients (46 ± 3 years old; three males, two females) undergoing cardiac transplantation surgery following a protocol approved by the Committee for the Protection of Human Subjects at Brigham and Women's Hospital, Boston, MA. These patients were diagnosed as having end-stage (NYHA IV) heart failure due to idiopathic dilated cardiomyopathy (Ejection Fraction = 0.12 ± 0.01 SEM). Excised hearts were washed of blood with chilled saline solution. Left-ventricular myocardium was dissected into 2–4 g sheets from the inner or outer walls of the left ventricle within 15 min of explantation.

Non-failing myocardium. With full approval from the University of Vermont Committee on Human Research and signed, informed consent, left-ventricular biopsies were obtained from nine patients (61 $\pm$ 3 years old, six males, three females) undergoing myocardial revascularization who had normal wall motion and ventricular function determined by contrast ventriculography (mean ejection fraction 0.62 $\pm$ 0.2). After placing the patients on cardiopulmonary bypass, cardiac arrest was obtained with cold, hyperkalemic (20 mM K^+/liter) blood cardioplegia (Fig. 1). The left ventricle was elevated and the myocardial fiber orientation was visually noted. Using a No.15 scalpel, two linear incisions were made in the anterior segment of the left ventricle parallel to the fibers 1.5 mm apart and 12 mm in length (Fig 1, circle). The myocardial biopsy was then elevated with forceps and the scalpel was used to separate the biopsy from the underlying tissue. This created a rectangular biopsy 1.5 mm in thickness. Mild electrocautery was applied to the trough followed by suture closure with one teflon pledgeted horizontal mattress suture of 4-0 polypropylene (see Fig. 1, bottom). There were no complications (bleeding, arrhythmias, etc.) in any patient participating in this study.

Protection of tissue from deterioration. All excised tissue was immediately submerged in BDM-Krebs protective solution (20) at room temperature and oxygenated by bubbling with 95 % O_2-5 % CO_2.

Tissue from explanted hearts was transported from Boston, Massachusetts, to Burlington, Vermont in continuously gassed jars within 5 h. Non-failing tissue arrived in the laboratory within 15 min of excision and was usually used within 2.5 h after excision. Control experiments indicate that differences in soak-time or aging do not effect the results.

Solutions. Krebs-Ringer solution (1) was modified by doubling the glucose concentration and adding insulin to improve viability and long-term survival of the myocardial

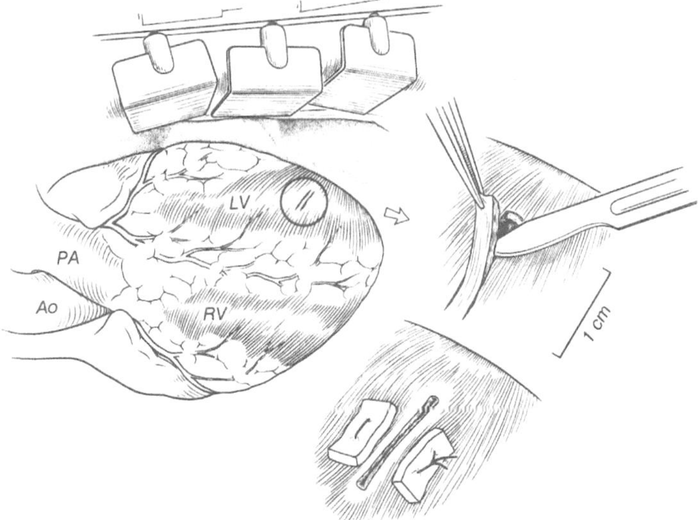

Fig. 1. Method of epicardial biopsy: The arrested heart is elevated and rotated to bring the anterior wall of left ventricle into view. Circle encloses the 1.5 mm deep by 12 mm long incisions separating the biopsy from adjacent myocardium. Right: biopsy specimen is cut from underlying myocardium. Bottom: suture closure of biopsy site.

tissue. The solution contained in mM: Na, 152; K, 3.6; Cl^-, 135; HCO_3^-, 25; Mg^{2+}, 0.6; $H_2PO_4^-$, 1.3; SO_4^{2-}, 0.6; Ca^{2+}, 2.5; glucose 11.2; and insulin, 10 IU/l. The BDM-Krebs protective solution consists of this Krebs-Ringer solution to which 30 mM of 2,3-butanedione monoxime (BDM) was added (20). BDM was obtained from Sigma. Glass distilled water was used for all solutions.

Muscle strip preparation. A biopsy strip (1.5 to 3mm wide) of myocardial tissue was clamped between the ends of plastic rods submerged in protective solution in a dissection chamber (Fig. 2A). The strip could be rotated axially to facilitate dissection, which was performed with microdissection scissors (6mm blade) and forceps under a 10X–20X binocular microscope. A strip of muscle tissue free of dense necrosis or fibrosis (see enlargement, Fig. 2A) was dissected from the main piece and transferred to a similar but smaller dissection chamber (Fig. 2B), also filled with protective solution. The endocardium or epicardium plus a few layers of underlying myocytes were dissected away from all strip preparations obtained from both failing and non-failing hearts. Skewed and damaged fiber bundles were carefully dissected away (Fig. 2B.1) prior to sculpting the strip to final dimensions (0.3–0.8mm diameter X 4–6mm length). In dissecting damaged or unwanted columns of myocytes away from the strip under preparation, scissor cuts were usually restricted to visible clefts formed where connective tissue layers occur between adjacent columns of myocytes. Lateral pulling on the unwanted tissue with forceps (Fig. 2B.1) was used during this cutting to exaggerate the width of the cleft and minimize the risk of cutting the surface of myocytes bordering the cleft. Cutting paths were chosen to maximize the possibility of having all columns of myocytes on the surface of the finished strip run undamaged along the entire length of the preparation. All finished strip preparations had parallel muscle fiber orientation as ascertained by 10-20X binocular observation and in some cases by 400X bright-field water immersion microscopy. Loops of 4-0 non-capillary, braided silk, previously wired with 25 µm diameter platinum stimulating electrodes (Fig. 2B.2) were attached to the ends of the preparation with silk ligatures (22). Upon completion of the preparation the muscle strip was allowed to rest for 15 to 30 min (Fig. 2B.3) and then it was transferred to the muscle chamber and submerged in normal oxygenated Krebs-Ringer solution to wash out the BDM.

Preparations were judged "intact" if they exhibited all-or-none twitch responses to electrical stimulation, maintained a steady threshold voltage (i.e., within 20 % of initial value) and did not exhibit "run down" of peak twitch tension by more than 20 % during the entire experiment (4–5 h).

Experimental Procedure

Isometric twitch tension was measured with a capacitance-type force gauge (Cambridge Technologies, Mod 400) and displayed, signal-averaged, and stored by a digital oscilloscope (Nicolet, Mod. 4094). Longitudinally propagated excitation was evoked by end-to-end stimulation of the muscle strips using 3-ms rectangular pulses 20 % above threshold voltage.

Muscle strip preparations were stimulated continuously at 1–2 Hz at 37 °C during a 60 min equilibration period after BDM washout. Then they were gradually stretched (in .05 mm increments) to lmax at which length active twitch tension was maximum. The steady-state force-frequency relation was obtained at 37 °C by

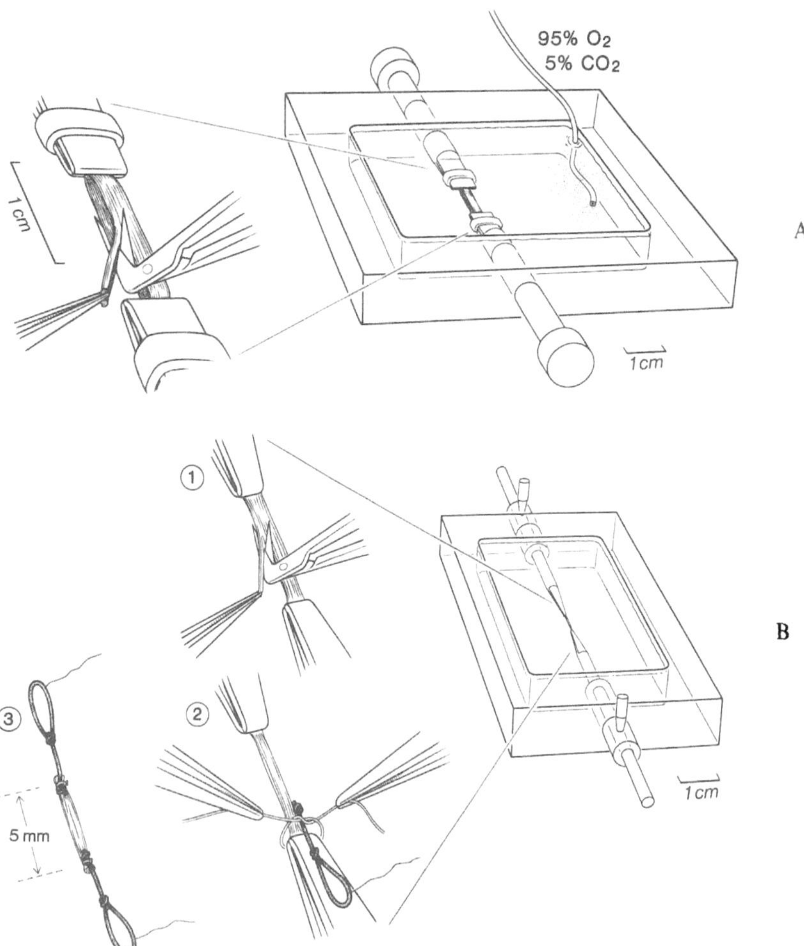

Fig. 2. Preparation of ventricular strips from myocardial biopsy tissue. A) gross dissection chamber with myocardial biopsy clamped in place. Enlargement shows method of dissecting a potential strip away from the bulk tissue. Chamber material is polished Lucite. Solution: Krebs-Ringer with 30 mM BDM. B) Fine dissection chamber with potential ventricular strip clamped in place. Enlargements show 1) method of dissecting columns of unwanted muscle tissue away from surface of strip, 2) method of attaching ligatures (with self-contained platinum stimulation wires) to finished strip, and 3) completed ventricular strip preparation.

recording and measuring the twitch after stimulation for 5 min at each frequency starting at 0.2 Hz and increasing in 0.2 Hz increments until the twitch declined by 30–40 % on the descending limb of the curve. Peak twitch tension and tension rate and timing parameters of the isometric myogram were measured by digital read-out from the oscilloscope.

Adequate oxygenation of the entire cross-section of the muscle strips was confirmed by applying the oxygen tension reduction test (23). This test was performed on each ventricular strip during stimulation at the highest frequency (f_o) giving maximal twitch tension. The gas bubbling through the muscle bath at 60 cc/min (four times normal rate) was changed from the normal 95 % O_2-5 % CO_2 to 80 % O_2, 15 % N_2, 5 % CO_2 for a period of 30 min with continuous stimulation at f_o (6-180 BPM). If the twitch force fell by 10 % or more by the end of this period the experiment was discarded. The average decline was 5 $\pm$ 2 % for all muscle strips included here.

After completion of the experiments, length and blotted weight of the portion of the muscle strip extending between the ligatures was measured. Cross-sectional areas used for normalization of twitch force were calculated by division of blotted weight by length at lmax.

Statistics

Mean values of twitch parameters from Class IV (dilated cardio-myopathy) hearts were compared with the means measured in tissue from the non-failing control hearts

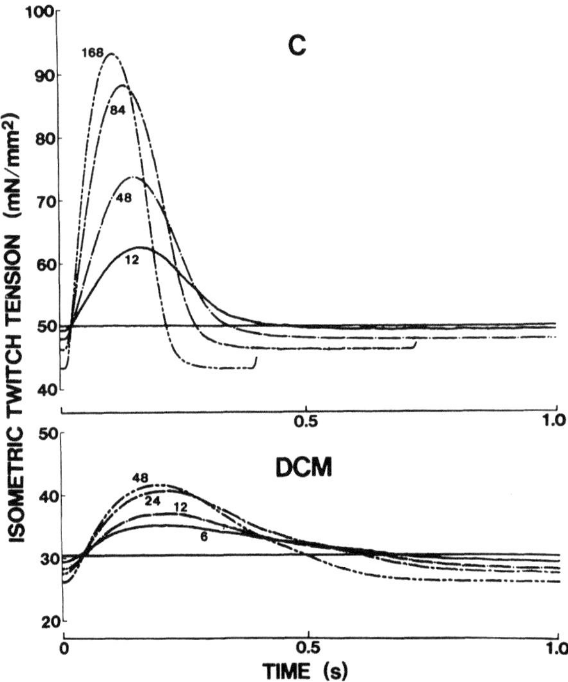

Fig. 3. Isometric myograms from non-failing (C) and failing (DCM) left-ventricular myocardial strips. Superimposed records on the ascending limb and peak of the tension-frequency relation showing simultaneous increase in peak tension and decrease in diastolic tension accompanying tachycardia. C = non-failing subepicardial strip, maximal twitch tension 50.6 mN/mm^2, cross-sectional area 0.14mm^2. DCM = failing subepicardial strip, maximal twitch tension 13.3 mN/mm^2, cross-sectional area 0.18mm^2. Horizontal lines reflect resting tension of the unstimulated muscle. Temperature 37 °C. Stimulation frequencies given in min^{-1}.

at each stimulation frequency using Students t-test. Group differences at the p < .05 level are considered significant and are starred on graphs and tables. Averages are given as mean $\pm$ standard error.

Results

The time-course and amplitude of the isometric twitch is extremely sensitive to stimulation frequency in both non-failing and failing myocardium. Figure 3 shows myograms at various stimulation frequencies on the ascending limb and at the peak of the tension-frequency relation. In both failing and non-failing tissue tachycardia causes the peak twitch tension (total tension minus end-diastolic) to rise (Fig. 3), while it causes the diastolic tension to fall (Figs. 3 and 6).

Individual tension-frequency relations for myocardial strips from both types of tissue are shown in Figure 4. In non-failing myocardium all but one of the peaks of the tension-frequency relations occur over a tight range of frequencies from 156 to 180 min^{-1}. In failing myocardium the positions of the peaks cover a wider range from 10 to 144 min^{-1} centered around 60 to 70 min^{-1}. Representative curves from the most widely separated members of the two groups are shown in Fig. 4A while some of the overlapping members are shown in Fig. 4B. The average optimal stimulation frequency for the failing myocardium was 56 % lower than in the non-failing myocardium (Table 1). However, for the five strips with lowest f_o in the failing subgroup in Fig. 4A, the average optimal stimulation frequency was 34 $\pm$ 11 min^{-1} while in the four strips in the sub-group in Fig. 4B it was 120 $\pm$ 13 min^{-1}.

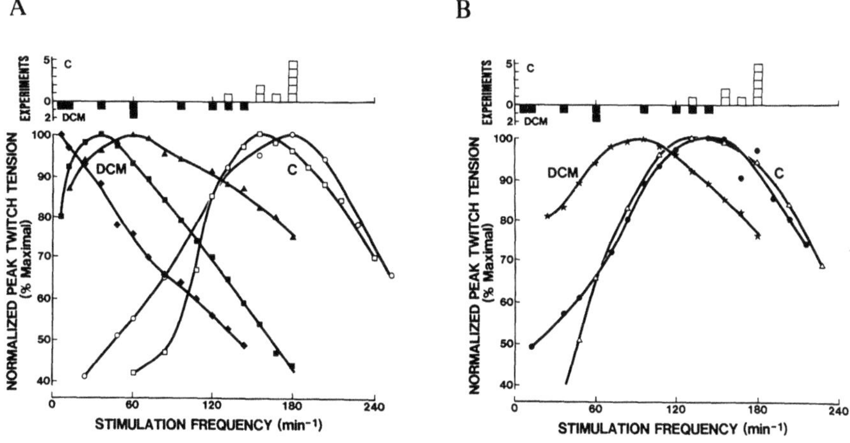

Fig. 4. Tension-frequency relations in non-failing and failing myocardial strips. A) Two representative curves from eight non-failing myocardial strips exhibiting the highest peak frequencies (open symbols) and three representative curves from the five failing myocardial strips exhibiting the lowest peak frequencies (filled symbols). Bar graph at top indicates the positions of the peaks of all myocardial strips studied (filled bars DCM, open bars C) and the number of preparations exhibiting peaks at each frequency. B) Same as A, but for three of the five myocardial strips (4 DCM, 1C) having their peaks occurring at intermediate values of frequency. 100% tension corresponds to 41.8 $\pm$ 10 mN/mm² in non-failing control strips and to 12.9 $\pm$ 1.6 mN/mm² in dilated cardiomyopathic strips; 37 °C.

Table 1. Isometric twitch parameters in human left-ventricular myocardial strips 37 °C

	Cross Section area (mm²)	Peak twitch tension (mN/mm²)	Optimum stimulus frequency (min⁻¹)	Diastolic tension (mN/mm²)
Non-failing control(n = 9)	0.19 ± 0.02	41.8 ± 10	163 ± 5	19.5 ± 4.9
Dilated Cardiomyopathy (n = 9)	0.27 ± 0.05	$12.9 \pm 1.6^{*}$	$72 \pm 17^{**}$	$22(n = 2)$

*2p < .02, **2p < .001

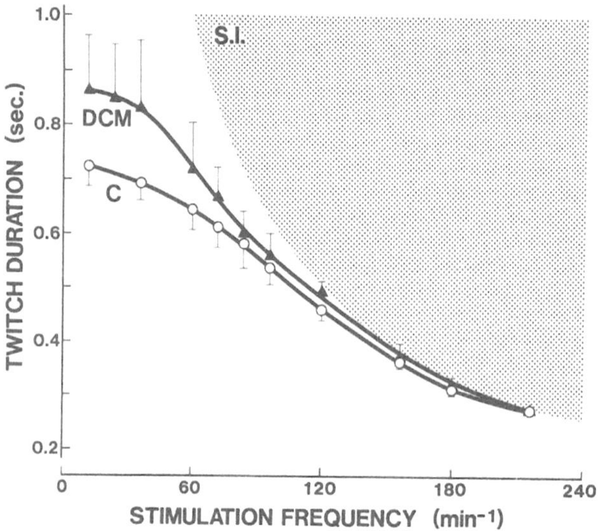

Fig. 5. Twitch duration vs stimulation frequency. Total duration of twitch from stimulus to complete relaxation in non-failing (C) and dilated cardiomyopathic (DCM) strips. Shaded area (delineated by S.I. duration $= f^{-1}$) indicates twitch durations which would result in incomplete relaxation of the twitch; 37 °C.

At optimal stimulation frequency the failing myocardium generated only 31 % (p < .02) as much twitch tension as did the non-failing myocardium (Fig. 3 and Table 1). The average twitch tension values for the two subgroups of failing myocardium in Figure 4 were not different (13.8 $\pm$ 2.9 mN/mm² in A vs. 14.1 $\pm$ 2 mN/mm² in B).

Three aspects of the frequency dependence of the isometric twitch relevant to the question of the role of diastolic tension build-up in contributing to the descending limb of the tension-frequency relations are shown in Figures 5, 6 and 7. The twitch duration (from stimulus to complete relaxation) is maintained shorter than the interval between stimulations for frequencies below 140 $\pm$ 16 min⁻¹ in the failing

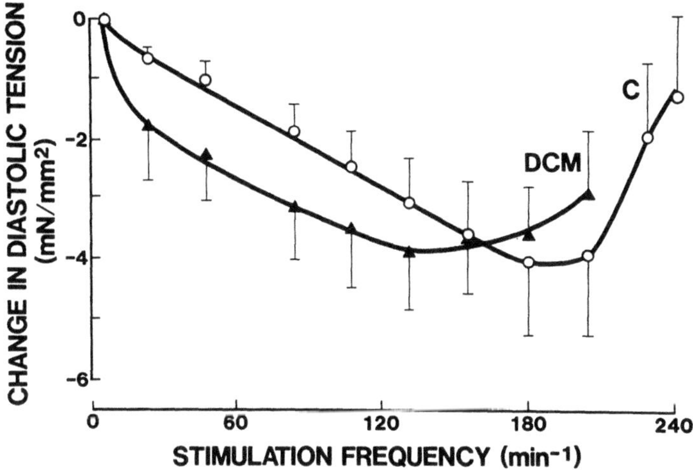

Fig. 6. Diastolic tension vs stimulation frequency. Values indicate change in diastolic tension from resting level at each frequency in non-failing (C) and dilated cardiomyopathic strips (DCM); 37 °C.

preparations and for frequencies below $161 \pm 14 \, \mathrm{min}^{-1}$ in the non-failing preparations. The ratio of the frequency at which the twitch duration equals the twitch interval (truncation frequency, f_T) to the frequency for maximal twitch tension generation (optimal frequency f_0) is 3.1 ± 0.96 in the failing myocardium while it is 1.0 ± 0.1 in the non-failing group.

The extent of decline in diastolic tension as stimulation frequency is increased is shown in Figure 6. In both groups the maximal decline is about $4 \, \mathrm{mN/mm^2}$, which amounts to 14 % and 37 % of the maximal peak twitch tension in the control and failing myocardium, respectively. The diastolic tension begins returning toward low-frequency values at a stimulation frequency of $192 \pm 10 \, \mathrm{min}^{-1}$ in control and at $165 \pm 18 \, \mathrm{min}^{-1}$ in the failing group. On an individual muscle basis this occurs at frequencies 19 ± 5 % above optimal in non-failing myocardium and 110 ± 50 % above optimal stimulation frequencies in the failing tissue.

The average amount of increase in diastolic tension that accompanies each decrease in peak twitch tension as stimulation frequency is raised above optimum frequency is shown in Figure 7 for the two groups. It amounts to 17 ± 6 % of the decrease in peak twitch tension in the non-failing group and 25 ± 6 % in the failing group.

Discussion

Use of the BDM protection method has enabled preparation of left-ventricular myocardial strips which are thin enough to be adequately oxygenated at normal physiological temperature and contraction frequencies (20). This allows questions concerning the mechanisms of myocardial failure to be addressed in isolated myocardium under conditions similar to those prevailing in vivo. Thus, values of peak

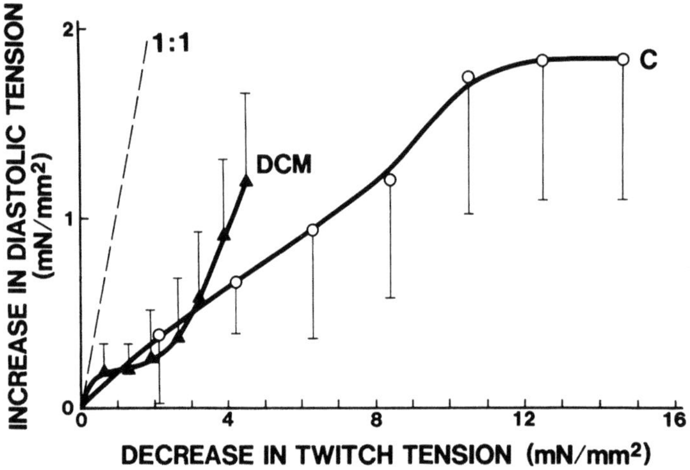

Fig. 7. Increase in diastolic tension vs decrease in peak twitch tension as stimulation frequency increases above f_o. Non-failing control (C) and dilated cardiomyopathic preparations (DCM). Dashed line indicates unity slope; 37 °C.

developed left-ventricular wall stress of $21.8\,mN/mm^2$ to $29\,mN/mm^2$ ($74\,min^{-1}$ heart rate) obtained by left-ventricular catherization of resting non-failing human subjects (12, 16) are similar to the present value from the non-failing strip preparations of about $25\,mN/mm^2$ (Fig. 4A). Similarly, from catherization measurements (5) in non-failing subjects peak rate of left-ventricular pressure development increased 30 % with ventricular pacing between 80 and $120\,min^{-1}$, which compares well with the 22 and 38 % increases in peak twitch tension over this same range of frequencies in the non-failing strips in Fig. 4A.

Of major interest are the present results obtained from failing myocardium indicating maximal twitch tension is depressed by 69 % (Table 1) and that the tension-frequency relation is shifted to lower frequencies. The deficient tension producing capacity of the failing myocardium may be a major mechanism of the 60–70 % reduction in left-ventricular ejection fraction observed in patients with dilated cardiomyopathy (5, 7).

An important effect of the shift of the tension-frequency relation to lower frequencies in dilated cardiomyopathy is that over the normal range of heart rates the normally positive inotropic effect of tachycardia becomes a negative one due to the presence of the descending limb of the tension-frequency curve in this frequency range (Fig. 4A). This defect of failing myocardium probably contributes to the blunting or disappearance of the normal tachycardia-induced increases in peak rate of left-ventricular pressure change and to the blunting of frequency dependent potentiation of circumferential fiber shortening velocity observed with atrial pacing in patients with dilated cardiomyopathy (5, 7).

These changes in myocardial performance limiting tension generating ability and frequency potentiation in dilated cardiomyopathy probably play a significant role in loss of both the ability to maintain normal cardiac output at rest and to increase cardiac output with exercise. Furthermore, the presence of a negative-sloped tension-frequency relation in the failing ventricle may contribute significantly to the dilation process since, in this case, accelerations in heart rate *per se* with exercise would cause the ventricle to undergo further loss of tension generating capacity at a time when venous return is rising thereby resulting in further increases in end-diastolic and end-systolic ventricular volumes. The ability to induce dilated cardiomyopathy in dogs by chronic ventricular pacing (2, 17) at high heart rates might result from shifting the "resting" heart rate into the negative portion of the normal tension-frequency relation.

It is also interesting to consider the possibility that the improved myocardial function and the regression of cardiomegaly in response to β-blocker therapy in patients with dilated cardiomyopathy (4, 25, 26) might result, in part, from its bradycardic effect *per se*. This might sufficiently lower the heart rate of a patient with a tension-frequency curve such as in Fig. 4B (stars) to move operation to a less negative-sloped, or even to a positive-sloped region of the curve.

The present results also suggest that a moderate shift in position of the tension-frequency relation to lower frequency does not in itself result in myocardial failure since one of the non-failing strips (LV ejection fraction = 0.75) exhibited a tension-frequency relation which peaked at $132 \, min^{-1}$ (open symbols in Fig. 4B). This is not different than three of the end-stage failure preparations which peaked between 120 and $144 \, min^{-1}$ (filled symbols in Fig. 4B). The non-failing myocardium had maintained a far superior tension generating capacity (optimal twitch tension $= 59.6 \, mN/mm^2$ compared with 13 to $17 \, mN/mm^2$ in the failing strips).

Since end-diastolic pressure and diastolic stiffness are increased in patients with dilated cardiomyopathy (7), it was of interest to examine the relation between diastolic tension and twitch tension in the present study. It is conceivable that the descending limb of the tension-frequency relation occurs at lower frequencies than normal in failing myocardium because of incomplete relaxation of the twitch as has recently been observed in hypertrophic cardiomyopathy (10). If so, we should expect to find the following conditions present: 1) the descending limb of the tension-frequency curve should begin at the same frequency (f_o) as the one (f_T) at which the stimulation interval becomes equal to the total duration of the twitch, 2) the reduction in diastolic tension that occurs with increasing stimulation frequencies should reverse and begin returning toward low frequency values also at the optimal stimulation frequency , and 3) the amount of increase in diastolic tension at each frequency above f_o should be equal to the amount of decrease in peak twitch tension associated with the frequency increase.

In the failing myocardium the premature decline in peak twitch tension does not appear to be caused by incomplete relaxation of the twitch before the next contraction begins, since the frequency at which this occurs is, on average, three times higher than the optimal stimulation frequency. In addition, the frequency at which end-diastolic tension begins rising from its lowest valve (see Fig. 6) is, on average, two times higher than the optimal stimulation frequency. Furthermore, the amount of twitch tension lost is about four to five times larger than the amount of rise in diastolic tension.

In the non-failing myocardium the descending limb of the twitch tension-frequency curve may be partly caused by incomplete relaxation, since the twitch duration is similar to the stimulus interval at optimal frequency and the diastolic tension begins its rise not much above (19 $\pm$ 5 %) this frequency. However, the contribution to the descending limb of the curve from this mechanism is only 17 % of the total decline in twitch tension on average. Hence, the results suggest that other mechanisms are dominant in controlling the shape of the tension-frequency relation in both failing and non-failing myocardium.

In view of reports of altered calcium handling in dilated cardiomyopathy, we consider the possibility that the tension-frequency relation is altered because excitation-contraction coupling is altered in the failing myocardium. Although the calcium sensitivity of the myofilaments (11), the Ca^{2+} release channel function (15) and the specific Ca^{2+} uptake by sarcoplasmic reticulum membrane (18) are normal in DCM, there is evidence for a reduced number Ca^{2+} pumps in the sarcoplasmic reticulum of DCM hearts (3). Further, in view of our recent finding of a 68 % reduction in heat liberation by the excitation-contraction coupling system (tension independent heat) of DCM ventricular strips (13), we suggest that reduced calcium release is a major intracellular defect leading to the reduced tension and altered tension-frequency performance observed here. This assumption is supported by the recent report of significant depression in peak intracellular Fura-2 transients in ventricular myocytes isolated from failing human hearts. (Beuckelman and Erdmann, this volume).

We propose that the shifts in position and shape of the tension-frequency curves in DCM are caused by differences in the way the intracellular and surface membrane calcium channels and/or pumps respond to changes in frequency. Note the tendency toward increased duration of isometric twitch in DCM in Figure 5. The report of reduced production of cyclic AMP in DCM hearts (6) suggests these differences may be mediated by differences in phosphorylation of Ca^{2+} handling proteins.

The present observation of the reduction in diastolic tension with increased frequency of stimulation may be a reflection of increase in calcium pumping rates in response to tachycardia. If so, the data of Figure 6 may be explained by assuming that a similar amount of increase occurs in DCM as in non-failing myocardium, but that in the former the ability to increase Ca^{2+} pumping rate saturates at a lower than normal stimulation frequency. This may contribute to the lowering of the optimal stimulation frequency in cardiomyopathy.

Appendix

We would like to call attention to the report of Feldman et al. (Feldman MD, Alderman JD, Aroesty JM, Royal HD, Ferguson JJ, Owen RM, Grossman W, McKay RG. (1988) Depression of Systolic and Diastolic Myocardial Reserve during Atrial Pacing Tachycardia in Patients with Dilated Cardiomyopathy. J Clin Invest 82:1661–1669.) Their Figures 6 and 7 H–M show that with tachycardia (from 74 to 134 min^{-1}) in patients with normal left-ventricular function at a given ventricular volume of 40 ml/m^2, left-ventricular diastolic pressure falls on average by 6.4 $\pm$ 0.5 mm Hg. In patients with IDCM (Fig. 5), increasing the heart rate from 86 to 141 min^{-1} caused a decline in diastolic pressure in four out of seven patients that averaged 1.10 $\pm$ 0.4 mm Hg (Of the remaining three, one had no change and two had increases of 1.4 or 3.6 mm Hg). We estimate that these changes in pressure correspond to changes in wall stress of about 1 and 0.2 mN/mm^2 in NF and IDCM hearts, respectively. In the NF and IDCM strips studied here (Fig. 6), increasing the frequency over these same ranges caused a fall in end-diastolic tension of 1.3 and 0.8 mN/mm^2, respectively.

Acknowledgements   We thank Richard Lachapelle for assistance with apparatus, Dr. Elizabeth Low and Tom Haviland for assistance with statistics and programming, Drs. Martin LeWinter, Frederick Schoen, Rosemarie Maddi, and John Fox for liaison with surgery and advice regarding use of human tissue. We thank Donna Kelley for preparing the manuscript.

References

1. Alpert NR, Mulieri LA (1982) Increased myothermal economy of isometric force generation in compensated cardiac hypertrophy induced by pulmonary artery constriction in the rabbit. Circ Res: 50:491–500
2. Coleman HN, Taylor RR, Pool PE, Whipple CH, Covell JW, Ross J, Braunwald E (1971) Congestive heart failure following chronic tachycardia. Am heart J: 81:790–798
3. Duc Ph, Lompre AM, Mercadier JJ, Wisnewsky C, Bouveret P, Fraysse JB, Komajda M, Allen P, Schwartz K (1989) Quantitative analysis of sarcoplasmic reticulum Ca^{2+} ATPase mRNA at the time of cardiac transplantation. J Clin Ivest: 21(III)S6
4. Engelmeier RS, O'Connell JB, Walsh R, Rad N, Scanlon PJ, Gunnar RM (1985) Improvement in symptoms and exercise tolerance by metoprolol in patients with dilated cardiomyopathy: a double-blind, randomized, placebo-controlled trial. Circulation 72(3):536–546
5. Feldman MD, Alderman JD, Aroesty JM, Royal HD, Ferguson JJ, Owen RM, Grossman W, McKay RG (1988) Depression of systolic and diastolic myocardial reserve during atrial pacing tachycardia in patients with dilated cardiomyopathy. J Clin Invest 82:1661–1669
6. Feldman MD, Copelas L, Gwathmey JK, Phillips P, Warren SE, Schoen FJ, Grossman W, Morgan JP (1987) Deficient Production of cyclic AMP; pharmacologic evidence of an important cause of contractile dysfunction in patients with end-stage heart failure. Circulation 75(2):331–339
7. Grossman W, McLaurin LP, Rolett EL (1979) Alterations in left ventricular relaxation and diastolic compliance in congestive cardiomyopathy. Cardiovas Res 13:514–522
8. Gwathmey JK, Slawsky MT, Hajjar RJ, Briggs GM, Morgan JP (1990) Role of intracellular calcium handling in force-interval relationships of human ventricular myocardium. J Clin Invest 85:1599–1613
9. Gwathmey JK, Copelas L, MacKinnon R, Schoen FJ, Feldman MD, Grossman W, Morgan JP (1987) Abnormal intracellular calcium handling in myocardium from patients with end-stage heart failure. Circ Res 61:70–76
10. Gwathmey JK, Warren SE, Briggs GM, Copelas L, Feldman MD, Phillips PJ, Callahan M, Schoen FJ, Grossman W, Morgan JP (1991) J Clin Invest 87:1023–1031
11. Hajjar RJ, Gwathmey JK, Briggs GM, Morgan JP. (1988) Differential effect of DPI 201-106 on the sensitivity of the myofilaments to Ca^{2+} in intact and skinned trabeculae from control and myopathic human hearts. J Clin Invest 82:1578–1584
12. Hasenfuss G, Holubarsch C, Blanchard EM, Mulieri LA, Alpert NR, Just Hj. (1989) Influence of isoproterenol on myocardial energetics. Experimental and clinical investigations. In Just Hj, Holubarsch Ch, Scholz H(ed) Inotropic Stimulation and Myocardial Energetics. Steinkopff, Darmstadt, pp 147–155
13. Hasenfuss G, Mulieri LA, Leavitt B, Ittleman F, Allen PD, Holubarsch Ch, Alpert NR (1990) Altered contractile function in failing dilated cardiomyopathy. Circulation 82:III-568
14. Higginbotham MB, Morris KG, Williams RS, McHale PA, Coleman RE, Cobb FR (1986) Regulation of stroke volume during submaximal and maximal upright exercise in normal man. Circ Res. 58:281–291
15. Holmberg SRM, Williams AJ (1989) Single channel recordings from human cardiac sarcoplasmic reticulum. Circ Res 65:1445–1449

16. Hood Jr. WP, Rackley CE, Rolett El (1968) Wall stress in the normal and hypertrophied human left ventricle. Am J Cardiol 22:550–558
17. Moe GW, Stopps TP, Howard RJ, Armstrong PW (1988) Early recovery from heart failure: insights into the pathogenesis of experimental chronic pacing-induced heart failure. J Lab Clin Med 112:426–432
18. Movsesian MA, Bristow MR, Krall J (1989) Ca^{2+} uptake by cardiac sarcoplasmic reticulum from patients with idiopathic dilated cardiomyopathy. Circ Res; 65:1141–1144.
19. Mulieri LA, Hasenfuss G, Leavitt B, Ittleman F, Allen P, Alpert NR (1990) Altered Tension and Force-Frequency Relation in Failing Human Myocardium. J Moll Cell Cardiol 22-S32
20. Mulieri LA, Hasenfuss G, Ittleman F, Blanchard EM, Alpert NR (1989) Protection of human left ventricular myocardium from cutting injury with 2,3-butanedione monoxime. Circ Res 65:1441–1444
21. Mulieri LA, Hasenfuss G, Ittleman F, Allen PD, Blanchard EM, Alpert NR (1989) Alterations in the force-frequency relation in stage III-IV failing human myocardium. Circulation 80(4):II-75
22. Mulieri LA, Luhr G, Trefry J, Alpert NR (1977) Metal-film thermopiles for use with rabbit right ventricular papillary muscles. Am J Physiol 233(5):C146–C156
23. Paradise NR, Schmitter JL, Surmitis JM (1981) Criteria for adequate oxygenation of isometric kitten papillary muscle. Am J Physiol 241:H348–H353
24. Plotnick GD, Becker LC, Fisher ML, Gerstenblith G, Renlund DG, Fleg JL, Weisfeldt ML. Lakatta EG. (1986) Use of the Frank-Starling mechanism during submaximal versus maximal upright exercise. Am J Physiol 251:H1101–H1105
25. Swedberg K, Hjalmarson A, Waagstein F, Wallentin I. (1980) Beneficial effects of long-term beta-blockade in congestive cardiomyopathy. British Heart Journal 44:117–133
26. Waagstein F, Hjalmarson A, Varnauskas E, Wallentin I. (1975) Effect of chronic beta-adrenergic receptor blockade in congestive cardiomyopathy. British Heart Journal 37:1022–1036

Author's address:
Louis A. Mulieri, Ph.D.
Dept. of Physiology and Biophysics
University of Vermont College of Medicine
Burlington, VT 05405, USA

Alterations of the force-frequency relationship in the failing human heart depend on the underlying cardiac disease

B. Pieske, G. Hasenfuss, Ch. Holubarsch, R. Schwinger[1], M. Böhm[1], H. Just.

Department of Cardiology, University of Freiburg and [1]Department of Cardiology, University of Munich, FRG

Summary: We investigated the force-frequency relationship (0.5–3 Hz) in non-failing human myocardium and in end-stage failing human myocardium due to dilated cardiomyopathy or subacute myocarditis. In non-failing myocardium, force of contraction increased with increasing stimulation frequency. In end-stage heart failure, the force-frequency relationship was inverse in myocardium from dilated cardiomyopathy, but was similar to control in myocardium from subacute myocarditis. After increasing extracellular Ca^{2+}-concentration from 2.5 to 7.2 mM, the shape of the force-frequency relationship was not changed in nonfailing myocardium. In dilated cardiomyopathy, the decline in force with increasing frequencies was even more pronounced at 7.2 mM compared to 2.5 mM extracellular Ca^{2+}. In subacute myocarditis, at Ca^{2+} 7.2 mM, increasing frequencies increased force in the lower frequency range (< 1.75 Hz) only, whereas at higher stimulation rates force declined again. These results indicate that (1.) alterations of the force-frequency relationship in the failing human heart depend on the underlying cardiac disease and/or the time-course of the disease, and (2.) an increase in the extracellular Ca^{2+}-concentration aggravates changes in the force-frequency relationship in the failing myocardium.

Key words: Force-frequency relation; human myocardium; dilated cardiomyopathy; myocarditis; excitation-contraction-coupling

Introduction

Bowditch (3) was the first to report that force of contraction in the frog heart increases with increasing heart rates. This "treppe"-phenomenon has also been shown in isolated myocardium (2, 19) and in whole-heart experiments (7, 12) from many mammalian species and was related to frequency-dependent increases in transmembrane Ca^{2+}-fluxes (1, 20). The situation in the human heart has recently been evaluated under in vitro conditions (6, 8, 16, 17). In these studies, it has been shown that in non-failing human myocardium, force of contraction increases with increasing stimulation frequencies. In contrast, this force-frequency relationship (FFR) was reported to be inverse in end-stage heart failure due to dilated and ischemic cardiomyopathy (6, 8, 16, 17).

It is not known, however, whether the inverse FFR is a common phenomenon in end-stage heart failure independent of its etiology or whether it depends on factors such as underlying cardiac disease or time-course of the disease.

We therefore investigated the FFR in non-failing human myocardium and in end-stage failing myocardium from patients with long-running dilated cardiomyopathy or with subacute viral myocarditis with a much shorter time-course. Since a defect in intracellular Ca^{2+} handling might be the underlying mechanism for the inverse FFR in heart failure (9, 10, 13, 14), experiments have also been performed in the presence of a high extracellular Ca^{2+}-concentration.

Materials and methods

Myocardial muscle strips

Human left-ventricular papillary muscle strips were obtained from end-stage (NYHA IV) failing hearts of six patients with dilated cardiomyopathy and three patients with subacute viral myocarditis, and from three non-failing hearts which were explanted from multiorgan donors and which could not be transplanted for technical reasons. Coronary heart disease or valvular disease has been excluded in all heart failure patients by angiography and echocardiography. All patients with viral myocarditis had a typical clinical course of the disease, and in two cases, myocarditis had been proven histologically from biopsies prior to transplantation. Time-course of the disease since diagnosis was 6.8 ± 1.5 years in the dilated cardiomyopathy group and 1.1 ± 0.4 years in the myocarditis group ($p < 0.05$). Mean age at the time of transplantation was 58 ± 3 years in the dilated cardiomyopathy group, 39 ± 7 years in the myocarditis group and 47 ± 5 years in the donor heart group. Mean ejection fraction was $22 \pm 1\%$ and $20 \pm 1\%$ in the dilated cardiomyopathy group and the myocarditis group, respectively (n.s.). No cardiac catheterization had been performed in the organ donors. Patients' data and hemodynamic parameters are listed in Table 1.

Preparation procedure

Papillary muscle strips were prepared and mounted to an isometric force transducer as described elsewhere (9, 15). Briefly, small pieces of left-ventricular papillary muscles were excised immediately after explantation and stored in a special cardio-protective solution containing 30 mM 2,3-butanedione monoxime, bubbled with carbogen (95% O_2, 5% CO_2, pH7.4) at room temperature. Upon arrival in the laboratory, small (cross-sectional area < 0.6 mm^2) muscle strips were dissected in exact longitudinal fiber orientation with the help of a stereomicroscope and specially designed preparation chambers (15). All preparation steps were carried out in the presence of the cardioprotective solution which minimizes cutting injury (15). The muscle strips were then mounted horizontally in an organ bath (Hugo Sachs Electronics, March-Hugstetten, FRG) and connected to an isometric force trans-ducer. Signals were recorded using a chart strip recorder (Graphtec Lineacorder, Yokohoma, Japan). Muscles were superfused with a carbogen-bubbled (pH 7.4) modified Krebs-Henseleit solution of the following composition (in mM): Na^+ 152, K^+ 3.6, Cl^- 135, Ca^{2+} 2.5, Mg^{2+} 0.6, HCO_3^- 25, $H_2PO_4^-$ 1.3, SO_4^{2-} 0.6, glucose 11.2 and insulin 10 IU/l at 37° C. After initially prestretching the muscle strips with 1 mN, they were stimulated by field stimulation with rectangular pulses of 5 ms duration, voltage 20% above threshold, at a frequency of 1 Hz and allowed to equilibrate for 1 h. At that time, the cardioprotective solution was completely washed out, and the muscles were stretched along their length-tension-curve in 0.05 mm-steps until force of contraction (FOC) was maximal. Then, the force-frequency relationship was obtained in each muscle by changing stimulation rates in 0.25 Hz-steps from 0.5 Hz to 3 Hz. Force-frequency relationships were obtained either at a basal Ca^{2+}-concentration of 2.5 mM, or after increasing force of contraction by raising the extracellular Ca^{2+}-concentration to 7.2 mM. At the end of each experiment, a test

Table 1. Age, time-course and hemodynamic data of the patients

Myocarditis	Age (y)	TT (y)	MRAP (mmHG)	LVEDP (mmHG)	LVEDV (ml)	EF (%)	CI (l/min/m^2)
1. SW	29	0.7	13	26	357	22	2.0
2. HJ	56	2.2	4	12	217	20	2.1
3. LH	33	0.5	8	44	480	18	2.5
Mean ± SEM	39 ± 7	1.1 ± 0.4	9 ± 2	27 ± 7	351 ± 62	20 ± 1	2.2 ± 0.2

DCM							
1. BR	45	2.0	6	21	235	18	1.7
2. DS	57	—	14	24	388	20	1.7
3. PE	67	10.0	11	25	289	25	1.8
4. FN	55	4.0	16	32	404	21	1.7
5. WF	62	8.0	11		—	21	1.6
6. SA	62	10.0	6	18	449	29	2.1
Mean ± SEM	58 ± 3	6.8 ± 1.5	11 ± 2	24 ± 2	353 ± 35	22 ± 1	1.8 ± 0.1

TT: time from first symptoms to transplantation; MRAP: mean right atrial pressure; LVEDP: left-ventricular end-diastolic pressure; LVEDV: left-ventricular end-diastolic volume; EF: ejection fraction; CI: cardiac index; y: years; DCM: dilated cardiomyopathy

for adequate oxygenation (18) was performed by changing the carbogen to 80% O_2, 5% CO_2 and 15% N_2 at the stimulation rate, where FOC was maximal. This procedure lowered oxygen tension significantly in the bathing solution and all muscle strips in which force of contraction declined by more than 10% during 30 min were considered hypoxic and discarded from the evaluation. By completion of the experiment, length and diameter of the muscle strip were measured by a scaled microscope ocular and wet weight of the portion of the muscle strip between the ligatures was measured. Cross-sectional area was determined by dividing wet weight by length.

Statistical analysis:

Average values are given as mean ± SEM. Mean values of peak twitch tension of the different groups at each stimulation frequency were compared using Student's t-test and the Bonferroni-Holms-procedure. Differences were considered significant if p was < 0.05.

Results

Figure 1 shows original recordings of typical experiments in papillary muscle-strip preparations from a non-failing heart (upper panel), a heart with subacute myocarditis (middle panel), and a heart with dilated cardiomyopathy (lower panel) at a Ca^{2+}-concentration of 2.5 mM. Stimulation frequency was increased stepwise from

FORCE - FREQUENCY RELATION
Ca^{2+} 2.5 mM

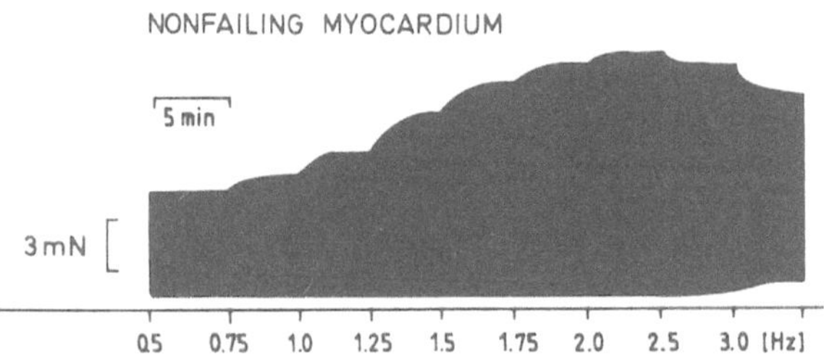

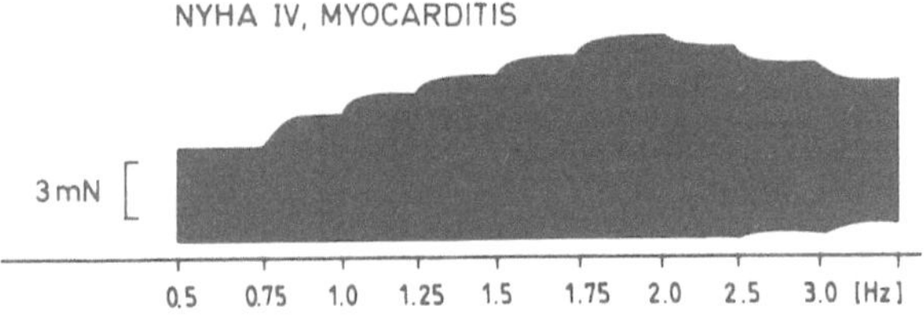

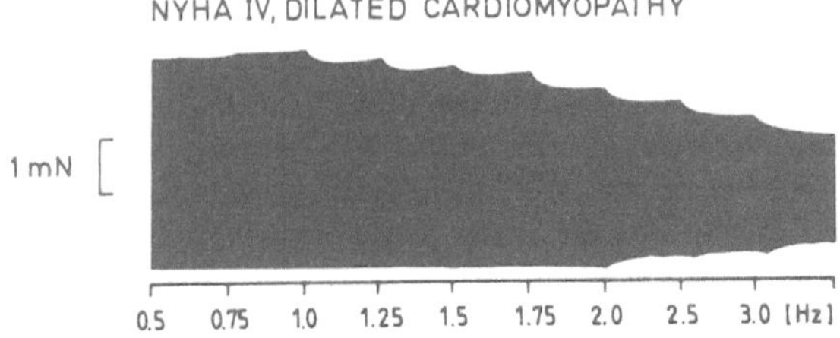

Fig. 1. Original tracings showing the influence of stimulation frequency on force of contraction in isolated human papillary muscle strips from a nonfailing heart (upper panel), and an end-stage failing heart due to subacute myocarditis (middle panel) and an end-stage failing heart due to dilated cardiomyopathy (lower panel). Stimulation frequency (Abscissa) was increased stepwise from 0.5 to 3 Hz.

0.5 to 3 Hz. In the muscle strip from the non-failing heart, force of contraction increased constantly with increasing stimulation frequency up to a frequency of 2 Hz at which FOC was maximal. In the muscle strip from a heart with myocarditis, FOC increased constantly up to a stimulation frequency of 1.75 Hz and then declined slightly. In contrast, in the muscle strip from a heart with DCM, FOC was maximal at a stimulation frequency of 0.75 Hz and then constantly declined with higher stimulation rates.

The average values are plotted in Fig. 2. Force of contraction is calculated as percent change from baseline (0.5 Hz corresponding to 100%). In the non-failing myocardium (left panel), force of contraction increased continuously up to 250 $\pm$ 24% at a stimulation frequency of 2 Hz and then declined slightly. In the myocarditis group (middle panel), the shape of the curve was similar. FOC increased up to a stimulation rate of 1.75 Hz, where FOC was 184 $\pm$ 16% from baseline, and then declined slightly. In both groups, FOC remained significantly elevated compared to baseline even at the highest stimulation frequencies (204 $\pm$ 27% and 149 $\pm$ 18%, respectively, at 3 Hz). In contrast, in dilated cardiomyopathy (right panel), FOC was maximal at 0.75 Hz (101 $\pm$ 2%) and then declined continuously with higher stimulation frequencies. At 3 Hz, FOC was 55 $\pm$ 9% from the baseline value at 0.5 Hz.

Figure 3 shows the results in the three types of myocardium after prestimulating muscle strips from the same hearts as in Fig. 2 with an extracellular Ca^{2+}-concentration of 7.2 mM. By this intervention, at a stimulation rate of 1 Hz, FOC increased significantly by 235 $\pm$ 29% in NF, by 139 $\pm$ 46% in myocarditis, and by 85 $\pm$ 12% in DCM. In the nonfailing myocardium (left panel) FOC increased constantly with increasing stimulation frequency and reached a maximum of 182 $\pm$ 16% at 2 Hz, compared to the value at 0.5 Hz. At 3 Hz, FOC had slightly declined to 168 $\pm$ 26% of the value at 0.5 Hz. Thus, after prestimulation with Ca^{2+} 7.2 mM, the force-frequency relation was parallel to the curve at Ca^{2+} 2.5 mM. In myocarditis (middle panel), FOC was maximal at a stimulation frequency of 1.75 Hz with 140 $\pm$ 6% of the baseline value at 0.5 Hz. However, at higher stimulation frequencies, FOC declined strikingly and was 95 $\pm$ 6% from baseline value at 3 Hz. In both groups, the frequency inotropism was significantly reduced (p < 0.05) in the Ca^{2+}-prestimulated state compared to Ca^{2+} 2.5 mM. In the DCM group (right panel), with 7.2 mM Ca^{2+}, FOC was maximal at the lowest frequency investigated (0.5 Hz, 100%) and declined constantly with higher stimulation frequencies. At 3 Hz, FOC was 57 $\pm$ 6% of the value at 0.5 Hz.

Discussion

The inversion of the FFR in end-stage human heart failure has been described recently (6, 8, 16, 17). However, the subcellular defect of this phenomenon still remains unknown. Furthermore, to date, no study has been performed that investigates FFR with respect to different etiologies of heart failure. End-stage heart failure is a final syndrome of a variety of diseases that may affect primarily the myocytes and/or result from an increase in fibrosis and scar tissue with reactive hypertrophy of the remaining myocardium.

The most striking finding of this study is that alterations of the force-frequency relationship in end-stage heart failure seem to depend on the underlying disease and/or the time-course of the disease. All patients under investigation suffered from end-stage (NYHA IV) heart failure according to clinical and hemodynamic data. However, as shown in Table 1, the time-course of the disease prior to transplantation was much shorter in the myocarditis group. Therefore, we cannot distinguish whether the observed differences between dilated cardiomyopathy and subacute myocarditis are primarily due to different subcellular defects, or whether the duration of the disease is the major factor.

There are now increasing reports on alterations in intracellular Ca^{2+}-handling in dilated cardiomyopathy (8–10, 13). As was shown recently in myocardium from dilated cardiomyopathic hearts, a significant reduction in isometric tension at 1 Hz goes

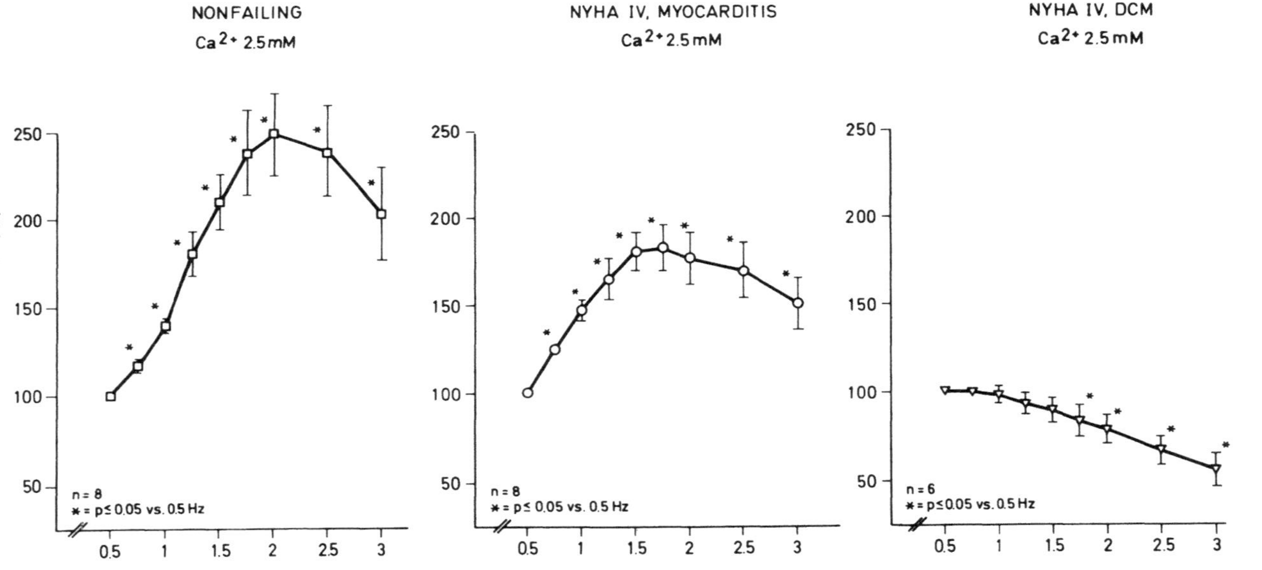

Fig. 2. FFR in isolated papillary muscle strips from non-failing hearts (left panel) and from end-stage failing hearts due to subacute myocarditis (middle panel) or dilated cardiomyopathy (right panel) at an extracellular Ca^{2+}-concentration of 2.5 mM. Ordinates: Force of contraction in % of the value at a stimulation frequency of 0.5 Hz. Basal force of contraction at a stimulation frequency of 1 Hz was 18.2 ± 3.6 mN/mm^2 in non-failing, 8.6 ± 0.7 mN/mm^2 in myocarditis and 9.2 ± 1.9 mN/mm^2 in dilated cardiomyopathy.

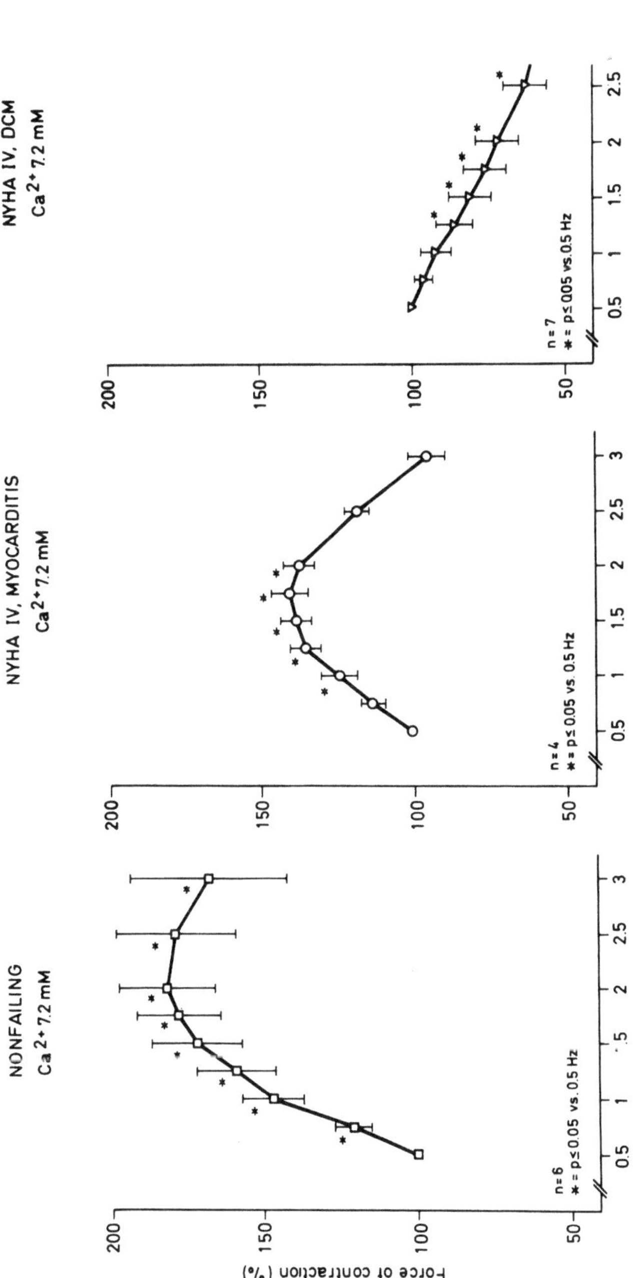

Fig. 3. FFR in isolated papillary muscle strips from non-failing hearts (left panel) and from end-stage failing hearts due to subacute myocarditis (middle panel) or dilated cardiomyopathy (right panel) at an extracellular Ca^{2+}-concentration of 7.2 mM. Ordinates: Force of contraction in % of the value at a stimulation frequency of 0.5 Hz. Basal force of contraction at a stimulation frequency of 1 Hz was 16.4 ± 5.1 mN/mm² in nonfailing, 12.5 ± 1.6 mN/mm² in myocarditis and 8.0 ± 1.6 mN/mm² in dilated cardiomyopathy. Ca^{2+} 7.2 mM at 1 Hz increased force of contraction by $235 \pm 29\%$, $139 \pm 46\%$ and $85 \pm 12\%$, respectively.

along with a significant reduction in intracellular Ca^{2+} cycling (9, 10). Disturbances of excitation-contraction-coupling may increase with increasing frequencies and may be the underlying mechanism for the inverse FFR in dilated cardiomyopathy.

To investigate whether increased extracellular Ca^{2+}-availability may normalize the inverse FFR in DCM, experiments were performed in the presence of 7.2 mM extracellular Ca^{2+}-concentration which increases Ca^{2+} influx into the myocytes (4). This intervention increased force of contraction significantly in all groups at 1 Hz. However, while the shape of the FFR at this high Ca^{2+} concentration was similar to the control experiments at Ca^{2+} 2.5 mM in nonfailing myocardium, the disturbance of the FFR was even more pronounced at Ca^{2+} 7.2 mM compared to Ca^{2+} 2.5 mM in myocardium from DCM hearts. This might suggest that intracellular Ca^{2+} handling rather than transsarcolemmal Ca^{2+} influx is related to the altered FFR in DCM. The disturbance of the intracellular Ca^{2+} handling may be more relevant if the degree of activation (i.e., extracellular Ca^{2+}-concentration) is high.

No information is available on subcellular defects in Ca^{2+}-handling in viral myocarditis. At Ca^{2+} 2.5 mM, the shape of the FFR in the myocardium from hearts with subacute viral myocarditis was similar to that of the non-failing myocardium. At the high extracellular Ca^{2+}-concentration, there was an attenuated frequency inotropism at intermediate stimulation frequencies and a lack of frequency inotropism at high stimulation frequencies. This might suggest that in subacute myocarditis, a manifestation of disturbances in intracellular Ca^{2+} handling is present only at higher degrees of activation and at high stimulation frequencies. These results may indicate that heart failure in subacute myocarditis is caused by defects distinct from those in dilated cardiomyopathy.

Recent investigations using immuno-histological methods have shown, however, that in up to 30% of DCM hearts, signs of chronic myocarditis can be demonstrated (11). This raises the possibility that in a significant number of patients, end-stage DCM may result from viral myocarditis. If this is the case, the duration of the disease may be a major factor in the development of disturbed FFR in those forms of DCM.

The shape of the force-frequency relationship in cardiac diseases may have significant implications for the therapy of congestive heart failure. Reducing heart rate may improve cardiac contractility in cases where the FFR is inverse. This might also explain the beneficial effects of β-blocker therapy in congestive heart failure due to dilated cardiomyopathy (5). However, our results indicate that this may not hold true for all forms of end-stage heart failure. It is challenging to speculate that, depending on the underlying cardiac disease, heart failure may improve by reducing heart rate in some cases, whereas in others, a higher heart rate may be necessary for adequate cardiac performance. To answer these questions, further studies on the force-frequency relationship have to be performed which differentiate between distinct underlying cardiac diseases.

References

1. Allen DG, Blinks JR (1978) Calcium-transients in aequorin-injected frog cardiac muscle. Nature Lond. 273:509–513
2. Blinks JR, Koch-Weser J (1961) Analysis of the effects of changes in rate and rhythm upon myocardial contractility. J Pharmacol Exp Ther 134:373–389
3. Bowditch HP (1971) Über die Eigenthümlichkeiten der Reizbarkeit, welche die Muskelfasern des Herzens zeigen. Ber Sächs Ges (Akad) Wiss 23:652–689

4. Carafoli E. (1987) Intracellular calcium homeostasis. Ann Rev Biochem 56:395–433
5. Engelmeier RS, O'Connell JB, Walsh R, Rad N, Scanlon PJ Gunnar RM (1985) Improvement in symptoms and exercise tolerance by metoprolol in patients with dilated cardiomyopathy. A double-blind, randomized, placebo-controlled trial. Circulation 72:536–546
6. Feldman MD, Gwathmey JK, Phillips P, Schoen F, Morgan JP (1988) Reversal of the force-frequency relationship in working myocardium from patients with end-stage heart failure. J Appl Cardiol 3:273–283
7. Freeman GL, Little WC, O'Rourke RA (1987) Influence of heart rate on left ventricular performance in conscious dogs. Circ Res 61:455–464
8. Gwathmey JK, Copelas L, MacKinnon R, Schoen FJ, Feldman MD, Grossman W, Morgan JP (1987) Abnormal intracellular calcium handling in myocardium from patients with end-stage heart failure. Circ Res 61:70–76
9. Hasenfuss G, Mulieri L, Leavitt B, Allen PD, Haeberle J, Alpert NR (1992) Alteration of contractile function and excitation-contraction coupling in dilated cardiomyopathy. Circ Res (in press)
10. Hasenfuss G, Mulieri LA, Leavitt B, Allen PD, Holubarsch Ch, Just H, Alpert NR (1992) Contractile protein function in failing and nonfailing human myocardium. In: Hasenfuss G, Holubarsch Ch, Just H, Alpert NR (eds) Cellular and molecular alterations in the failing human heart. Steinkopff Verlag, Darmstadt
11. Kühl U, Daun B, Seeberg B, Schultheiß HP (1991) High incidence of myocarditis by endomyocardial biopsy in patients with dilated cardiomyopathy. Circulation 84 (Suppl II) 2
12. Mitchell JH, Wallace AG, Skinner NS Jr (1963) Intrinsic effects of heart rate on left ventricular performance Am J Physiol 205:41–48
13. Morgan JP, Erny RE, Allen PD, Grossman W, Gwathmey JK (1990) Abnormal intracellular calcium handling, a major cause of systolic and diastolic dysfunction in ventricular myocardium from patients with heart failure. Circulation 81 (Suppl. III) 21–32
14. Morgan JP (1991) Abnormal intracellular modulation of calcium as a major cause of cardiac contractile dysfunction. N Engl J Med 325:625–632
15. Mulieri LA, Hasenfuss G, Ittleman F, Blanchard EM, Alpert NR (1989) Protection of human left ventricular myocardium from cutting injury with 2,3-Butanedione Monoxime. Circ Res 65:1441–1444
16. Mulieri LA, Hasenfuss G, Leavitt B, Ittleman F, Allen PD, Alpert NR (1990) Altered tension and force-frequency relation in failing human myocardium. J Mol Cell Cardiol 22(I) 32
17. Mulieri LA, Leavitt B, Hasenfuss G, Allen PD, Alpert NR (1992) Contraction frequency dependence of twitch and diastolic tension in human dilated cardiomyopathy. In: Hasenfuss G, Holubarsch Ch, Just H, Alpert NR (eds) Cellular and molecular alterations in the failing human heart. Steinkopff Verlag, Darmstadt
18. Paradise NF, Schmittler JL, Surmitis JM (1981) Criteria for adequate oxygenation of isometric kitten papillary muscle. Am J Physiol 241:H348–H353
19. Parmley WW, Sonnenblick EH (1969) Relation between mechanics of contraction and relaxation in mammalian cardiac muscle. Am J Physiol 216:1084–1091
20. Winegard S, Shanes AM (1962) Calcium flux and contractility in guinea pig atria. J Gen Physiol 45:371–394

Author's address:
Dr. B. Pieske
Klinikum der Albert-Ludwigs Univ.
Abt. Innere Medizin III
Hugstetterstr. 55
W.-7800 Freiburg, FRG

Pathophysiology of cardiac hypertrophy and failure of human working myocardium: abnormalities in calcium handling

A.J. Meuse, C.L. Perreault, J.P. Morgan

Charles A. Dana Research Institute of the Harvard-Thorndike Laboratory, Department of Medicine (Cardiovascular Division), Beth Israel Hospital and Harvard Medical School, Boston, Massachusetts, USA

Summary: Abnormal intracellular calcium ($[Ca^{2+}]_i$) handling appears to be a major cause of both systolic and diastolic dysfunction in animals and human beings with hypertrophy and/or heart failure. We utilized the bioluminescent calcium indicator aequorin to examine the cyclical variations in intracellular calcium levels during isometric contractions. Studies of ventricular muscle from patients with end-stage heart failure exhibited three physiologic findings not seen in preparations taken from normal hearts including: 1) abnormalities in calcium handling; 2) deficient production of cyclic AMP; and 3) a reversed force-frequency relationship. These observations have important implications with regard to the pathogenesis and therapeutics of heart failure in man.

Key words: Calcium; heart failure; diastolic dysfunction; systolic dysfunction

Excitation-contraction coupling

This review summarizes some of the studies our laboratory has performed since 1979 to delineate the role played by abnormal intracellular calcium handling in producing the systolic and diastolic dysfunction that characterizes hypertrophied and failing human myocardium (7, 8, 12, 15, 29). Because the calcium ion (Ca^{2+}) plays a central role in the process of excitation-contraction coupling in the heart, alteration from normal in the subcellular handling of Ca^{2+} may provide the basis for cardiac contractile failure. The process of cardiac excitation-contraction coupling is outlined in Figure 1; details are presented in the legend. Note the central role played by the Ca^{2+} ion as a second messenger. Disturbance of any of these multiple steps in the excitation-contraction coupling process could produce significant contractile or relaxant dysfunction (5, 6, 19, 20, 28, 39-42, 43).

Cyclic adenosine monophosphate (cAMP) is now well-established as a significant second messenger in the heart (27, 30, 32, 33, 37). Cyclic AMP is generated by the action of adenylate cyclase on ATP; it is eventually degraded through the activity of cardiac phosphodiesterases. With regard to excitation-contraction coupling, of particular importance is cAMP's ability to activate protein kinase A, which phosphorylates various sites within the heart. Within the context of this discussion, three sites of

* Supported in part by grant HL-31117 and a Research Career Development Award from the National Institutes of Health, Bethesda, Maryland, USA (HL-01611); DA-05171 from the National Institute of Drug Abuse, Bethesda, Maryland, USA, and a Grant-In-Aid from the American Heart Association.

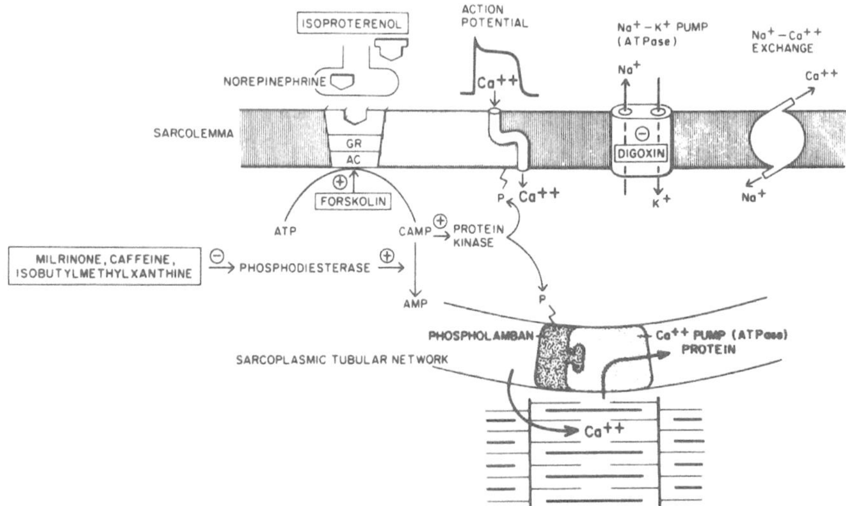

Fig. 1. In the mammalian heart, excitation-contraction coupling is initiated when depolarization permits trigger Ca^{2+} to enter the myoplasm through calcium channels located in the sarcolemma. This trigger Ca^{2+} releases a much larger quantity of activator Ca^{2+} from intracellular stores contained in the sarcoplasmic reticulum (SR). The released Ca^{2+} interacts with troponin C of the regulatory complex on the myofilaments to initiate cardiac contraction; relaxation occurs when Ca^{2+} dissociates from the contractile apparatus and is resequestered by the energy-dependent Ca^{2+} pump of the SR (an ATPase).

Intracellular Ca^{2+} homeostasis is maintained by sarcolemmal mechanisms that extrude Ca^{2+} into the extracellular space, including a sodium-calcium exchange mechanism and an energy-dependent Ca^{2+} pump (not shown). The initial excitatory event of membrane depolarization is also associated with intracellular entry of Na^+, which is ultimately extruded by the energy dependent Na^+, K^+ pump (an ATPase) or by the sodium-calcium exchanger, which operates on the basis of concentration gradients and does not appear to require ATP.

Each of these steps represents a potential site of action for inotropic drugs. The figure also shows that Ca^{2+} entry via the voltage-dependent Ca^{2+} channels and resequestration of Ca^{2+} by the SR are governed by their degree of phosphorylation (indicated by P). See text for additional details. (From 7)

phosphorylation are particularly important (Fig. 1): the first site is in or near the voltage-dependent, L-type calcium channels. Phosphorylation at this site increases conductance of calcium so that more can enter the cell during each depolarization and leads to an increase in intracellular stores of activator calcium. Phospholamban, the regulatory subunit of the sarcoplasmic reticulum Ca^{2+} pump is a second important site of phosphorylation. Phosphorylation of phospholamban significantly increases the rate at which the sarcoplasmic reticulum can resequester Ca^{2+}, thereby enhancing relaxation. A third important site of phosphorylation (not shown in Fig. 1) is at the level of the regulatory subunits themselves, where one of the components of the troponin-tropomyosin regulatory subunit (Troponin-I) can be phosphorylated by a cAMP-dependent protein kinase, which decreases the affinity of troponin-C for calcium. This increases the rate of dissociation of calcium from troponin-C and enhances relaxation.

Intracellular calcium measurement

Several approaches are available for monitoring $[Ca^{2+}]_i$ within myocytes, including the bioluminescent Ca^{2+} indicator aequorin (1). Aequorin is a naturally occurring photoprotein that emits light when it combines with free ionized $[Ca^{2+}]_i$. When loaded into the cytoplasm of cardiac myocytes, aequorin can be used to study intracellular Ca^{2+} handling (1-3, 22). For example, Figure 2B shows the light (an index of intracellular $[Ca^{2+}]_i$) and tension responses recorded from a ferret right-ventricular papillary muscle. Similar signals have been recorded in myocardium from a variety of mammalian species, including man. We have also applied the aequorin technique to the study of $[Ca^{2+}]_i$ in the isolated, saline perfused whole heart (Fig. 2A; 21), blood perfused whole heart (23) and isolated cardiac myocytes (Fig. 2C; 25), as well as vascular smooth muscle and skeletal muscle preparations. The techniques for obtaining and purifying aequorin, and for loading the purified protein into cardiac myocytes, have been described in detail (3, 22, 24).

In the experiments that will be described below, the muscles were maintained in a physiologic salt solution containing 1–2.5 mM $[Ca^{2+}]_o$ at 30 °C and stimulated to contract at 0.1–2 Hz. In most experiments, the muscles were maintained at the length at which maximal isometric tension developed (L_{max}) and light and tension were simultaneously recorded. Typical experimental results are shown in Figure 2, which

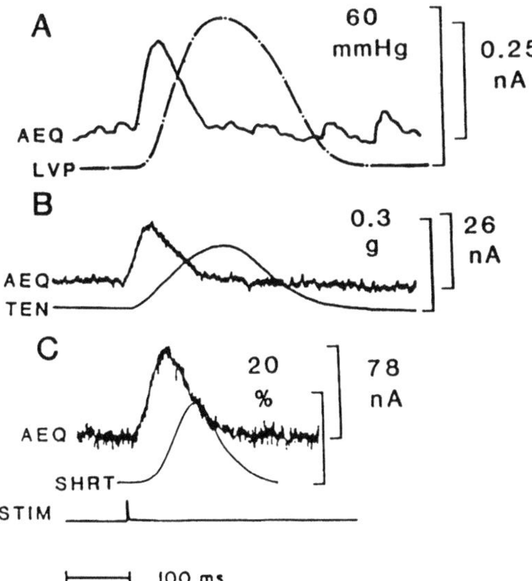

Fig. 2. Relationship between contractile force development and intracellular Ca^{2+} transients in ferret working myocardium. Panel 2A, isolated perfused whole heart preparation; panel 2B, papillary muscle preparation; panel 2C, a suspension of isolated cardiac myocyte (approximately 300 cells). In each panel, upper trace is aequorin (AEQ) light signal expressed in nanoamperes (nA) of current recorded from photoanode; lower trace is stimulus (STIM) artifact. The middle trace in Panel 2A is left ventricular pressure (LVP) in millimeters of mercury (mm Hg); in 2B, isometric force in grams (g) and 2C, shortening (SHRT), expressed as percent of control length. (From 37).

demonstrates papillary muscles, isolated whole heart and single cell preparations from the ferret (37). Note that in each of the three models shown in Figure 2, the aequorin light signal consists of a single component that increases in amplitude much more rapidly than does tension pressure or peak shortening and then declines toward basal levels by the time peak tension pressure or peak shortening is reached. Because the Ca^{2+} released from the intracellular stores during a single cardiac contraction is many times greater than that entering the sarcoplasm from the extracellular space, it is probable that, under normal circumstances, the amplitude of the aequorin light response in mammalian ventricular muscle predominantly reflects the quantity of Ca^{2+} released by the sarcoplasmic reticulum (SR). The time-course of the light transient can be influenced by several factors, including the rate of entry of Ca^{2+} into the sarcoplasm, the rate of Ca^{2+} diffusion and binding to the myofilaments, and the rate of Ca^{2+} reaccumulation from the sarcoplasm and myofilaments by the SR during relaxation. The SR is the major site of both Ca^{2+} release and reaccumulation during the normal cardiac cycle; therefore, the time-course of the light signal in mammalian working myocardium predominantly reflects Ca^{2+} handling by this organelle. Approaches to quantifying the aequorin signal have previously been described (1).

Subcellular mechanisms in human hypertrophy and failure

We have studied a variety of animal models of cardiac hypertrophy and failure and correlated abnormal changes in diastolic and systolic mechanical performance with abnormalities in $[Ca^{2+}]_i$ handling (13, 14, 21, 23, 24, 44). Researchers in our laboratory have been fortunate to obtain, for experimental study, myopathic and control human myocardium during the past several years (8, 9, 12, 15, 17, 18). These hearts were obtained from recipients of cardiac transplants or from donors whose tissue was not suitable for transplantation. From each of these hearts, experimental preparations were removed and returned to the laboratory for study. Figure 3 compares the $[Ca^{2+}]_i$ transients recorded with aequorin and mechanical contractions of trabeculae carneae isolated from a control patient without heart failure, a patient with end-stage dilated cardiomyopathy, and a patient with end-stage hypertrophic cardiomyopathy (12). Note that the light signal recorded from the control muscle in Figure 3 consists of a single component that rises to a peak and then declines towards baseline before peak tension is reached. Monophasic aequorin signals appear to be typical for both animal and human working myocardium (14, 16). In contrast, the calcium transients recorded from myopathic muscles were not only prolonged as compared with the controls, but also consisted of two temporally distinct components (L_1 and L_2 in the figure). Moreover, the plateau phase of the cardiac action potential is prolonged in myopathic muscle vs the controls. These data indicate that the prolonged contractions of myopathic muscles in vitro, as well as the myocardial relaxation abnormalities seen in patients with cardiomyopathy, appear to correlate with changes in $[Ca^{2+}]_i$ handling (11, 29). Additional studies indicated that the muscles from myopathic hearts have a lesser capacity to maintain calcium homeostasis in the presence of normal and increased transsarcolemmal gradients. In addition, studies with selective antagonists indicated that L_2 in myopathic muscle reflects dysfunction of both the sarcolemma and sarcoplasmic reticulum; the former can cause increased Ca^{2+} entry, possibly by voltage-dependent calcium channels,

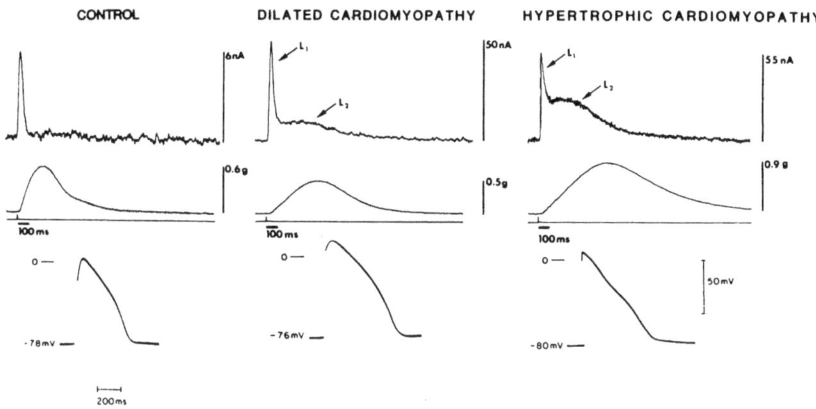

Fig. 3. Representative light tracing (upper noisy trace) and tension tracing (middle trace) in control and myopathic human muscles; lower trace is stimulus artifact. Light expressed in nanoAmperes (nA) of anode current; tension in g. The cross-sectional areas of the muscles are (in mm^2) control, 1.3; dilated cardiomyopathy, 0.5; hypertrophic cardiomyopathy, 0.5. Action potentials recorded in muscle from these same hearts are also shown. (From 12).

and the latter can cause slowed restoration of low resting tone during diastole due to a decreased rate of $[Ca^{2+}]_i$ resequestration (12). The inability of myopathic muscle to maintain $[Ca^{2+}]_i$ homeostasis may be a primary cause of contractile dysfunction in heart failure. Similar abnormalities in $[Ca^{2+}]_i$ handling have been identified in a variety of animal models of cardiac hypertrophy and failure (for example, see (25, 45)).

Deficient production of cAMP in human hypertrophy and failure

Data displayed in Figure 4 show the effects of milrinone which increases intracellular cAMP concentration through inhibition of phosphodiesterases (7). Note that in contrast to control muscle, fibers from the heart failure group showed little response to this agent. In contrast, in the presence of a minimally effetive dose of forscolin to produce activation of adenylate cyclase, milrinone was capable of producing a marked positive inotropic response in the heart failure group. Similar observations were obtained with other phosphodiesterase inhibitors like caffeine and isobutyl-methylxanthine (7). These results support the working hypothesis that a primary defect in human heart failure is deficient production of cAMP (7, 9, 26). This hypothesis can account for many of the abnormalities we have described in myopathic human myocardium (35). For example, if cAMP (located centrally in Fig. 1) levels are depressed, we would expect to have a lower level of activity of protein kinase A. Therefore, decreased phosphorylation of phospholamban could decrease the rate at which calcium is resequestered by the sarcoplasmic reticulum and account for the prolonged calcium transients and delayed relaxation seen in the myopathic hearts.

The possible mechanism accounting for depressed cAMP levels remains speculative. As shown in Figure 1, adenylate cyclase can probably function normally since forskolin produces responses similar to the controls in the myopathic tissue. This also

                                                      A.J. Meuse, C.L. Perreault, J.P. Morgan

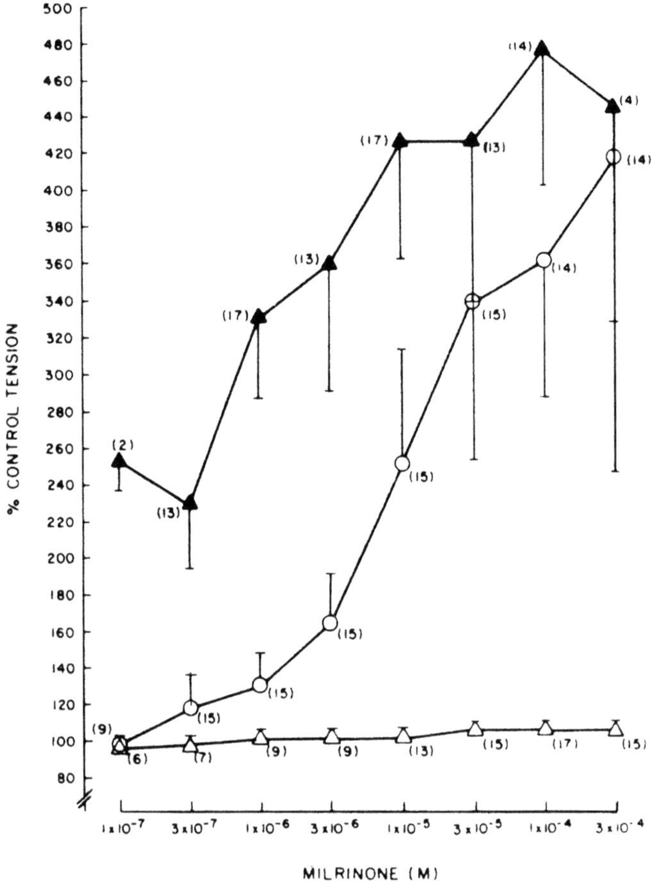

Fig. 4. Dose-response curves for milrinone expressed as percent changes in isometric tension from baseline. Open circles represent control values, open triangles the values from the heart failure group, and filled triangles are values from the heart failure group in the presence of a minimally effective dose of forskolin. (From 7).

may indicate that the substrate for adenylate cyclase, ATP, is present in adequate quantities. It is unlikely that intrinsic phosphodiesterase activity is enhanced since under these conditions, maximally effective concentration of phosphodiesterase inhibitors would be expected to produce a greater than normal response. It seems unlikely that down-regulation of beta-adrenergic receptors could account for these findings since propranolol in control cardiac muscle isolated in vitro does not produce the prolonged Ca^{2+} transient, action potential, or isometric twitch that characterize muscle from patients with heart failure. In recent years, it has become clear that major changes occur in the regulatory G-proteins that govern adenylate cyclase activity in the heart that would tend to diminish adenylate cyclase activity and could produce a diminished production of cAMP (4, 10, 36).

The frequency-response relationship of control versus myopathic human muscle demonstrates the potential clinical significance of these changes. The upper panel (A) of Figure 5 shows the $[Ca^{2+}]_i$ transients and twitches recorded from a myopathic muscle (17); the lower tracing (B) shows similar responses in a control muscle. As the rate of stimulation is increased, the control myocardium shows increased levels of $[Ca^{2+}]_i$ and enhanced force development. Moreover, the calcium level and end-diastolic tension return to normal between contractions. In myopathic muscle, on the other hand, the prolonged calcium transients and twitches fuse at higher pacing rates

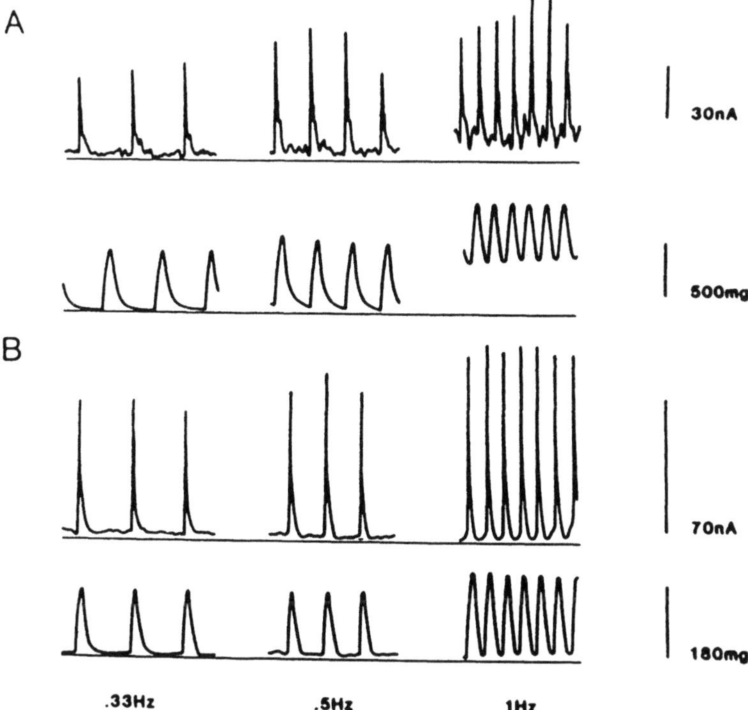

Fig. 5. Light and tension responses to varying frequencies of stimulation from 0.33 to 1 Hz in a hypertrophic trabecular strip (Panel A) from a patient with hypertrophic obstructive cardiomyopathy and a control trabecular strip (Panel B). Light intensity expressed in nanoAmperes (nA), tension in milligrams (mg). Temperature = 30 °C; $[Ca^{++}]_0 = 16$ mM. (From 17).

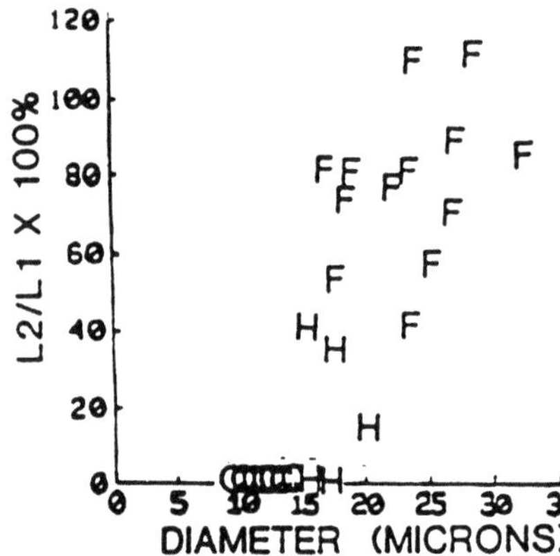

Fig. 6. Plot of $L_2/L_1 \times 100$ versus myocyte diameter in 16 mM $[Ca^{2+}]_o$. See text for details. (From 28).

producing elevated end-diastolic calcium and end-diastolic tension in these muscles. As a result, even though total developed tension is increased, the active tension generated during each systole is much less than that which occurs at lower pacing rates. These changes are enhanced by agents that increase $[Ca^{2+}]_i$ levels further, such as digitalis, and are ameliorated by agents that enhance relaxation, such as drugs that increase cAMP generation (38). Extrapolations of these experimental findings to the clinical setting in man may explain why patients with end-stage heart failure tolerate tachycardia so poorly and manifest marked decreases in cardiac compliance.

Calcium handling in human hypertrophy and failure

Three lines of evidence suggest that the changes we have described may be related to the degree of hypertrophy that is present with or without clinical signs and symptoms of heart failure. First, all of the myopathic hearts studied showed significant degrees of compensatory hypertrophy (12). Second, we have studied one muscle from a patient with hypertrophic cardiomyopathy and hyperdynamic systolic function who underwent septal resection for outflow tract obstruction (31). Although this patient demonstrated hyperdynamic systolic function, the calcium transients showed the two components characteristic of human cardiomyopathy. Third, there is a close correlation between the presence of hypertrophy and the L_2/L_1 ratio, which increases with the severity of the patient's clinical signs and symptoms of heart failure (27; Fig. 6). It remains to be determined why these changes in Ca^{2+} handling processes cease being adaptive and reach the point where they result in significant degrees of systolic and diastolic dysfunction.

References

1. Blinks JR (1986) Intracellular $[Ca^{++}]$ measurements. In: Fozzard HA, Haber E, Jennings RB (eds) The Heart and Cardiovascular System. New York Scientific Foundations, pp 671–701

2. Blinks Jr, Endoh M (1986) Modification of myofibillar responsiveness to Ca^{2+} as an inotropic mechanism. Circulation 76 (suppl III):85
3. Blinks JR, Wier WG, Hess P, Prendergast FG (1982) Measurement of Ca^{2+} concentrations in living cells. Prog Biophys Mol Biol 40:1
4. Böhm M, Gierschik P, Jakobs KH, Pieske B, Schnabel P, Ungerer M, Erdmann E (1990) Increase in G_i alpha in human hearts with dilated but not ischemic cardiomyopathy. Circulation 82:1249–1265
5. Brutsaert DL, Sys SU (1989) Relaxation and diastole of the heart. Physiol Rev 69:1228–1315
6. Erdmann E, Bohm M (1989) Positive inotropic stimulation in the normal and insufficient human myocardium. Basic Res Cardiol 84:125–133
7. Feldman MD, Copelas L, Gwathmey JK, Phillips P, Warren SE, Schoen FJ, Grossman W, Morgan JP (1987) Deficient production of cyclic AMP: pharmacologic evidence of an important cause of contractile dysfunction in patients with end-stage heart failure. Circulation 75:331–339
8. Feldman MD, Gwathmey JK, Phillips P, Schoen F, Morgan JP (1988) Reversal of the force-frequency relationship in working myocardium from patients with end-stage heart failure. J Appl Cardiol 3:272–283
9. Ginsburg R, Bristow MR, Billingham ME, Stinson EB, Schroeder JS, Harrison DC (1983) Study of the normal and failing isolated human heart: Decreased response of failing heart to isoproterenol. Am Heart J 3:535-540
10. Grossman W, Lorell B (eds) (1988) Diastolic Relaxation of the Heart. Martinus Nijhoff, Boston pp. 1–305
11. Grossman W, McLaurin LP, Rolett EL (1979) Alterations in left ventricular relaxation and diastolic compliance in congestive cardiomyopathy. Cardiovasc Res 13:514–52
12. Gwathmey JK, Copelas L, Mackinnon R, Schoen FJ, Feldman MD, Grossman W, Morgan JP (1987) Abnormal intracellular calcium handling in myocardium from patients with end-stage heart failure. Circ Res 61:70–76
13. Gwathmey JK, Morgan JP (1985) Altered calcium handling in experimental pressure-overload hypertrophy in the ferret. Circ Res 57:836–843
14. Gwathmey JK, Morgan JP (1990) Calcium handling in myocardium from amphibian, avian and mammalian species: the search for two components. J Comp Physiol 161:19–25
15. Gwathmey JK, Slawsky MT, Briggs GM, Morgan JP (1988) The role of intracellular sodium in the regulation of intracellular calcium and contractility. Effects of DPI 201–106 on excitation-contraction coupling in human ventricular myocardium. J Clin Invest 82:1592–1605
16. Gwathmey JK, Slawsky MT, Hajjar RJ, Briggs GM, Morgan JP (1990) The role of intracellular calcium handling in force-interval relationships of human ventricular myocardium. J Clin Invest 85:1599–1613
17. Gwathmey JK, Warren SF, Briggs GM, Copelas L, Feldman MD, Phillips PJ, Callahan M Jr., Schoen FJ, Grossman W, Morgan JP (1991) Diastolic dysfunction in hyptertrophic cardiomyopathy: Effect on active force generation during systole. J Clin Invest 87:1023–1031
18. Hajjar RJ, Gwathmey JK, Briggs GM, Morgan JP (1988) Differential effect of DPI 201–106 on the sensitivity of myofilaments to Ca^{2+} in intact and skinned trabeculae from control and myopathic human hearts. J Clin Invest 82:1578–1584
19. Hathaway DR, March KL (1989) Molecular cardiology: new avenues for the diagnosis and treatment of cardiovascular disease. J Am Coll Cardiol 13:265–282
20. Katz AM (1989) Changing strategies in the management of heart failure. J Am Coll Cardiol 13:513–523
21. Kihara Y, Grossman W, Morgan JP (1989) Direct Measurement of changes in intracellular calcium transients during hypoxia, ischemia, and reperfusion of the intact mammalian heart. Circ Res 65:1029–1044

22. Kihara Y, Morgan JP (1989) A comparative study of three methods for intracellular loading of the calcium indicator aequorin in ferret papillary muscles. Biochem Biophys Res Commun 162:402–407
23. Levine M, Meuse AJ, Watanabe J, Bentivegna L, Morgan JP (1990) Intracellular [Ca^{2+}]$_i$ during ischemia in the blood perfused dog heart. Biophys J 57:2:173
24. MacKinnon R, Gwathmey JK, Allen PD, Briggs GM, Morgan JP (1988) Modulation by the thyroid state of intracellular calcium and contractility in ferret ventricular muscle. Circ Res 63:1080–1089
25. Meuse AJ, Perreault CL, Grossman W, Morgan JP (1989) An experimental procedure for obtaining aequorin-loaded isolated mammalian cardiac myocytes. J Gen Physiol 94:46
26. Morgan JP (1988) Intracellular calcium in heart failure. Cardiovasc Drugs Ther 1:621–624
27. Morgan JP (1989) Mechanism of action of inotropic drugs. In: Braunwald E (ed) Heart Disease Update. W.B. Saunders, Philadelphia, pp 136–144
28. Morgan JP, Bentivegna LA, Perreault CL, Meuse AJ, Allen PD, Ransil BJ, Grossman W, Gwathmey JK (1989) Molecular Biology of the Cardiovascular System. In: Roberts R, Sdineidej M (eds) UCLA Symposia on Molecular and Cellular Biology, New Series, Volume 131. Wiley Liss, New York, NY, pp 241–248
29. Morgan JP, Chesebro JH, Pluth JR, Puga FJ, Schaff HV (1984) Intracellular calcium transients in human working myocardium as detected with aequorin. J Am Coll Cardiol 3:410–418
30. Morgan JP, Erny RE, Allen PD, Grossman W, Gwathmey JK (1990) Abnormal intracellular calcium handling: A major cause of systolic and diastolic dysfunction in ventricular myocardium from patients with heart failure. Circulation 81 (Suppl III):21–32
31. Morgan JP, Morgan KG (1984) Calcium and cardiovascular function: Intracellular calcium levels during contraction and relaxation of mammalian cardiac and vascular smooth muscle as detected with aequorin. Am J Med 77 (Suppl 5A):33–46
32. Morgan JP, Morgan KG (1989) Intracellular calcium and cardiovascular function in heart failure: effects of pharmacologic agents. Cardiovasc Drugs Ther 3(Suppl 3):959–970
33. Morgan JP, Perreault CL, Morgan KG (1991) The cellular basis of contraction and relaxation in cardiac and vascular smooth muscle. Am Heart J 121:961–968
34. Morgan JP, Wier WG, Hess P, Blinks JR (1983) Influence of Ca^{2+} channel blocking agents on calcium transients and tension development in isolated mammalian heart muscle. Circ Res 52 (Suppl I):47–52
35. Näbauer M, Böhm M, Brown L, Diet F, Eichhorn MN, Kemkes B, Pieske B, Erdmann E (1988) Positive inotropic effects in isolated ventricular myocardium from non-failing and terminally failing human hearts. Eur J Clin Invest 18:600–606
36. Neumann J, Schmitz H, von Meyerinck L, Döring V, Kalmar P (1988) Increase in myocardial Gi-proteins in heart failure. Lancet 2:936–946
37. Perreault CL, Meuse AJ, Bentivegna L, Morgan JP (1990) Abnormal intracellular calcium handling in acute and chronic heart failure: Role in systolic and diastolic dysfunction. Eur Heart J 11:8–21
38. Phillips PJ, Gwathmey JK, Feldman MD, Schoen FJ, Grossman W, Morgan JP (1990) Post-extrasystolic potentiation and the force-frequency relationship: Differential augmentation of myocardial contractility in working myocardium from patients with end-stage heart failure. J Molec Cell Cardiol 22:99–110
39. Poole-Wilson PA (1989) Relevance of the aetiology of heart failure to drug therapy. Eur Heart J 10 (Suppl B):64–68
40. Rüegg JC (ed) (1988) Calcium in Muscle Activation. Springer, New York, pp 1–285
41. Shub C (1989) Heart failure and abnormal ventricular function. Pathophysiology and clinical correlation (Part 1). Chest 96:636–640
42. Shub C (1989) Heart failure and abnormal ventricular functiton. Pathophysiology and clinical correlation (Part 2). Chest 96:906–914
43. Swynghedauw B, Schwartz K, Lauer B, Lompre AM, Mercadier JJ, Samuel JL, Rappaport L (1988) Striated muscle overload. Eur Heart J 9(Suppl E):1–6

44. Warren SE, Hague NL, Morgan JP (1989) Normal intracellular calcium availability in the hypertrophic Syrian hamster. Clin Res 37:305A
45. Wikman-Coffelt J, Steffanelli T, Wu ST, Parmley WW, Jasmin G (1991) Intracellular calcium transients in the cardiomyopathic hamster heart. Circ Res 68:45–51

Author's address:

James P. Morgan, M.D., Ph.D.
Cardiovascular Division
Beth Israel Hospital
330 Brookline Avenue
Boston, MA 02215, USA

Ca^{2+}-currents and intracellular [Ca^{2+}]$_i$-transients in single ventricular myocytes isolated from terminally failing human myocardium

D.J. Beuckelmann, E. Erdmann

Department of Medicine I, University of Munich, FRG

Summary: The purpose of the present study was to test the hypothesis that steps between the excitation of the cell membrane and contraction are altered in cardiac failure. Ca^{2+}-currents and [Ca^{2+}]$_i$-transients were measured in single ventricular myocytes isolated from explanted hearts of patients with terminal heart failure undergoing transplantation, or from donors whose organs could not be used for technical reasons.

Peak Ca^{2+}-current densities were unchanged, as was the current-voltage relation. However, in myocytes isolated from severely failing hearts resting [Ca^{2+}]$_i$-levels were elevated, peak [Ca^{2+}]$_i$-transients were significantly smaller, and the diastolic decline of [Ca^{2+}]$_i$ was markedly slowed.

As the trigger for the release of Ca^{2+} from the sarcoplasmic reticulum is unchanged and the systolic [Ca^{2+}]$_i$-transient is reduced, severe heart failure can be described as partial electro-mechanical uncoupling.

Key words: Human ventricular myocyte; heart failure; [Ca^{2+}]$_i$-handling, Ca^{2+}-current; fura-2

Introduction

The pathophysiological basis of heart failure is still controversial. Maximal force developed by electrically driven papillary muscles of patients with terminal heart failure has been shown to be unchanged as compared to normal controls (3, 7). Therefore, the alterations of systolic contraction and diastolic relaxation in vivo in these patients cannot be caused simply by a reduction of contractile proteins, but rather by functional and therefore potentially reversible changes. Other authors have found that the Ca^{2+}-sensitivity of the myofilaments in chemically skinned fibers of these hearts was comparable to healthy controls (15). From these results it seemed likely that intracellular [Ca^{2+}]$_i$-handling might be altered in heart failure. Gwathmey et al. were the first to describe an abnormal intracellular [Ca^{2+}]$_i$-handling in myocardium from patients with cardiac contractile failure. They performed experiments in isometrically contracting papillary muscles from patients with terminal heart failure using the bioluminescent photoprotein aequorin as a [Ca^{2+}]$_i$-indicator. An additional [Ca^{2+}]$_i$-signal in this tissue (L$_2$) indicated a slowed diastolic decline of [Ca^{2+}]$_i$ (8). From these results and pharmacological interventions they postulated a combined defect of sarcolemmal Ca^{2+}-influx and Ca^{2+}-reuptake by the sarcoplasmic reticulum (SR).

Sarcolemmal Ca^{2+}-currents can only be quantified in single myocytes. The following study was performed to investigate whether Ca^{2+}-influx via the Ca^{2+}-

---

Supported by the Deutsche Forschungsgemeinschaft (Be 1113/2–1) current or Ca^{2+}-release from the sarcoplasmic reticulum was altered in heart failure. Single ventricular myocytes were isolated from explanted human hearts and Ca^{2+}-currents were registered using common voltage-clamp techniques (9), intracellular $[Ca^{2+}]_i$-transients were recorded through the use of the fluorescent $[Ca^{2+}]_i$-indicator, fura-2.

Methods

Cell isolation

Myocytes were isolated from hearts of patients with end-stage heart failure due to dilated cardiomyopathy or ischemic heart disease undergoing transplantation. Cells isolated from organ donors whose hearts could not be transplanted for technical reasons were used as controls. Details of the isolation procedure have been described before (1).

Solutions

Cells were superfused with a modified Tyrode solution containing 2.0 mM $CaCl_2$; 138 mM NaCl; 10 mM CsCl; 1 mM $MgCl_2$; 10 mM glucose; 10 mM Hepes; pH was 7.3 with addition of NaOH. Cs^+ was included to block K^+ currents that might interfere with the measurement of Ca^{2+}-current.

Loading of cells with fura-2 and internal perfusion

Cells were loaded with fura-2 (penta-potassium salt) by internal perfusion through diffusion from the micropipette electrode over 5 to 15 min. Electrodes had resistances of 2.0 to 3.0 Mohm and were filled with: 0.050 mM Fura-2 (Molecular Probes); 120 mM Cs-glutamate; 10 mMCsCl; 1 mM $MgCl_2$; 5 mM NaCl; 10 mM Hepes (cesium salt); and 2 mM Mg-ATP; pH was 7.2 with addition of CsOH. The calibration technique has been described in detail elsewhere (2).

Recording technique

The experimental apparatus was constructed around a Zeiss Axiovert 35 microscope with a MPM 201 photometer attachment (Fig. 1). UV light, emitted from a 75-watt Xenon arc lamp was passed through 10-nm interference filters (340-nm or 380-nm wavelengths) and was reflected by a dichroic mirror centered at 405 nm into the fluorescence objective (Zeiss NeoFluar, n.a. 1.3, 100x) for excitation of the Ca^{2+}-indicator in the cell. Selection of excitation filters was done by a computer-controlled filter switcher. A computer-controlled shutter limited the illumination to periods of interest. Fluorescence emitted from the cell passed through the objective and a beam splitter reflected 30% of the light into the oculars for visual observation. 70% of the fluorescence was directed into a photomultiplier tube (PMT) after passing through a 510–540 nm bandpass filter and a short pass filter to eliminate interfering red and

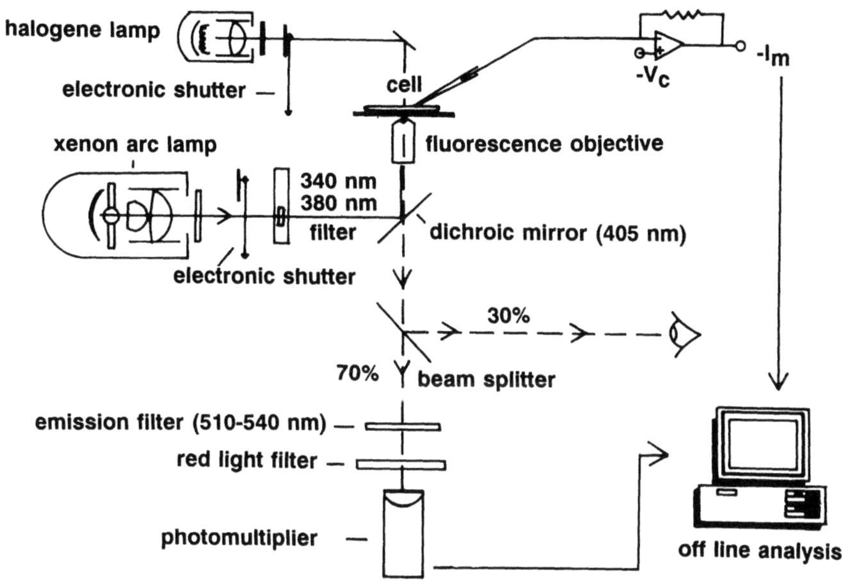

Fig. 1. Experimental apparatus for intracellular [Ca^{2+}]$_i$-measurements in single ventricular myocytes with the fluorescent indicator, fura-2. Details are given in the text.

infrared light. An aperture at the image plane in front of the PMT selected light from a circular region (diameter approximately 10 µm) of the cell for fluorescence recordings.

Whole-cell currents were recorded by standard techniques (9) using a patch-clamp amplifier (List Instruments, model EPC-7) with a 100-MΩ feedback resistor. Microelectrodes were pulled from borosilicate glass and had resistances of 2.0 to 3.0 MΩ.

Fluorescence recordings were filtered with a cut-off frequency of 120 Hz (Bessel) and, together with the current recordings from the patch clamp amplifier, digitized (1-5 KHz, Indec IDA interface) and stored for off-line analysis.

Results

Cell isolation

Cells were prepared from 14 hearts of patients with end-stage heart failure due to dilated cardiomyopathy (DCM, n = 9) or ischemic cardiomyopathy (ICM, n = 5) undergoing transplantation. Mean age of patients was 51 years, cardiac index in all cases was < 2.7 l/min/m^2, ejection fraction was < 35%. Results were compared with cells isolated from four normal human hearts without cardiac disease that could not be transplanted for technical reasons. The living cell yield was approximately 5%. Figure 2 depicts an exemplar cell, isolated from an explanted heart of a patient with terminal heart failure due to ischemic cardiomyopathy.

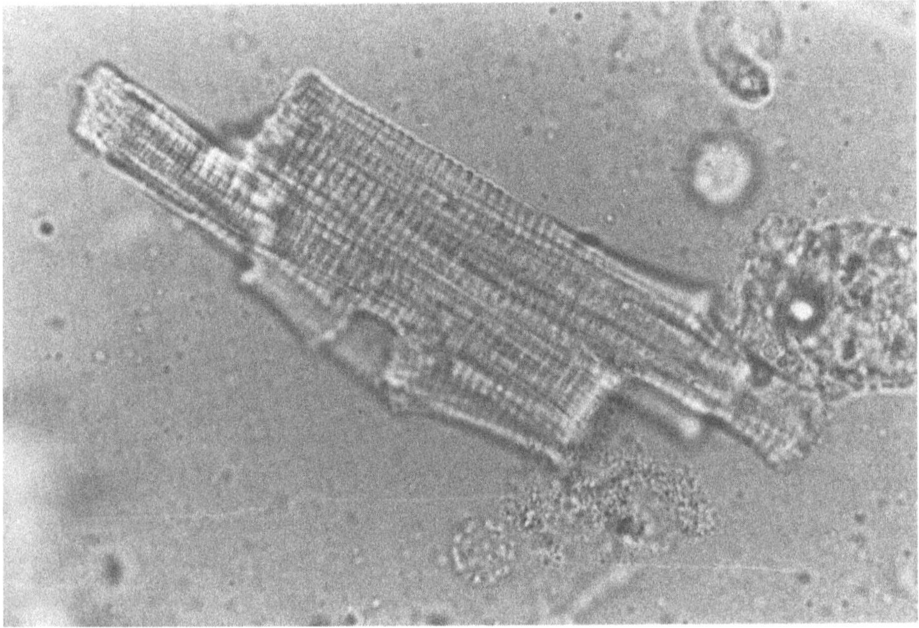

Fig. 2. Single human ventricular myocyte isolated from an explanted heart of a patient with dilated cardiomyopathy (Objective 100x).

Characteristics of L-type Ca^{2+}-currents

In seven diseased hearts and in three healthy control hearts the characteristics of the Ca^{2+}-current and the effect of isoprenaline were investigated. The dependence of L-type Ca^{2+}-currents on clamp-pulse potential over the range from -30 to $+80$ mV is shown in a cell isolated from a control heart and from myocardium of a patient with dilated cardiomyopathy. Interfering K^+-currents were blocked by adding Cs^+ to the intra- and extracellular solutions. Figure 3 shows an original voltage-clamp record upon depolarization to $+10$ mV under control conditions and after addition of isoprenaline (10^{-7} M). Cells were isolated from an undiseased heart and from a patient with terminal heart failure due to dilated cardiomyopathy. In both cells the current-voltage relation was similar with a maximum of the peak current at $+10$ mV and an apparent reversal potential of $+65$ mV. Cells were comparable in size (cell capacity: control 216 pF, DCM 229 pF). The size of the peak Ca^{2+}-current was similar (0.61 vs. 0.63 nA). However, β-adrenergic stimulation of the current increased its size to a much higher degree in normal myocytes than in cells isolated from failing hearts.

The average Ca^{2+}-current densities under control conditions and after incubation in isoprenaline (10^{-7} M) are shown in Table 1. I_{Ca}-densities were taken as the peak Ca^{2+}-current per cm^2 cell surface, assuming a surface area of 10^{-6} cm^2/pF

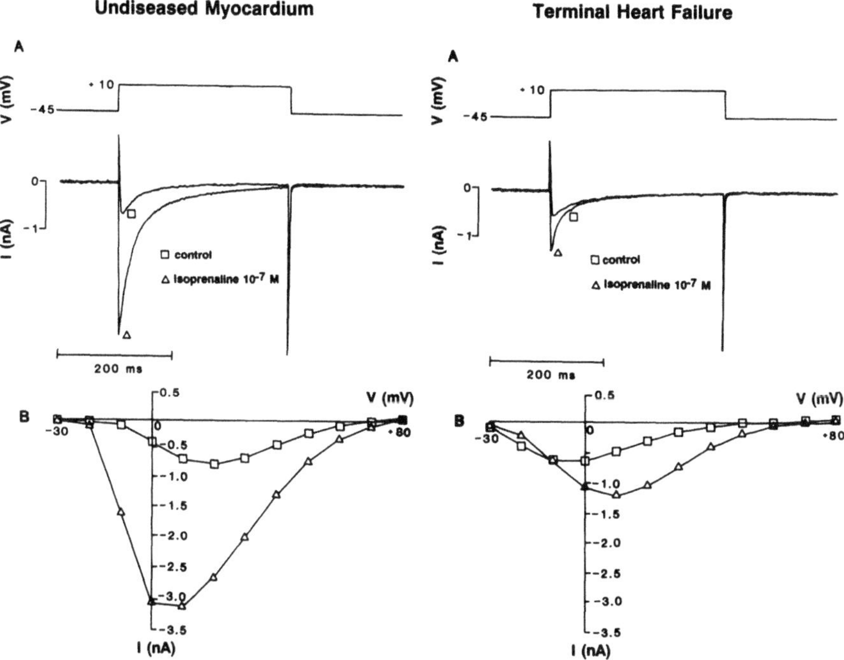

Fig. 3. A) Current records in single myocytes under voltage-clamp isolated from an undiseased heart and from a heart with dilated cardiomyopathy. Depolarization (300 ms) from a holding potential of -45 mV to $+10$ mV under control conditions and after incubation in isoprenaline 10^{-7} M. Interfering K$^+$-currents were blocked with Cs$^+$. B) Current-voltage relation. Values were obtained by subtracting the average current during the last 10 ms of the pulse from the peak current.

Table 1. Ca^{2+}-current densities under control conditions and after β-adrenergic stimulation with isoprenaline 10^{-7} M. Current densities were calculated assuming a surface area of 10^{-6} cm^2/pF membrane capacity. Mean values and standard deviations are shown. The t-test for unpaired data was used for statistical analysis.

	I$_{Ca}$-density (μA/cm^2) control	I$_{Ca}$-density (μA/cm^2) isoprenaline (10^{-7} M)
Heart failure	-2.9 ± 1.0	-4.6 ± 0.9
Control	-2.6 ± 1.4	-7.6 ± 4.6
	n.s.	p < 0.05

membrane capacity. Unstimulated Ca^{2+}-currents were unchanged in cells from organs with heart failure as compared to undiseased controls. However, the Ca^{2+}-current increase after stimulation with isoprenaline elicited a significantly smaller response in diseased cells as compared to controls.

Intracellular $[Ca^{2+}]_i$-transients during voltage clamp pulses

To investigate whether intracellular $[Ca^{2+}]_i$-handling may be altered in severe heart failure, myocytes loaded with the $[Ca^{2+}]_i$-indicator fura-2 were stimulated repetitively to $+10$ mV. Upon stimulation to $+10$ mV the maximum of the $[Ca^{2+}]_i$-transient could be elicited. Stimulus duration was 300 ms, stimulation frequency was 0.5 Hz.

Figure 4 shows the intracellular $[Ca^{2+}]_i$-transient in exemplar cells isolated from an undiseased heart and from an explanted heart of a patient with heart failure (ICM). Under normal conditions there was a fast rise of $[Ca^{2+}]_i$ from a resting level

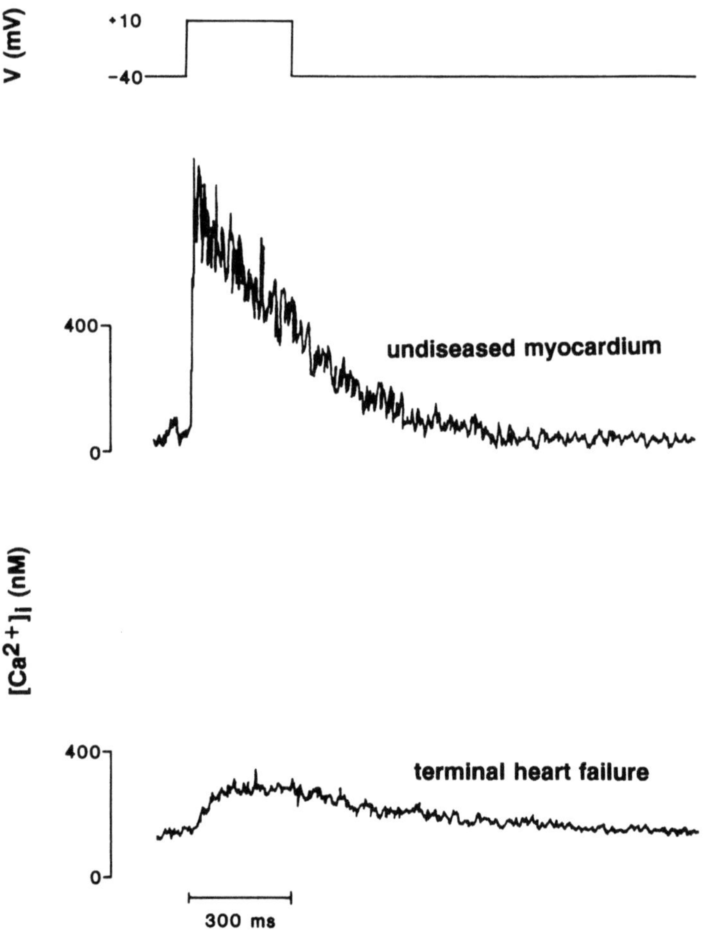

Fig. 4. Intracellular $[Ca^{2+}]_i$-transients. Cells were stimulated through repetitive depolarizations to $+10$ mV. Pulse duration was 300 ms, stimulation frequency was 0.5 Hz. Intracellular fura-2 was 50 μM.

of $\sim$ 70 nM to a maximum of 900 nM. Decay of [Ca^{2+}]$_i$ started already during depolarization. The time necessary for [Ca^{2+}]$_i$ to decay to its half maximum concentration (t$_{1/2}$) was 320 ms. Conversely, when the same protocol was applied to a cell from a patient with terminal heart failure the [Ca^{2+}]$_i$-resting level was 150 nM and the peak [Ca^{2+}]$_i$-transient was only 330 nM after a markedly slower rise. Furthermore, the decline was significantly slower as compared to the control cell (t$_{1/2}$: 560 ms).

Discussion

In mammalian heart muscle the link between excitation of the cell membrane and release of Ca^{2+} from intracellular stores has been shown to be through Ca^{2+}-induced Ca^{2+}-release (2, 5). It is the initial fast component of the Ca^{2+}-current that triggers the release of a much larger amount of Ca^{2+} from the sarcoplasmic reticulum. Figure 3 and Table 1 show that the Ca^{2+}-current density of the peak Ca^{2+}-current is unchanged in terminal heart failure. This is in line with results that indicated an unaltered number of dihydropyridine binding sites in these hearts (14). Therefore, the trigger signal for the release of Ca^{2+} from the sarcoplasmic reticulum does not seem to be modified in heart failure.

It has been shown by many groups that the positive inotropic response of myocardium from severely failing hearts to β-adrenergic stimulation is significantly reduced as compared to normal myocardium. This is thought to be due to down-regulation of β-adrenergic receptors (7, 13) and an increase of the inhibitory regulatory guanine binding protein G$_i$ (4, 6). The results shown in Fig. 3 and Table 1 demonstrate that at least part of this reduced inotropic response to isoprenaline may be explained by a significantly smaller increase of the sarcolemmal Ca^{2+}-current in these hearts.

One of the main results of the present study was that the intracellular [Ca^{2+}]$_i$-handling in myocytes from severely failing hearts was markedly altered. Systolic [Ca^{2+}]$_i$-transients were significantly reduced and diastolic [Ca^{2+}]$_i$-decline markedly slowed. As the Ca^{2+}-current of the sarcolemmal membrane is unchanged and the affinity of the contractile proteins to Ca^{2+} is unaffected the reduced systolic [Ca^{2+}]$_i$-increase can only be explained by an altered release of Ca^{2+} from the sarcoplasmic reticulum. Several mechanisms may theoretically account for this finding:

– the Ca^{2+}-release mechanism of the sarcoplasmic reticulum may be altered and, therefore, a trigger signal is not adequately answered. However, recently, other authors have shown that the electrophysiological properties of single Ca^{2+}-release channels of the sarcoplasmic reticulum in hearts from patients with heart failure are unchanged (10).
– elevation of the diastolic [Ca^{2+}]$_i$ and its slow diastolic decline may lead to Ca^{2+}-dependent inactivation of the Ca^{2+}-release channel of the sarcoplasmic reticulum.
– reduction of the Ca^{2+}-reuptake rate of the sarcoplasmic reticulum may lead to a reduced availability of Ca^{2+} to be released during the next beat. Results concerning the Ca^{2+}-reuptake in human heart failure have been controversial. Some authors found the Ca^{2+}-uptake to be reduced in heart failure (11), while others showed an unchanged Ca^{2+}-uptake rate (12).

A reduced Ca^{2+}-reuptake would explain the slow $[Ca^{2+}]_i$-decline shown in Fig. 4. It would also explain the beneficial effect of slowing the heart rate in patients with heart failure. As diastole is primarily shortened during tachycardia, a slow heart rate would provide more sufficient time to refill the intracellular stores, even when the Ca^{2+}-ATPase of the sarcoplasmic reticulum would have a reduced reuptake rate.

Further studies will be necessary to clarify which of the above mechanisms is responsible for the altered $[Ca^{2+}]_i$-handling in heart failure. However, accumulating data from other groups and from the experiments shown above indicate that the attention for pharmacologic interventions should be primarily directed to increasing the diastolic Ca^{2+}-reuptake of the SR.

References

1. Beuckelmann DJ, Näbauer M, Erdmann E (1991) Characteristics of calcium-current in isolated human ventricular myocytes from patients with terminal heart failure. J Mol Cell Cardiol 23:929–937
2. Beuckelmann DJ, Wier WG (1988) Mechanism of release of calcium from sarcoplasmic reticulum of guinea-pig cardiac cells. J Phys (Lond) 405:233–255
3. Böhm M, Beuckelmann D, Brown L, Feiler G, Lorenz B, Näbauer M, Kemkes B and Erdmann E. (1988) Reduction of beta-adrenoceptor density and evaluation of positive inotropic responses in isolated, diseased human myocardium. Eur Heart J 9:844–852
4. Böhm M, Gieschik P, Jakobs KH, Pieske B, Schnabel P, Ungerer M and Erdmann E. (1990) Increase of Gia in human hearts with dilated but not ischemic cardiomyopathy. Circulation 82:1249–1265
5. Fabiato A (1985) Simulated calcium current can both cause calcium loading in and trigger calcium release from the sarcoplasmic reticulum of a skinned canine cardiac purkinje cell. J Gen Physiol 85:291–320
6. Feldman AM, Cates AE, Bristow MR and Van Dop C (1989) Altered expression of α-subunits of G proteins in failing human hearts. J Mol Cell Cardiol 21:359–365
7. Feldman MD, Copelas L, Gwathmey JK, Phillips P, Warren SE, Schoen FJ, Grossman W and Morgan JP (1987) Deficient production of cyclic AMP:pharmacologic evidence of an important cause of contractile dysfunction in patients with end-stage heart failure. Circulation 75 (2):331–339
8. Gwathmey JK, Copelas L, Mackinnon R, Schoen FJ, Feldman MD, Grossman W and Morgan JP (1987) Abnormal intracellular calcium handling in myocardium from patients with end-stage heart failure. Circ Res 61:70–76
9. Hamill OP, Marty A, Neher E, Sakmann B and Sigworth FJ (1981) Improved patch-clamp techniques for high-resolution current recording from cells and cell-free membrane patches. Pflugers Arch 391:85–100
10. Holmberg SRM and Williams AJ (1989) Single channels recordings from human cardiac sarcoplasmic reticulum. Circ Res 65:1445–1449
11. Limas CJ, Olivari M-T, Goldenberg IF, Levine TB, Benditt DG and Simon A (1987) Calcium uptake by cardiac sarcoplasmic reticulum in human dilated cardiomyopathy. Cardiovasc Res 21:601–605
12. Movsesian MA, Bristow MR and Krall J (1989) Ca^{2+} uptake by cardiac sarcoplasmic reticulum from patients with idiopathic dilated cardiomyopathy. Circ Res 65:1141–1144
13. Näbauer M, Böhm M, Brown L, Diet F, Eichhorn M, Kemkes B, Pieske B and Erdmann E (1988) Positive inotropic effects in isolated ventricular myocardium from non-failing and terminally failing hearts. Eur J Clin Invest 18:600–606
14. Rasmussen PR, Minobe W and Bristow MR (1990) Calcium antagonist binding sites in failing and nonfailing human ventricular myocardium. Biochem Pharmacol 39:691–696

15. Wankerl M, Böhm M, Morano I, Rüegg JC, Eichhorn M and Erdmann E (1990) Calcium sensitivity and myosin light chain pattern of atrial and ventricular skinned cardiac fibers from patients with various kinds of cardiac disease. J Mol Cell Cardiol 22:1425–1438

Author's address:
Dr. Dirk J. Beuckelmann, M.D.
Medizinische Klinik I
der Universität München
Klinikum Großhadern
Marchioninistraße 15
D-8000 München 70, FRG

Dynamic calcium requirements for activation of human ventricular muscle calculated from tension-independent heat

E.M. Blanchard, B.J. Leavitt, L.A. Mulieri, N.R. Alpert

Department of Physiology and Biophysics, University of Vermont, Burlington, USA

Summary: The heat and tension generated by strips of human left ventricle taken from nonfailing hearts were measured at 30 °C before and after partial inhibition of ATP splitting by the contractile proteins. We used 2, 3-butanedione monoxime (BDM) (4mM) as the chemical inhibition agent and alterations in solution calcium concentration and stimulus frequency to estimate the heat associated with calcium cycling for a wide range of activation levels. Tension-independent heat (TIH) was used to calculate the total calcium cycled per twitch by assuming that two-thirds of TIH was due to ATP splitting by the sarcoplasmic reticulum Ca^{2+} ATPase with a coupling ratio of 2 Ca^{2+}/ATP split and that one-third of TIH was due to ATP splitting by the sarcolemmal $Na^+ - K^+$ ATPase supporting the $Na^+ - Ca^{2+}$ exchanger (1 Ca^{2+}/ATP). The enthalpy of creatine phosphate hydrolysis buffering ATP was taken as -34 KJ/mol. There was a highly positive correlation between TIH and mechanical activation during steady-state and nonsteady-state stimulation. The estimated total calcium turnover per twitch at 39 % activation (0.3 Hz pacing rate and 2.5 mM Calcium) was ~ 0.17 nmol/g wet weight. This estimate is less than that calculated from biochemical data describing the cellular content and Ca^{2+} affinity of major Ca^{2+} buffers, but is similar to values calculated from recent electron probe microanalysis experiments.

Key words: Dynamic calcium requirements; tension-independent heat; human ventricular muscle

Introduction

An abnormality in calcium homeostasis has been implicated as an important mechanism for the performance deficit measured for human hearts with end-stage failure (10). Changes in the magnitude and kinetics of the free Ca^{2+} transient detected with photoproteins or fluorescent indicators provide support for this idea. While this aspect of calcium flux into and out of the cytoplasm with each twitch is vitally important due to the close relation between free Ca^{2+} and the saturation of the regulatory sites of troponin C (TnC), the detailed mechanisms leading to changes in the free Ca^{2+} transient cannot be uncovered by measuring free Ca^{2+} alone. This is true because the peak free Ca^{2+} concentration achieved in a twitch is the result of a dynamic interaction among (1) the total amount of calcium released, (2) all the intracellular Ca^{2+} buffers exchanging Ca^{2+} on the time scale of the muscle twitch, and (3) the rate of Ca^{2+} removal from the cytosol.

We have taken a biophysical approach to evaluating calcium homeostasis in ventricular preparations that focuses on the requirements that the total amount of calcium released into the cytosol from extracellular and sarcoplasmic reticular (SR) compartments has to be actively transported out of the cytosol for the preparation to twitch in the steady-state (1, 3). We measure the heat (tension independent heat, TIH) associated with this active transport of Ca^{2+} by SR and sarcolemmal (SL) pumps and exchangers and estimate total calcium cyled per twitch on the basis of reasonable assumptions about the coupling between Ca^{2+} transport and ATP hydrolysis. Steady-state measurements where mechanical activation is varied over a wide range indicate that the total calcium required for activation is less than the value derived from calculations based on the likely content and Ca^{2+} affinity of known calcium buffers. An extra stimulation imposed within a steady pacing regime elicits a substantial increase in mechanical output in the next regular stimulation (post extrasystolic potentiation, PESP) which then decays over several twitches. This PESP protocol elicits a similar potentiation and decay of TIH for these human ventricular preparations.

Methods

Muscle strips were prepared from epicardial samples taken from nonfailing (angiocardiographic ejection fraction $> 60\%$) human left ventricles after cardioplegia and incubated in a protective Krebs-Ringer solution as previously described (18, 19). Briefly, the ventricular samples were immediately placed in preoxygenated (95 % O_2-5% CO_2) Krebs-Ringer solution containing 30 mM 2,3-butanedione monoxime (BDM) which reversibly suppresses all contractile activity, particularly contracture zones associated with dissection injury. After 30 min of incubation at room temperature ($\sim 22\,°C$) the strip preparations were dissected from the sample while incubating in the same BDM-Krebs protective solution at room temperature. Strips were cut along the fiber direction under a x7–10 binocular microscope. After tying silk ligatures containing platinum wires to the strip, the preparation was mounted vertically against the thermopile (17) with one end attached to a stationary hook and the other end attached to an isometric force transducer. After initial equilibration ($30\,°C$, 0.3 Hz pacing) in normal Krebs-Ringer solution for about 1 h, the muscle strip was stretched to the optimal length for isometric force production. The protocol for measuring heat and mechanical output over a wide range of activation at $30\,°C$ was then executed. After incubation of the strip in normal Krebs-Ringer containing 2.5 mM $CaCl_2$, steady-state heat and tension were recorded at 0.3, 0.2, and 0.08 Hz pacing rates. These measurements were then repeated after incubating the strip in solutions containing 4, 6, and 0.3 mM $CaCl_2$. The protocol employing variations in solution calcium concentration and stimulus frequency was then repeated except that the incubation solutions contained 4 mM BDM. In some of the experiments heat and tension were recorded during a post extrasystolic potentiation (PESP) protocol where a single extra stimulus was interposed 800 msec after a regular stimulus. Control pacing (0.3 Hz) resumed after the extra stimulus. To obtain heat, the temperature difference between the evolving temperature record and the falling baseline that would have occurred if stimulation had ceased after the previous twitch was corrected for heat loss by the single time constant method of Hill (11). The temperature record corrected for heat loss was multiplied by the heat loss coefficient, the cool-off time constant (determined from "Joule" heating) (1), and the thermal capacity of the muscle and adhering solution to convert temperature to heat (mJ/g wet weight). An example of the human temperature data is shown in Figure 1.

Tension independent heat (TIH) was estimated by fitting the initial heat (Y) and tension-time integral (x) values measured with and without 4 mM BDM with a straight line. The heat intercept at zero tension-time integral was taken as TIH for

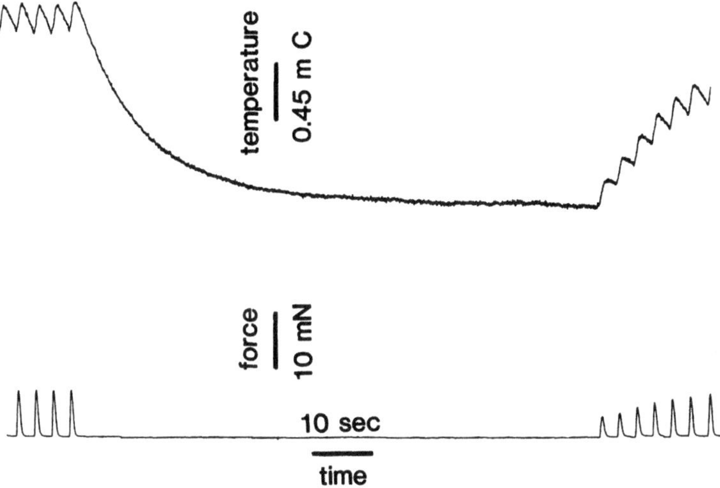

Fig. 1. The time-course of isometric force (below) and temperature (above) from a human left-ventricular strip paced at 0.3 Hz after incubation in normal Krebs solution at 30 °C. The muscle cools to the resting temperature when stimulation in stopped.

each condition. We converted TIH (mJ/g) to total calcium cycled in the following way: nmol calcium cycled/g = TIH (mJ/g) $\times$ $1/\Delta H_{PCr}$ (μmol/mJ) $\times$ 1.67 $\times$ 1000 where ΔH_{PCr} = $-$ 34 KJ/mol (6) and 1.67 represents a Ca^{2+}/ATP coupling ratio based on the assumption that two-thirds of Ca^{2+} transport is achieved by the SR with a coupling ratio of 2 Ca^{2+}/ATP (22), and that one-third of transport is achieved by the Na^+–Ca^{2+} exchanger supported by the Na^+–K^+ ATPase (net 1 Ca^{2+}/ATP) (5). The choice of 1.67 to represent the combination of the two types of transport systems is somewhat arbitrary as the exact proportions have not been experimentally defined. We use phosphocreatine enthalpy because adenosine diphosphate is probably rapidly rephosphorylated to ATP by the Lohmann reaction. No correction for the fraction of TIH unrelated to Ca^{2+} cycling was made, although this idea is discussed. The choice of BDM as the chemical partitioning agent is supported by data demonstrating direct effects of BDM on myofilaments (13, 15, 23), but minimal effects on the free Ca^{2+} transient (4, 12) at low concentrations.

The Krebs-Ringer solution contained in mM: Na^+, 152; K^+, 3.6; Cl^{2-}, 135; HCO_3^-, 25; Mg^{2+}, 0.6; $H_2PO_4^-$, 1.3; SO_4, 0.6; Ca^{2+}, 2.5; glucose, 11.2; insulin, 10 IU/L. Solutions were gassed with 95%O_2–5%CO_2. Some muscle strips were tetanized after incubation at the end of the above protocol with a PO_4-free Krebs solution containing 16 mM $CaCl_2$, 2 mM caffeine, and 10 μM ryanodine in order to estimate the maximum tension a strip could generate. 2,3-butanedione monoxime was purchased from Sigma.

Results

Steady-state: The geometric, mechanical, and thermal characteristics of the strip preparations with the control conditions of 0.3 Hz pacing, 30 °C, and 2.5 mM $CaCl_2$ are outlined in Table 1.

Table 1. Characteristics of human left-ventricular strips; n = 15 (30 °C, 2.5 mM $CaCl_2$, 0.3 Hz)

	Length (mm)	Wet weight (mg)	Cross-sec. area (mm^2)	Peak tension (mN/mm^2)	Tension integral (mNs/mm^2)	Twitch time (msec)	TPT* (msec)
Mean	5.86	2.18	0.38	22.44	15.09	1539	508
SEM	0.21	0.29	0.05	2.81	1.90	60	20

	T$\frac{1}{2}$ relax (msec)	Initial heat (mJ/g)	TIH* (mJ1g)	TIH ÷ initial heat	TDH* (mJ/g)	TDH ÷ tension integral	Twitch ÷ tetanus
Mean	834	3.03	0.34	0.10	2.85	0.42	0.39
SEM	22	0.40	0.08	0.02	0.36	0.04	0.02

*TPT = time to peak tension; TIH = tension independent heat; TDH = tension dependent heat.

The protocol employing variations in solution $CaCl_2$ concentration and in stimulus frequency produced a four-fold increase in peak twitch tension and a three-fold increase in twitch tension-time integral (TTI) when these parameters were raised from the lowest to the highest values. The maximum tension achieved by tetanizing the preparations after incubation in solutions with high levels of calcium and agents chosen to diminish Ca^{2+} removal from the cytosol was 68.21 ± 8.03 mN/mm^2. Thus, the average ratio of twitch tension under control conditions (0.3 Hz, 2.5 mM $CaCl_2$) to maximum tetanic tension was 0.39. This value was then used to normalize the average TTI values obtained with all the other conditions so that mechanical activation could be expressed relative to maximum activation. TIH (mJ/g) associated with each level of mechanical activation (ranging from 16–52 % of maximum) was determined as described in Methods and this relation is plotted in Figure 2. There was a high (r = 0.945) positive correlation between TIH and TTI and the relation appears to be adequately described by a linear equation (TIH = 0.01[TTI] + 0.06). The positive heat intercept of the TIH-TTI relation indicates that there is a small amount of residual TIH remaining at the threshold for mechanical activation.

The TIH values shown in Figure 2 were converted to calcium cycled (nmol/g wet weight) as described in Methods and are plotted vs the associated mechanical activation levels in Figure 3. The squares represent the estimate based on the myothermal data and the diamonds represent values taken from Fabiato (7). The slopes of the two relations (1.84 and 1.70, respectively) are similar but the myothermal relation is shifted to the left of the relation derived from Fabiato's calculations. The position of the myothermal data suggests that less total calcium is required to cycle per twitch to achieve a specific level of mechanical activation in the human ventricular strips compared to the amount calculated on the basis of biochemical measurements on cardiac tissue from species other than man.

The effect of the extra stimulation protocol (PESP) on heat and mechanical output in the twitches following the extra stimulation are depicted in Figure 4. The magnitude of twitch TTI (solid bar) and the associated TIH (hatched bar) in the first

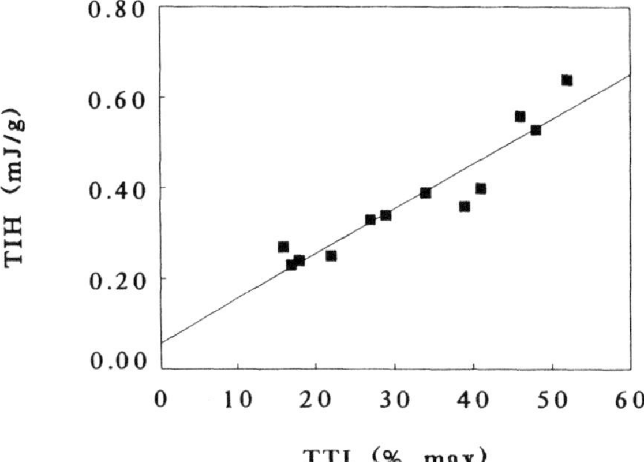

Fig. 2. The relation between tension independent heat (TIH) and the integral of isometric twitch tension (TTI). The variation in TTI was achieved by varying solution $CaCl_2$ concentration (0.3, 2.5, 4.0, 6.0 mM) and stimulus frequency (0.3, 0.2, 0.08 Hz) at each $CaCl_2$ level. TTI was normalized relative to the maximum level of activation that could be obtained by tetanizing the preparation as described in the text.

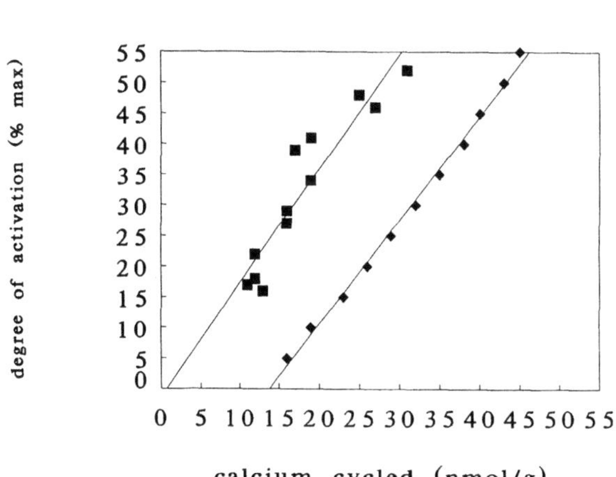

Fig. 3. The relation between the degree of mechanical activation and the amount of calcium cycled (nmol/g wet weight). The relation derived by converting TIH to calcium cycled (squares) is plotted along with the relation taken from the review of Fabiato (7) (diamonds).

four regular twitches following (post) the extra "systole" are expressed relative to the value measured for the last regular twitch before (pre) the extra systole. Both parameters were substantially potentiated ($\sim$ 45 %) in the first post twitch and then the potentiation decayed toward zero in the next three twitches. These are preliminary experiments (n = 3) and the apparent difference between TTI and TIH in the time constant of the decay of potentiation suggested in Figure 4 is not statistically significant.

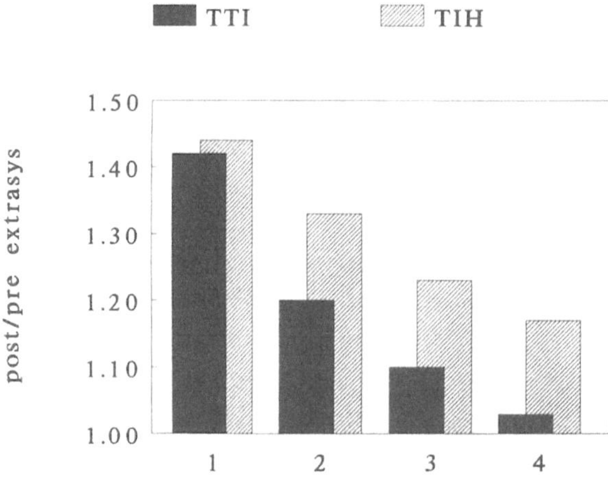

Fig. 4. The effect of a single extra stimulation interposed within a 0.3 Hz pacing regime on the tension-time integral (TTI) and tension independent heat (TIH) associated with the twitches in the immediate post "extrasystolic" period. The magnitude of TTI and TIH is plotted on the ordinate as a ratio of the post extrastimulation value to the pre extrastimulation value.

Discussion

The significant results of this myothermal study with ventricular strips from non-failing human hearts are: 1) The amount of total calcium cycling required to achieve $\sim 50\%$ level of activation of the myofilaments is $\sim 28\,\text{nmol/g}$ wet weight as derived from the tension independent heat data. This value is 35% smaller than the $\sim 43\,\text{nmol/g}$ value that can be extracted from Fabiato's biochemical analysis (7). 2) For the first time, we report that a change in an index of total calcium cycling, TIH, parallels the potentiation and decay of potentiation observed for mechanical output. This parallel between TIH and mechanical output is similar to the parallel between the intracellular free Ca^{2+} transient and mechanical output (24) following a single extra stimulation interposed within a regular pacing regime. These data are consistent with the idea that the extra stimulation either increases SR Ca^{2+} loading or the behavior of the SR release site so that total calcium release during the immediate post "extrasystollic" period is transiently larger than the steady-state value (24).

A complete biochemical description of the total calcium required for mechanical activation on a twitch-to-twitch basis requires information about the concentrations, equilibrium binding constants for Ca^{2+}, and the rates of Ca^{2+} exchange for the physiologically relevant buffers and pumps. This information is largely unavailable for human cardiac muscle, either normal or diseased. Early measurements of Ca^{2+} binding by dog myofibrils (21) provided a minimum estimate of $\sim 90\,\text{nmol/g}$ of calcium that would be needed to bind to troponin C (TnC) to fully activate the myofilaments. Rapid kinetic measurements (20), however, revealed that two-thirds of the TnC sites (Ca^{2+}-Mg^{2+} sites) are not physiologically relevant because they exchange Ca^{2+} too slowly on the time scale of the twitch. That correction then brings the calcium requirements to $\sim 30\,\text{nmol/g}$ to saturate the Ca^{2+} regulatory sites of

TnC. Obviously, more calcium will be needed to the extent that other rapidly exchanging buffers compete with TnC; Fabiato's analysis includes calmodulin and outer sites on the SR (8) as competitors with TnC. Figure 5 shows the equilibrium Ca^{2+} saturation of these three buffers as f (pCa). It is important to note that significant calcium will bind to the calmodulin and SR sites as intracellular free Ca^{2+} rises from rest ($\sim$ pCa 7.1) to the threshold of mechanical activation ($\sim$ pCa 6.7). This requirement for Ca^{2+} binding to buffers before the mechanical threshold is reached is the basis for the displacement of the activation-calcium cycled relation derived from Fabiatio's review along the calcium cycled axis toward higher amounts of calcium seen in Figure 3. The difference between the relation derived from human thermal data and the relation derived from Fabiato's analysis appears to derive from the absence of the x axis displacement for the former. We speculate that the requirement for calcium cycling before the mechanical threshold is reached, which is present in the Fabiato analysis, is absent from the relation in Figure 3 derived from human myothermal data because of two reasons: 1) The resting pCa of the cells in the human preparation is probably lower ($\sim$ pCa 6.64) (26) than the Fabiato value (pCa 7.1), and 2) the Ca^{2+} sensitivity of the human myofilaments at 30°C ($pCa_{50} \sim 5.9$) (25) is higher than the value ($pCa_{50} \sim 5.6$) used by Fabiato derived from 22°C data. The result of these two adjustments for the biochemical analysis would be to remove the requirement that significant amounts of calcium be cycled and bound to the major intracellular buffers before mechanical threshold is reached. Calcium buffering during mechanical activation is consistent with the biochemical prediction as seen by the similarity in slopes. The lower estimate of calcium cycling compared to the biochemical estimate of Fabiato is also consistent with a recent electron probe microanalysis study (16) measuring the amount of calcium released from the junc-

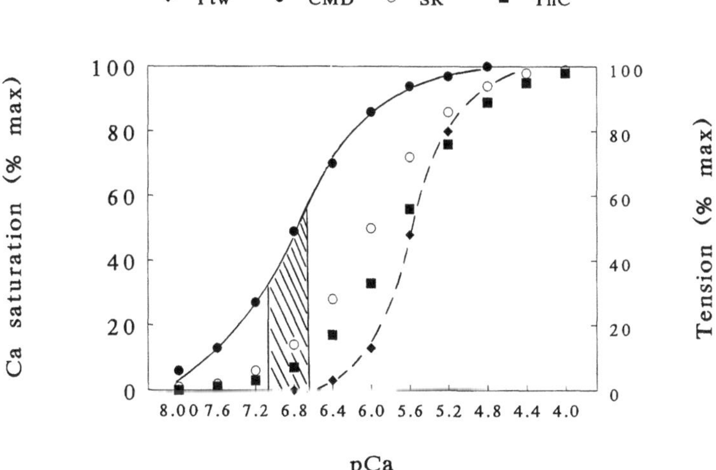

Fig. 5. The equilibrium Ca^{2+} saturation-pCa relations for the major calcium buffers in the Fabiato (7) review. The K_{Ca} (M^{-1}) used for calmodulin (CMD), sarcoplasmic reticulum high affinity site (SR), and troponin C (TnC) was 6×10^6, 1×10^6, and 5×10^5, respectively. The anticipated tension ($\blacklozenge$ % max) at each pCa (P_{tw}) from the review is also plotted. The shaded area represents the pCa range from the resting pCa of 7.1 to an estimate of the mechanical threshold, $\sim$ pCa 6.7.

tional SR of hamster papillary muscles. Maximum calcium release was calculated to correspond to a value of ~ 39 µmol/1 H_2O (or about 16 nmol/g wet weight).

The conversion of TIH to calcium cycling requires some comment regarding the fraction of TIH that is truly due to calcium cycling, as opposed to heat from other ion transport such as Na^+ pumping by the Na^+-K^+ ATPase associated with the action potential. Values for the heat due to electrical activation vary widely, but this "Ca^{2+}-independent" heat could represent a substantial fraction of TIH and, thus, could cause calcium cycling to be overestimated (9, 14). Our initial approach to partitioning TIH into "Ca^{2+}-dependent" and "Ca^{2+}-independent" fractions is to examine the value of TIH remaining when twitch tension is zero. Figure 2 indicates that there is only a small (~ 0.04 mJ/g) TIH signal expected at zero twitch tension-time integal—the threshold for mechanical activation. This result contrasts sharply with much higher values (~ 0.4 mJ/g) observed for rabbit papillary muscles at 21 °C by us (unpublished data) and others (9). Preliminary experiments with the Ca^{2+} channel blocker nifedipine (25 µM) indicate that elimination of most of the inward calcium current leaves a heat value, presumably due to Na^+ transport, that is similar to the TIH intercept seen in Figure 2. Thus, TIH remaining at the threshold of mechanical activation of the human ventricular strips is more likely to be due to the heat of Na^+ transport than to Ca^{2+} cycling between resting sites and high affinity Ca^{2+} buffers.

We previously reported that TIH and mechanical activation were dissociated when twitch tension of rabbit papillary muscles was altered by nonsteady-state interventions. Thus, TIH did not change during the sharp increase in mechancial activation (treppe) when stimulation resumed after a rest period (1). In addition, no potentiation of TIH appears to occur when an extra stimulation is interposed within a regular pacing regime (2). Explanations have focused on the possibility that the interventions cause poorly understood changes in calcium exchange with buffers rather than the expected alterations in total calcium cycling. The post extrasystolic potentiation data from the human experiments, however, indicate that no such hypotheses are required to explain the relation between calcium cycling and mechanical activation for these preparations.

References

1. Alpert NR, Blanchard EM, Mulieri LA (1989) Tension-independent heat in rabbit papillary muscle. J Physiol (Lond) 414:433–453
2. Blanchard EM, Alpert NR (1990) Dissociation between post extrasystolic potentiation and tension independent heat in rabbit papillary muscles. Biophys J 57:176a
3. Blanchard EM, Mulieri LA, Alpert NR (1990) Dynamic calcium requirements for activation of rabbit papillary muscle calculated from tension-independent heat. Amer J Cardiol 65:8G–11G
4. Blanchard EM, Smith GL, Allen DG, Alpert NR (1990) The effects of 2,3-butanediene monoxime on initial heat, tension, and alquo light output of ferret papillary muscles. Pflugers Arch 406:219–221
5. Carafoli E (1985) The homeostasis of Ca in heart cells. J Mol Cell Cardiol 17:203–212
6. Curtin NA, Woledge R (1978) Energy changes and muscular contraction. Physiol Rev 58:690–761
7. Fabiato A (1983) Calcium-induced release of calcium from the cardiac sarcoplasmic reticulum. Am J Physiol 245:C1–C14
8. Feher JJ, Briggs FN (1979) The effect of Calcium load on the calcium permeability of sarcoplasmic reticulum. J Biol Chem 257:10191–10199

9. Gibbs CL, Vaughan P (1968) The effect of calcium depletion upon the tension-independent component of cardiac heat production. J Gen Physiol 52:532–549
10. Gwathmey JK, Copelas L, MacKinnon R, Schoen FJ, Feldman MD, Grossman W, Morgan JP (1987) Abnormal intracellular calcium handling in myocardium from patients with end-stage heart failure. Circ Res 61:70–76
11. Hill AV (1939) Recovery heat in muscle. Proc R Soc London, B, 127:297–307
12. Kurihara S, Saeking Y, Hongo K, Tanaka E, Sudo N (1990) Effects of length change on intracellular Ca^{2+} transients in ferret ventricular muscle treated with 2,3-butanedione monoxime (BDM). Jap J Physiol 40:915–920
13. Lenart TD, Tanner JW, Goldman YE (1989) 2,3-butanedione monoxime (BDM) suppresses cross-bridge reattachment following laser Photolysis of caged ATP. Bio Phys J 55:260a
14. Langer GA (1967) Sodium exchange in dog ventricular muscle. J Gen Physiol 50:1221–1239.
15. Li T, Sperelakis N, Teneick RE, Solaro RJ (1985) Effects of diacetyl monoxime on cardiac excitation-contraction coupling. J Pharmacol Exp Ther 232:688–695
16. Moravec CS, Bond M (1991) Calcium is released from the junctional sarcoplasmic reticulum during cardiac muscle contraction. Amer J Physiol 260:H989–H997
17. Mulieri LA, Luhr G, Trefry J, Alpert NR (1977) Metal-film thermopiles for use with rabbit right ventricular papillary muscles. Am J Physiol 233:C146–C156
18. Mulieri LA, Hasenfuss G, Ittleman F, Blanchard EM, Alpert NR (1989) Protection of human left ventricular myocardium from cutting injury with 2,3-butanedione monixime. Circ Res 65:1441–1444
19. Mulieri LA, Leavitt BJ, Hasenfuss G, Allen PD, Alpert NR (1992) Contractile frequency dependence of twitch and diastolic tension in human dilated cardiomyopathy. In: Hasenfuss G, Holubarsch Ch, Just H, Alpert NR (eds) Cellular and molecular alterations in the failing human heart. Steinkopff, Darmstadt
20. Robertson SP, Johnson JD, Potter JD (1981) The time-course of Ca^{2+} exchange with calmodulin, troponin, parvalbumin, and myuosin in response to transient increases in Ca^{2+}. Biophys J 34:559–569
21. Solaro RJ, Wise AM, Shiner JS, Briggs FN (1974) Calcium requirements for cardiac myofibrillar activation. Circ Res 34:525–530
22. Tada M, Imei M (1983) Regulations of calcium transport by the ATPase phospholamban system. J Mol Cell Cardiol 15:565–575
23. West JM, Stephenson DG (1989) Contractile activation and the effects of 2,3-butanedione monoxime (BDM) in skinned cardiac preparations from normal and dystrophic mice (129/ReJ). Pflugers Arch 413:546–552
24. Wier WG, Yue DT (1986) Intracellular calcium transients underlying the short-term force-interval relationship in ferret ventricular myocardium. J Physiol 376:507–530
25. Gwathmey JK, Hajjar RJ (1990) Relation between steady-state force and intracellular $[Ca^{2+}]$ in intact human myocardium. Index of myofibrillar responsiveness to Ca^{2+}. Circulation 82:1266–1278.
26. Gwathmey JK, Slawsky MT, Hajjar RJ, Briggs GM, Morgan JP (1990) Role of intracellular calcium handling in force-interval relationships of human venticular myocardium. J Clin Invest 85:1599–1613

Author's address:
E. M. Blanchard
Department of Physiology and Biophysics
University of Vermont
USA-Burlington
VT 05405

The calcium-release channel from cardiac sarcoplasmic reticulum: Function in the failing and acutely ischaemic heart

S.R.M. Holmberg[1], A.J. Williams[2]

[1]Department of Cardiology, Royal Sussex County Hospital, Brighton, East Sussex
[2]Department of Cardiac Medicine, National Heart & Lung Institute, Dovehouse Street, London, England

Summary: Junctional SR membrane vesicles have been isolated from chronically failing human hearts explanted at transplant operations. Vesicles have been incorporated into artificial planar phospholipid bilayers and the activity of single calcium-release channels investigated under voltage-clamp conditions. The properties of these channels are similar to those previously reported from normal animal tissue and do not provide evidence that the function of individual calcium-release channels is altered in the failing heart. Using radio-labelled ryanodine binding as a specific marker for the calcium-release channel, we demonstrate that, in the sheep heart, ischaemia results in the degradation of the calcium-release channel. The activation of proteases and oxidant stress in the ischaemic and re-perfused post-ischaemic myocardium are likely mediators of cell injury. Using the protease trypsin and the photosensitisation of rose bengal to generate the reactive oxygen species (ROS) singlet oxygen and superoxide radicals we demonstrate a direct effect on the calcium-release channel in vitro. Exposure of junctional SR vesicles to trypsin or oxidant stress resulted in the progressive loss of specific ryanodine binding and the degradation of high molecular weight proteins identified by polyacrylamide gel electrophoresis. The activity of single channels was also modified during exposure to proteolysis or oxidant stress; an initial increase in channel opening was observed followed by irreversible loss of channel function. Degradation of specific proteins, such as the calcium-release channel, may contribute to contractile dysfunction in the ischaemic and reperfused post-ischaemic myocardium.

Key words: Sarcoplasmic reticulum; Ca^{2+} release; failing heart; ischaemia

Introduction

In the mammalian heart, contraction is initiated by an elevation of the cytosolic calcium concentration. The bulk of this calcium is stored in and released from the sarcoplasmic reticulum (SR). The site for calcium release is localised to terminal cisternal regions of the membrane network, where the SR forms triadic junctions with the t-tubule system of the sarcolemma. Fabiato (7) had initially suggested that the principal stimulus for SR calcium release was the influx of calcium through sarcolemmal ion channels – so called calcium-induced calcium-release. On the basis of experiments with skinned myocytes, he proposed (6) that the SR calcium efflux pathway was an ion channel gated principally by calcium and modulated additionally by other agents including ATP, magnesium and caffeine.

Using canine hearts, Rousseau et al. (26) prepared vesicles enriched with membranes from junctional regions of the SR. Vesicles were fused with artificially constructed lipid bilayers to permit the recording of calcium currents through single ion channels under voltage-clamp conditions. Using this technique, a high conductance ligand-operated channel was characterised with properties consistent with those predicted by the work of Fabiato. Channel open probability was primarily increased by raising the calcium concentration at the cytoplasmic face of the channel and was further increased by the addition of adenine nucleotides and caffeine (25). Open probability was reduced in the presence of magnesium and ruthenium red. Rousseau et al. (27) also demonstrated that the channel was the site of action of the plant alkaloid ryanodine, which uncouples cell excitation from contraction. Ryanodine abolishes the normal pattern of channel gating, which involves bursts of brief opening events and instead locks the channel in an open state which has a reduced conductance relative to the normal open state. It was proposed that higher concentrations of ryanodine might fully close the channel.

SR function in the failing human heart is abnormal both with impaired calcium release (8) and uptake (17). Calcium transients in the acutely ischaemic myocardium are also abnormal, with an early rise in cytosolic calcium (1, 15, 28). This increase precedes the elevation of total cell calcium (4, 5) and is, therefore, due to a redistribution of intracellular calcium, suggesting the possibility of SR dysfunction.

In the present study, we report the results of recordings from single calcium-release channels of junctional SR membrane vesicles isolated from chronically failing human myocardium. We also report the results of experiments on animal tissue which investigate channel function in the ischaemic myocardium. The calcium-release channel protein is composed of four identical sub-units, each with a molecular weight of between 330 and 400 kDa, and it is associated with the specific binding site for ryanodine (13, 14); we have investigated [^{3}H]ryanodine binding to samples of ischaemic myocardium to demonstrate that the calcium-release channel is a target for early degradation during ischaemia.

The activation of proteases and oxidant stress – particularly in the reperfused, post-ischaemic myocardium – are likely mediators of protein degradation. Using singlet oxygen and superoxide free radicals generated by the photosensitisation of rose bengal (16), Hearse et al. (9) have shown that oxidant stress is arrhythmogenic in Langendorff heart preparations. In isolated papillary muscles, similar oxidant stress has been shown to induce transient inotropy followed by contractile failure (19). These effects appear to be mediated through cytosolic calcium overload (3) and it seems likely that this calcium originates from the SR (10). In the present study, we show that proteolysis and oxidant stress both result in degradation of the calcium-release channel in vitro and that this process may be associated with functional changes which could contribute to contractile dysfunction in the ischaemic and post-ischaemic myocardium.

Materials and methods

Materials

[^{3}H]ryanodine was purchased from New England Nuclear, Hertfordshire, UK. Unlabelled ryanodine was purchased from Progressive Agri-Systems, Wind Gap, Pennsylvania, USA. Phosphatidylethanolamine (bovine heart) was purchased from Avanti Polar Lipids, Pelham, Alabama, USA. Aqueous counting scintillant was purchased from Amersham International, Buckinghamshire, UK. All other chemicals were of AnalaR or best available grade from BDH Ltd or Sigma Chemical Co Ltd.

Vesicle Preparation

Human cardiac muscle was obtained from six hearts explanted at transplantation. Clinical details are shown in Table 1. Specimens were taken from areas of the left ventricular free wall, which did not show gross scarring, and were preseved in an ice-cold cardioplegic solution within 5–10 min of explant for transportation to the laboratory. Animal tissue (sheep heart) was obtained from a local abattoir and preserved in a similar manner. In experiments in which we investigated the effects of ischaemia, sheep hearts were immediately dissected into two to four portions and maintained at 37 ± 2° C in an air-excluded environment for between 0–120 min prior to preservation in cardioplegia. Using a modification (11) of the method of Meissner and Henderson (20), a mixed microsomal preparation was prepared from muscle homogenate by differential centrifugation and from this a fraction enriched with junctional SR membrane vesicles was isolated by a further centrifugation on discontinuous sucrose density gradients.

Single-Channel Recordings

Experiments were performed at 22° C. Planar phospholipid bilayers, composed of phosphatidylethanolamine dispersed in n-decane at a concentration of 30 mg/ml, were painted across a 200 µm-diameter hole in the septum between two experimental chambers containing choline chloride solutions as previously described (12). Junctional SR vesicles were added to the designated cis chamber and the solution fortified with choline chloride to produce a 7:1 concentration gradient across the membrane. Vesicle fusion was detected by the appearance of a chloride selective conductance. Calcium channels were observed after perfusion of the cis chamber, which is equivalent to the cytoplasmic face of the channel (22, 29), with 250 mM HEPES/-TRIS, pH 7.4, containing a controlled free calcium concentration and the trans chamber (SR luminal face of the channel) with 250 mM glutamic acid and 10 mM HEPES titrated to pH 7.4 with $Ca(OH)_2$ ($[Ca^{2+}]$ 67 mM). The composition of the chambers was then modified as detailed in Results. In the experiments described, the

Table 1. Summary of clinical data of patients from whom cardiac explants were obtained at time of heart or heart-lung* transplantation.

Age	Sex	NYHA Class	Diagnosis
53	M	IV	Chronic mitral insufficiency
34	F	III	Eisenmenger VSD*
45	M	IV	Idiopathic dilated cardiomyopathy
45	M	III	Ischaemic cardiomyopathy
56	M	IV	Ischaemic cardiomyopathy
66	M	IV	Ischaemic cardiomyopathy

cis chamber was voltage-clamped at 0 mV relative to the trans chamber, which was held at ground.

Channel opening results in a flow of ions across the bilayer. This current was converted to a voltage, amplified (23) and recorded on FM tape. Data were displayed on a Hewlett-Packard 7475A plotter after digitisation at 2kHz, using a PDP 11/73 based computer system (Indec, Sunnyvale, Calif.)

[^3H]ryanodine Binding Assay

SR vesicles (approx. 100 µg protein) were incubated at 37 °C with 5 nM [^3H]ryanodine in 1 ml of a buffered medium containing 1 M KCl, 5 µM phenylmethylsulphonyl fluoride, 100 µM $CaCl_2$, 25 mM PIPES/KOH, pH 7.4. Other additions are described for individual experiments. At the completion of incubation, the medium was diluted with 5 ml of ice cold buffer and filtered through Whatman GF-B filters which had been pre-soaked in buffer. Filters were washed with 3×5 ml aliquots of buffer and counted in 10 ml of aqueous counting scintillant. All incubations were performed in triplicate. Non-specific binding was measured using matching control media to which 2.5 µM unlabelled ryanodine had been added; these counts were subtracted from total binding to produce specific binding which is shown in Results.

Reactive oxygen species

Singlet oxygen and superoxide free radicals were generated by the illumination of the photosensitive dye rose bengal with visible light from a fiber-optic system.

Gel Electrophoresis

Samples of junctional SR (approx. 150 µg protein) were resolved on continuous gradient (4–10%) polyacrylamide gels and stained with Coomasie Blue. Molecular weight standards were employed to estimate the M_r of specific protein bands. Laser densitometric scans (LKB, Sweden) were performed on the high molecular weight portion of the gel ($M_r > 120$ kDa).

Protein assays were performed using the modification of the Lowry method described by Markwell et al (18).

Results

Figure 1 shows recordings from a single calcium-release channel from human myocardium and illustrates increasing channel opening frequency with increasing calcium concentration at the cytosolic face of the bilayer. In this figure data are displayed on a long time base to demonstrate channel activity over a prolonged period. The diagrams show current fluctuations apparently occurring to levels less than that of the fully open channel conductance, but less severe low-pass filtering (data not shown) permits full resolution of these opening events and demonstrates that, within the limits of resolution(> 1 ms), all openings are to the fully open state

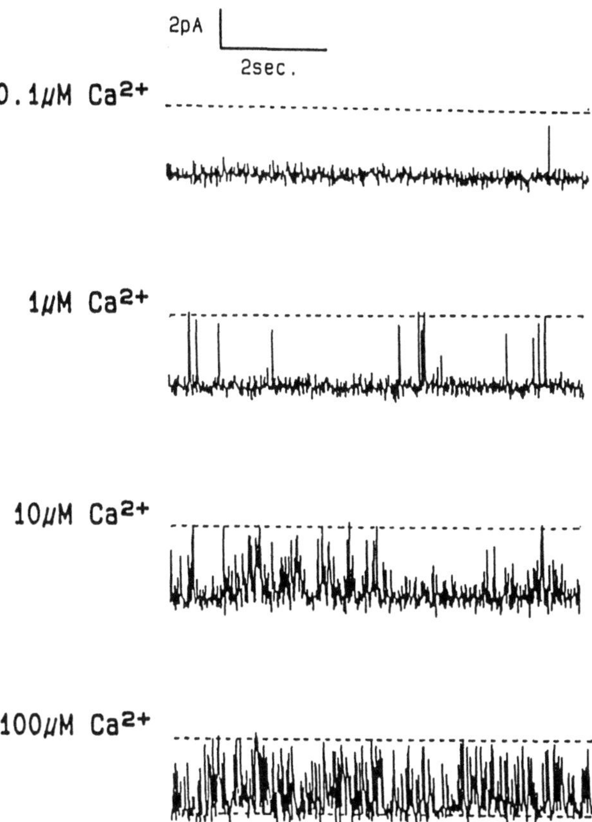

Fig. 1. Activation of the Ca^{2+}-release channel by Ca^{2+}. Current fluctuations from a bilayer containing a single calcium-release channel with 250 mM HEPES/TRIS, pH 7.4 cis and 250 mM glutamic acid/ 10 mM HEPES titrated to pH 7.4 with $CaOH_2$ (67 mM), trans. The free Ca^{2+} concentration of the solution in the cis chamber has been raised from 0.1 μM to 100 μM as indicated. Upward deflections of current represent channel opening. Recordings were made with the cis chamber clamped at 0 mV relative to ground.

without evidence of any subconductance levels. Using calcium as the sole activating ligand, channel open probability reached a maximum value of approximately 0.4 in the presence of 100 μM calcium. Further increases in the calcium concentration above 100 μM did not result in increased open probability and were sometimes associated with decreased openings and irreversible loss of channel function.

The channel recordings in Figure 2 demonstrate the effects of millimolar magnesium and ATP added at the cytoplasmic face of the channel. Magnesium reduces open probability and this effect can be overcome by raising the calcium concentration. When channel opening frequency is increased, current fluctuations above the single channel conductance level are seen, demonstrating the presence of at least two channels in the bilayer. ATP increases channel opening frequency in the presence of calcium and typically raises open probability to near unity, ie. full channel activation.

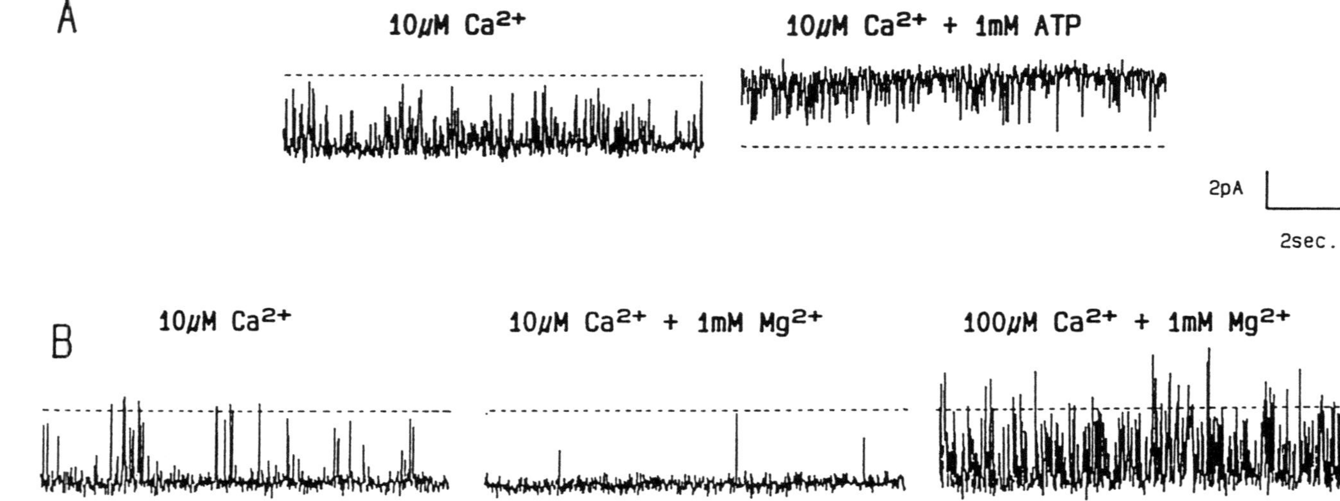

Fig. 2. Modulation of the activity of the Ca^{2+}-release channel by physiological ligands. A) Effect of ATP. Current fluctuations from a bilayer containing a single Ca^{2+}-release channel with 250 mM HEPES/TRIS, pH 7.4 cis and 250 mM glutamic acid/ 10 mM HEPES titrated to pH 7.4 with $CaOH_2$ (67 mM), trans. The free Ca^{2+} concentration of the solution in the cis chamber was 10 μM. Upward deflections indicate channel opening. Current levels of the open and closed states are indicated by the dashed lines. Recordings were made at a holding potential of 0 mV. B) Effect of magnesium. Current fluctuations from a bilayer containing at least 2 Ca^{2+}-release channels with 250 mM HEPES/TRIS, pH 7.4 cis and 250 mM glutamic acid/ 10 mM HEPES titrated to pH 7.4 with $CaOH_2$ (67 mM), trans. The Ca^{2+} and Mg^{2+} concentrations of the cis solution have been varied as indicated. Current levels of the single open and fully closed channel states are indicated by the dashed lines. The holding potential was 0 mV.

The effects of caffeine, ruthenium red and ryanodine on channel gating when added to the cytosolic face of the channel are shown in Figure 3. Channel activation by caffeine was reversible following drug washout, but the effects of ruthenium red and ryanodine were irreversible. Block by ryanodine did not occur immediately

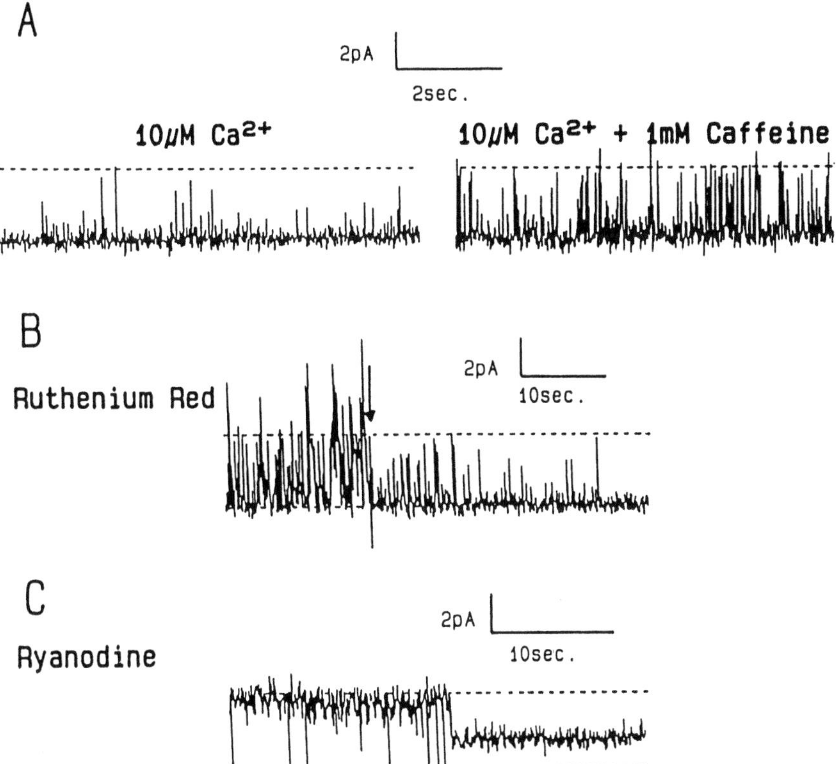

Fig. 3. Modulation of the activity of the Ca^{2+}-release channel by pharmacological ligands. A) Effect of caffeine. Current fluctuations from a bilayer containing a single Ca^{2+}-release channel in 250 mM HEPES/TRIS, pH 7.4 cis and 250 mM glutamic acid/ 10 mM HEPES titrated to pH 7.4 with $CaOH_2$ (67 mM), trans, before and following the addition of 1 mM caffeine to the solution in the cis chamber. Upward deflections indicate channel opening. Current levels of the open and closed states are indicated by the dashed lines. B) Effect of ruthenium red. Current fluctuations from a bilayer containing at least 2 Ca^{2+}-release channels with 250 mM HEPES/-TRIS, pH 7.4 cis and 250 mM glutamic acid/ 10 mM HEPES titrated to pH 7.4 with $CaOH_2$ (67 mM), trans. The free Ca^{2+} concentration of the solution in the cis chamber was 10 µM 1 µM ruthenium red was added to the solution in the cis chamber at the point indicated by the vertical arrow. Upward deflections indicate channel opening, dashed lines indicate the single channel open and fully closed current levels. C) Effect of ryanodine. Current fluctuations from a bilayer containing a single Ca^{2+}-release channel in 250 mM HEPES/TRIS, pH 7.4 cis and 250 mM glutamic acid/ 10 mM HEPES titrated to pH 7.4 with $CaOH_2$ (67 mM) trans. The free Ca^{2+} concentration of the solution in the cis chamber was 100 µM and this solution additionally contained 5 µM ryanodine. Upward deflections indicate channel opening. Current levels of the open and closed states are indicated by the dashed lines. In all cases the holding potential was 0 mV.

following drug addition but occurred more rapidly under conditions in which channels had a high open probability. Block always appeared to develop from the open channel state.

In Figure 4 the effects of ischaemia on [³H]ryanodine binding to mixed microsomal fractions of sheep myocardium are shown. Seven separate experiments were performed – each heart contributing a control and 1–3 ischaemic samples. Ischaemia reduces ryanodine binding but does not affect the total yield of mixed microsomal membrane protein (data not shown) demonstrating that this effect is not due to an artefact of the isolation technique.

Samples of mixed microsomal membranes from tissue subjected to 60 and 120 min ischaemia were further fractionated to produce junctional SR membranes. The yield of these membranes was greatly reduced relative to similar preparations from control, non-ischaemic tissue. Recordings of single calcium-release channels from ischaemic junctional SR vesicles showed no obvious differences from those of non-ischaemic tissue. It seems likely that these results demonstrate that membrane vesicles containing degraded calcium-release channels do not sediment with the junctional SR membrane fraction and, thus, such channels are not amenable to observation by this technique.

In Figure 5a the effects of trypsin on [³H]ryanodine binding to sheep junctional SR membrane vesicles are shown. Proteolysis degrades the calcium-release channel complex, resulting in a reduction in ryanodine binding; this effect is attenuated by the addition of trypsin inhibitor to the pre-incubation medium (data not shown). Data from the same experiment (illustrated in Fig. 5b) show that proteolysis results in degradation of high molecular weight proteins on gel electrophoresis which correspond to the monomer of the calcium-release channel protein complex.

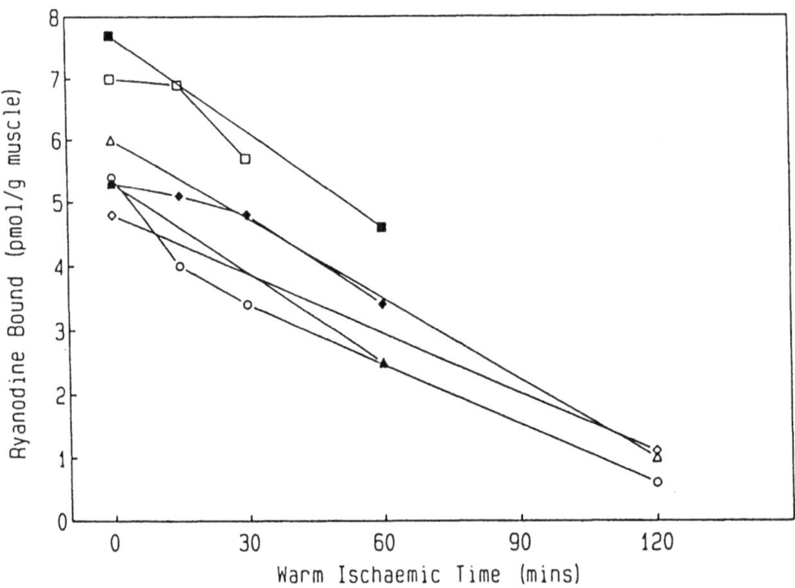

Fig. 4. The effect of normothermic ischaemia on [³H]ryanodine binding to mixed microsomal membranes from sheep heart. Data from the same preparation are shown joined by solid lines.

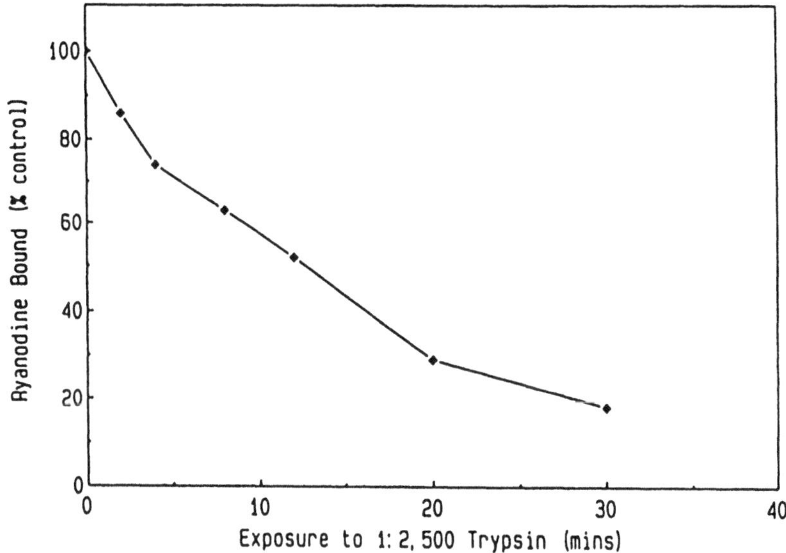

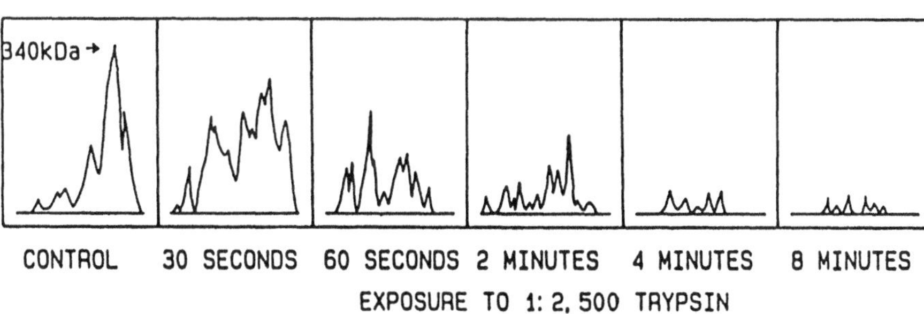

Fig. 5. The effect of trypsin on [³H]ryanodine binding and high molecular weight proteins from heavy SR membrane vesicles. The data shown are from an individual experiment. Heavy SR membrane vesicles were incubated in media containing 400 mM sucrose, 5 mM HEPES/-TRIS, pH 7.4 + 1 : 2500 w/w trypsin : membrane protein. Subsequently, matched samples were used either for [³H]ryanodine binding (upper panel) or placed into gel buffer for SDS polyacrylamide gel electrophoresis. The lower panel shows laser densitometric scans of the high molecular weight (> approx. 160 kDa.) portion of the gel.

Similar data are shown in Figure 6 for the effects of prior exposure to oxidant stress on subsequent [³H]ryanodine binding and gel electrophoresis of junctional SR membrane vesicles. In these experiments the addition of sodium azide, as a quenching agent for singlet oxygen (24), attenuates the effects of oxidant stress.

Figure 7 shows typical recordings of calcium-release channels exposed to oxidant stress by the addition of rose bengal to both cis and trans chambers in the presence of bright illumination. In 83% of experiments (n = 15) channel open probability showed an initial increase, continued exposure to oxidant stress resulted in eventual

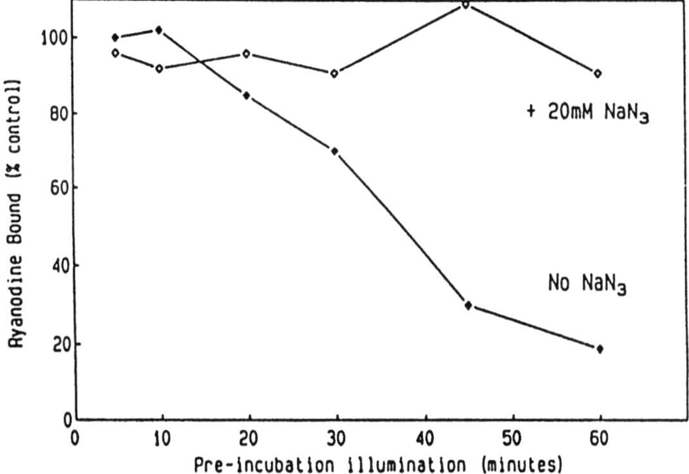

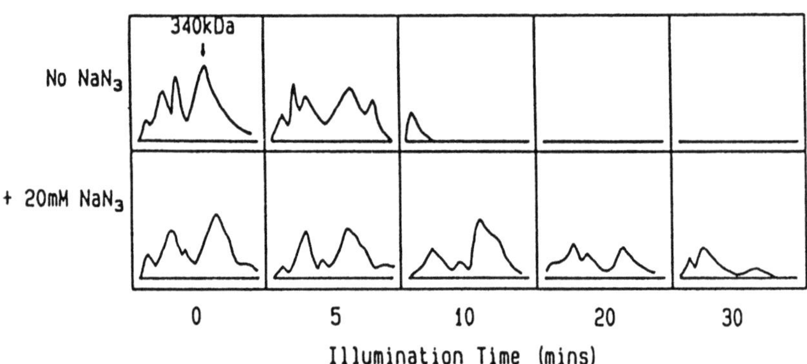

Fig. 6. The effect of ROS on [^{3}H]ryanodine binding and high molecular weight proteins from heavy SR membrane vesicles. Data shown are from an individual experiment. Heavy SR membrane vesicles were incubated in media containing 400 mM sucrose, 5 mM HEPES/TRIS, pH 7.4 + 1:1000 w/w rose bengal:membrane protein, ± 20 mM sodium azide. After pre-incubation, matched samples were used for either [^{3}H]ryanodine binding (upper panel) or placed into gel buffer for SDS polyacrylamide gel electrophoresis. The lower panel shows laser densitometric scans of the high molecular weight (> approx. 160 kDa.) portion of the gel. (Figure reproduced with permission from Holmberg et al. (1991) Cardioscience 2:19–25.)

irreversible loss of channel function which preceded disruption of the bilayer membrane. Sub-conductance states, possibly indicative of damage to the channel conduction pathway and which were not seen with intact channels were observed frequently during the period of transient channel activation. In these experiments 50 mM histidine as a quenching agent for singlet oxygen (9) retarded both the period of channel activation and the eventual loss of channel function. Experiments studying the effects of cis trypsin on sheep cardiac calcium-release channel function in bilayers

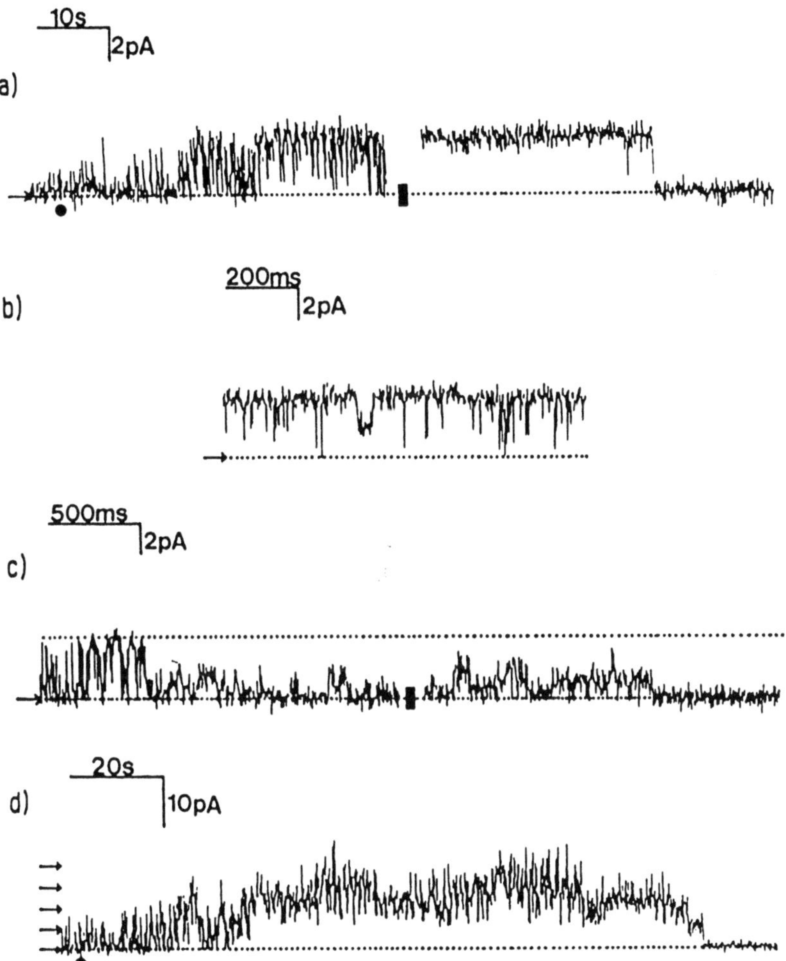

Fig. 7. The effect of ROS on Ca^{2+}-release channel activity. All traces show current fluctuations from bilayers containing Ca^{2+}-release channels with 250 mM HEPES/TRIS, pH 7.4 cis and 250 mM glutamic acid/ 10 mM HEPES titrated to pH 7.4 with $CaOH_2$ (67 mM) trans. The free Ca^{2+} concentration of the solution in the cis chamber was 10 μM and 50 nM rose bengal was present in both chambers. Upward deflections indicate channel openings and the holding potential was 0 mV. a) Activity from a single Ca^{3+}-release channel. Illumination was started at the time indicated by the filled circle. Approximately 20 s elapses between the two portions of the recording. Increased channel opening is followed by the loss of channel function. b) Ca^{2+}-release channel activity during illumination showing the occurrence of a sub-conductance level. c) Ca^{2+}-release channel activity during illumination. Multiple opening events to incomplete or sub-conductance levels are seen prior to complete loss of channel function. The filled symbol indicates a 2 s break in recording. d) Channel activity from a bilayer containing at least 5 Ca^{2+}-release channels. Illumination commenced at the time indicated by the filled circle. Current increases and then falls as individual channels are inactivated. The arrows indicate multiples of the single-channel conductance. (Figure reproduced with permission from Holmberg et al. (1991) Cardioscience 2:19–25).

have shown a similar pattern of increased open probability followed by loss of channel function (data not shown) and similar activation and subsequent loss of activity has been reported following the trypsinisation of rabbit skeletal muscle SR calcium-release channels in bilayers (21).

Discussion

Channel function from failing human hearts

The data shown in this study for human SR calcium-release channel activity are all recorded from failing myocardium resulting from a number of different aetiologies. No 'normal' control hearts were available to allow comparison with these findings. Nevertheless, the evidence suggests that abnormalities of single-channel behaviour do not underlie the previously reported impairment of SR function in the chronically failing heart. First, human channel recordings were similar in all the hearts studied regardless of the aetiology of cardiac dysfunction. Second, the data in this report show that the properties of human calcium-release channels are virtually identical with those previously reported for sheep (2, 30) and canine (26) cardiac channels.

This suggests that other factors may contribute to abnormalities of calcium release from the SR of failing myocardium. It is possible that these factors may include a reduction in the number of channels, altered cytosolic ligand concentrations, or amounts of available calcium stored within the SR. The experiments on ischaemic sheep myocardium raise the possibility that a subpopulation of abnormally functioning channels may exist within the failing heart, but that these channels do not sediment with the junctional SR fraction. This possibility cannot definitely be discounted on present evidence.

Channel Function in acute ischaemia

The results of experiments on the effects of ischaemia, summarised in Figure 4, demonstrate that the calcium-release channel is indeed degraded during ischaemia, but because the isolation process for junctional SR membranes appears to select for residual, normally functioning channels, these experiments do not give any information regarding the functional effects of this process at the single channel level.

Figures 5 and 6 demonstrate that both proteolysis, possibly occurring during ischaemia and oxidant stress, possibly also occurring during ischaemia, but most probably in the period of post-ischaemic reperfusion, result in the degradation of the calcium-release channel complex as seen both by the effects on [3H]ryanodine binding and protein electrophoresis. A further interesting observation is that complete loss of high molecular weight proteins preceeded loss of [3H]ryanodine binding sites. These data are consistent with previous reports that ryanodine binding activity is localised to a portion of the calcium-release channel protein which may remain intact despite partial channel degradation. The effect of this would be that [3H]ryanodine binding underestimates the extent of calcium-release channel degradation.

The probable functional consequences of channel degradation are shown in Figure 7. The initial phase of increased channel opening would result in increased permeability of the SR membrane to calcium which would contribute to the cytosolic calcium overload seen in the ischaemic myocardium. Irreversible loss of channel function in the myocardium surviving non-lethal ischaemia might deplete the amount of calcium which could be released by the SR during excitation-contraction coupling and factors such as this may contribute to the phenomenon of myocardial stunning.

Acknowledgements. We are grateful to the British Heart Foundation for providing funds to support this work.

References

1. Allen DG, Lee JA, Smith GL (1988) The effects of simulated ischaemia on intracellular calcium and tension in isolated ferret ventricular muscle. J Physiol 401:81P (Abstract)
2. Ashley RH, Williams AJ (1990) Divalent cation activation and inhibition of single calcium release channels from sheep cardiac sarcoplasmic reticulum. J Gen Physiol 95:981–1005
3. Bernier M, Vandeplassche G, Kusama Y, Borgers M, Hearse DJ (1988) Singlet oxygen induced ultrastructural injury, calcium shifts and arrhythmias in the aerobic isolated perfused rat heart during the photosensitization of rose bengal. J Mol Cell Cardiol 20 (Suppl. 5):S24(Abstract)
4. Burton KP, Hagler HK, Templeton GH, Willerson JT, Buja LM (1977) Lanthanum probe studies of cellular pathophysiology induced by hypoxia in isolated cardiac muscle. J Clin Invest 60:1289–1302
5. Chien KR, Reeves JP, Buja LM, Bonte F, Parkey RW, Willerson JT (1981) Phospholipid alterations in canine ischaemic myocardium. Circ Res 48:711–719
6. Fabiato A (1983) Calcium-induced release of calcium from the cardiac sarcoplasmic reticulum. Am J Physiol 245:Cl–C14
7. Fabiato A, Fabiato F (1972) Excitation-contraction coupling of isolated cardiac fibers with disrupted or closed sarcolemmas. Circ Res 31:293–307
8. Gwathmey JK, Copelas L, MacKinnon R, Schoen FJ, Feldman MD, Grossman W, Morgan JP (1987) Abnormal intracellular calcium handling in myocardium from patients with end-stage heart failure. Circ Res 61:70–76
9. Hearse DJ, Kusama Y, Bernier M (1989) Rapid electrophysiological changes leading to arrhythmias in the aerobic rat heart: photosensitization studies with rose bengal-derived reactive oxygen intermediates. Circ Res 65:146–153
10. Holmberg SRM, Cumming D, Kusama Y, Hearse DJ, Poole-Wilson PA, Shattock MJ, Williams AJ (1991) Reactive oxygen species modify the structure and function of the cardiac sarcoplasmic reticulum calcium-release channel. Cardioscience 2:19–25
11. Holmberg SRM, Williams AJ (1989) Single Channel Recordings from Human Cardiac Sarcoplasmic Reticulum. Circ Res 65:1445–1449
12. Holmberg SRM, Williams AJ (1990) Patterns of interaction between anthraquinone drugs and the calcium release channel from cardiac sarcoplasmic reticulum. Circ Res 67:272–283
13. Imagawa T, Smith JS, Coronado R, Campbell KP (1987) Purified ryanodine receptor from skeletal muscle sarcoplasmic reticulum is the Ca^{2+}-permeable pore of the Ca release channel. J Biol Chem 262:16636–16643
14. Inui M, Saito A, Fleischer S (1987) Purification of the ryanodine receptor and identify with feet structures of junctional terminal cisternae of sarcoplasmic reticulum from fast skeletal muscle. J Biol Chem 262:1740–1747
15. Lee H-C, Mohabir R, Smith N, Franz MR, Clusin WT (1988) Effect of ischemia on calcium-dependent fluorescence transients in rabbit hearts containing Indo-1. Circulation 78:1047–1059

16. Lee PCC, Rodgers MAJ (1987) Laser flash photokinetic studies of rose bengal sensitized photodynamic interactions of nucleotides and DNA. Photochem Photobiol 45:79–86
17. Limas CJ, Olivari MT, Goldenberg IF, Levine TB, Benditt DG, Simon A (1987) Calcium uptake by cardiac sarcoplasmic reticulum in human dilated cardiomyopathy. Cardiovas Res 21:601–605
18. Markwell MAK, Haas SM, Bieber LL, Tolbert NE (1978) A modification of the Lowry procedure to simplify protein determination in membrane and lipoprotein samples. Anal Biochem 87:206–210
19. Matsuura H, Shattock MJ (1989) Functional and electrophysiological effects of reactive oxygen intermediates on isolated rat ventricular muscle. J Mol Cell Cardiol 21 (Suppl. II):S129 (Abstract)
20. Meissner G, Henderson JS (1987) Rapid Ca release from cardiac sarcoplasmic reticulum vesicles is dependent on Ca^{2+} and is modulated by Mg^{2+}, adenine nucleotide and calmodulin. J Biol Chem 262:3065–3073
21. Meissner G, Rousseau E, Lai FA (1989) Structural and functional correlation of the trypsin-digested Ca^{2+} release channel of skeletal muscle sarcoplasmic reticulum. J Biol Chem 264:1715–1722
22. Miller C (1978) Voltage-gated cation conductance channel from fragmented sarcoplasmic reticulum: Steady-state electrical properties. J Memb Biol 40:1–23
23. Miller C (1982) Open-state substructure of single chloride channels from Torpedo electroplax. Phil Trans Roy Soc Lon B 299:401–411
24. Pooler JP, Valenzeno DP (1979) The role of singlet oxygen in photooxidation of excitable cell membranes. Photochem Photobiol 30:581–584
25. Rousseau E, Meissner G (1989) Single cardiac sarcoplasmic reticulum Ca^{2+}-release channel: activation by caffeine. Am J Physiol 256:H328–H333
26. Rousseau E, Smith JS, Henderson JS, Meissner G (1986) Single channel and $^{45}Ca^{2+}$ flux measurements of the cardiac sarcoplasmic reticulum calcium channel. Biophys J 50:1009–1014
27. Rousseau E, Smith JS, Meissner G (1987) Ryanodine modifies conductance and gating behaviour of single Ca^{2+} release channel. Am J Physiol 253:C364–368
28. Steenbergen C, Murphy E, Levy L, London RE (1987) Elevation in cytosolic free calcium concentration early in myocardial ischemia in perfused rat heart. Circ Res 60:700–707
29. Tomlins B, Williams AJ, Montgomery RAP (1984) The characterization of a monovalent cation selective channel of mammalian cardiac muscle sarcoplasmic reticulum. J Memb Biol 80:191–199
30. Williams AJ, Ashley RH (1989) Reconstitution of cardiac sarcoplasmic reticulum calcium channels. Ann NY Acad Sci 560:163–173

Author's address:
Alan Williams,
Cardiac Medicine,
National Heart and Lung Inst.,
Dovehouse St.,
London SW3 6LY
England

Immune-mediated modulation of sarcoplasmic reticulum function in human dilated cardiomyopathy

C. J. Limas, C. Limas

Departments of Medicine (Cardiovascular Division) and Laboratory Medicine and Pathology, University of Minnesota School of Medicine and the Department of Veterans Affairs Medical Center, Minneapolis, Minnesota, USA.

Summary: Calcium transport by the cardiac sarcoplasmic reticulum is depressed in human dilated cardiomyopathy, but the mechanisms involved are not clear. The possible involvement of immunological mechanisms was explored by evaluating the effect of sera from 49 patients with dilated cardiomyopathy on oxalate-facilitated Ca^{2+} uptake. In 14 of these patients, serum or IgG induced a time- and concentration-dependent decline ($29 \pm 4\%$ at 100-fold serum dilution) in Ca^{2+} transport. In 14 patients, autoantibodies against the β_1-adrenoceptor were also demonstrated by a ligand binding inhibition assay. Serum from these patients inhibited the isoproterenol-mediated stimulation of Ca^{2+} uptake in permeabilized cardiac myocytes, but did not prevent the effect of protein kinase A. Anti-β-receptor antibodies were present in 50% of the sera inhibiting Ca^{2+} uptake compared to 20% of those without inhibitory activity, ($p < 0.01$). There was a strong correlation between the inhibition of sarcoplasmic reticulum Ca^{2+} transport and the HLA-DR4 phenotype (78% compared to 30% in patients with no inhibitory effect). These results suggest that immunological mechanisms play an important role in modifying sarcoplasmic reticulum function in about a third of the patients with detailed cardiomyopathy.

Key words: Sarcoplasmic reticulum; dilated cardiomyopathy; autoantibodies; calcium transport; HLA antigens

Introduction

Abnormal calcium transport is a consistent finding in the failing human heart (4) and is thought to be an important determinant of the decline in systolic and diastolic performance (3). Decreased sarcoplasmic reticulum (SR) function has been reported in end-stage heart failure (5, 10, 13), but its pathogenesis remains unclear. Possible mechanisms involved include changes in the functional properties of the sarcoplasmic reticulum, quantitative decline in the density of Ca^{2+} pump units (14) without change in the ability of SR to take up and release Ca^{2+} (6, 17) and abnormalities in cyclic AMP-dependent regulation. It is also likely that the pathogenesis of SR dysfunction is heterogeneous and may be, in part, influenced by the etiology of heart failure. We have explored this hypothesis by taking advantage of the observation that immunological factors play an important role in the pathogenesis of dilated cardiomyopathy. Our results suggest that cardiac SR function may be modulated through immunological mechanisms at two levels: a) the β-receptor-mediated phosphorylation of phospholamban, and b) Ca^{2+} uptake by the sarcoplasmic reticulum vesicles.

Materials and Methods

Serum was obtained from 49 patients with dilated cardiomyopathy (37 males and 12 females) aged 25–64 (mean: 47) years. The severity of cardiac dysfunction in these patients is reflected in depressed left-ventricular ejection fraction (25.8 $\pm$ 4%), cardiac output (4.51 $\pm$ 0.7 L/min), and elevated pulmonary capillary wedge pressure (20 $\pm$ 6 mmHg). Two groups served as controls: a) 26 patients with noncardiomyopathic heart disease (20 with ischemic heart disease, four with valvular heart disease, and two with myocarditis) and comparable hemodynamic severity of dysfunction (ejection fraction 30.1 $\pm$ 5%, cardiac output 9.9 $\pm$ 0.8 L/min and capillary wedge pressure of 19.2 $\pm$ 4 mmHg); b) 25 normal subjects. In 12 dilated cardiomyopathy patients, IgG was also prepared from their serum, utilizing previously described methodology (11).

Cardiac sacroplasmic reticulum was isolated from adult male sprague-Dawly rats as previously described (9). Briefly, the hearts were washed twice with a solution containing 10 mM $NaHCO_3$ – 5 mM NaN_3 and was then homogenized in the same solution with a Polytron PT-20 homogenizer three times at rheostat setting of 3 for 5 s with 15-s rest intervals. The homogenates were processed for sarcoplasmic reticulum (SR) isolation as described by Harigaya and Schwartz (5). The extent of mitochondrial contamination as reflected in cytochrome oxidase activity of the SR fraction was less than 8% of the activity in the mitochondrial fraction. Na^+-K^+-ATPase activity was also low in the absence of added detergents, reflecting minimal contamination with sarcolemmal membranes.

For studies of Ca^{2+} uptake, microsomes were suspended in 1 ml of reaction mixture consisting of 40 mM histidine-HCl buffer (pH 6.8), 5 mM $MgCl_2$, 0.1 M KCl, 0.005 M sodium oxalate, 0.05 μ Ci^{45} $CaCl_2$ (Ca^{2+}-EGTA buffer containing 390 μM EGTA and 125 μM $CaCl_2$, giving a free Ca^{2+} concentration of 1 μM by the use (20) of 4.4×10^5 M^{-1} as the association constant for the Ca^{2+}-EGTA complex), 5 mM ATP, and 40-60 μg microsomal protein. The mixture was preincubated with varying dilutions of serum (1:25–1:200) or IgG (10^{-7}–10^{-4} M) for 15 min 37°C before starting the reaction with the addition of $^{45}CaCl_2$, and the recation was stopped by filtering through a Millipore filter (HA 0.45 μm).

For studies of isoproterenol or protein kinase A mediated stimulation of Ca^{2+} uptake, cardiac myocytes were first isolated by enzymatic dissociation and then permeabilized with saponin as described by Miyakoda et al. (15). Isoproterenol was added at a final concentration of 0.05 mM and protein kinase A (catalytic subunit, Sigma Chemical) at 50 μg/ml. The incubation medium contained in addition to the Ca^{2+} uptake buffer, 10 μM ruthenium red. After preincubation at 37°C for 15 min with or without serum, the reaction was initiated with the addition of $^{45}CaCl_2$ and was terminated by filtration, as described above.

The presence of anti-β-receptor antibodies was determined using a ligand binding inhibition assay (11). Briefly, rat cardiac membranes were preincubated with diluted (1:100) serum from patients for 2 h at 4°C. The dilutions were made in assay buffer containing 50 mM Hepes-4mM $MgCl_2$, pH 8.0. After the preincubation, membranes were washed twice to remove the serum and were resuspended in the same buffer. For beta-adrenoceptor assay, the ligand [^{3}H] dihydroalprenolol (New England Nuclear Co, sp. act. 105 Ci/mmol) was used (0.2–20 nm) and the incubation carried out at 37°C for 30 minutes. At the end of the incubation, 4 ml of ice-cold buffer was added to each tube and the samples were filtered through Whatman GF/C filters. The filters were washed three times with 5 ml of cold buffer and counted in 5 ml Aquasol-2 (New England Nuclear). Nonspecific binding was assessed in the presence of 1µM propranolol. Binding in the absence of serum was taken as 100%.

The distribution of the Class II HLA-DR antigens in antibody-positive and -negative dilated cardiomyopathy patients was studied as previously described (10).

Results and Discussion

Serum from a subset of dilated cardiomyopathy patients was found to inhibit calcium uptake in a time-dependent manner (Fig. 1). This inhibitory effect was reproduced by IgG from the dilated cardiomyopathy patients, suggesting that it was due to the presence of autoantibodies (Fig. 2). Furthermore, no effect was noted when skeletal, rather than cardiac, sarcoplasmic reticulum was used (Fig. 3), indicating a rather restricted distribution of antigenic epitopes. The influence of the etiology of heart failure on the presence of anti-sarcoplasmic reticulum antibodies is shown in Fig. 4; a significant difference in the prevalence and titer of these antibodies was noted between cardiomyopathic and noncardiomyopathic patients.

We have previously reported (11) that a substantial proportion of dilated cardiomyopathy patients have antibodies directed against the cardiac β_1-adrenoceptor. Indeed, 14 of the 49 patients in this series had such antibodies as determined by the ligand binding inhibition assay. It might be expected that these antibodies would prevent the isoproterenol-mediated stimulation of Ca^{2+} uptake, since we have demonstrated that they block stimulation of adenylate cyclase in cardiac membrane preparations (13). This prediction was indeed borne out (Fig. 5). In permeabilized cardiac myocytes, the effect of isoproterenol, but not the catalytic subunit of protein kinase A was prevented by sera known to have anti-β-receptor antibodies. The fact that protein kinase A was still effective suggests that the

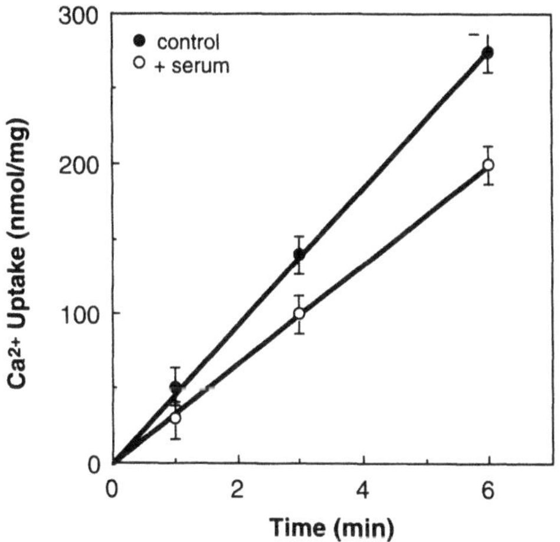

Fig. 1. Time-course of Ca^{2+} uptake by cardiac sarcoplasmic reticulum in the presence ($\bigcirc$) or absence ($\bullet$) of a 100-fold dilution of serum from cardiomyopathic patients. Results represent mean $\pm$ SE for 14 comparisons.

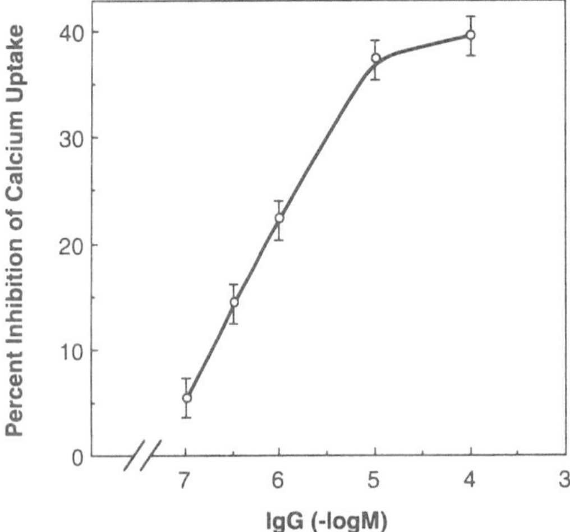

Fig. 2. Effect of IgG from sera of dilated cardiomyopathy patients on Ca^{2+} uptake by rat cardiac sarcoplasmic reticulum. Results represent mean $\pm$ SE for eight experiments.

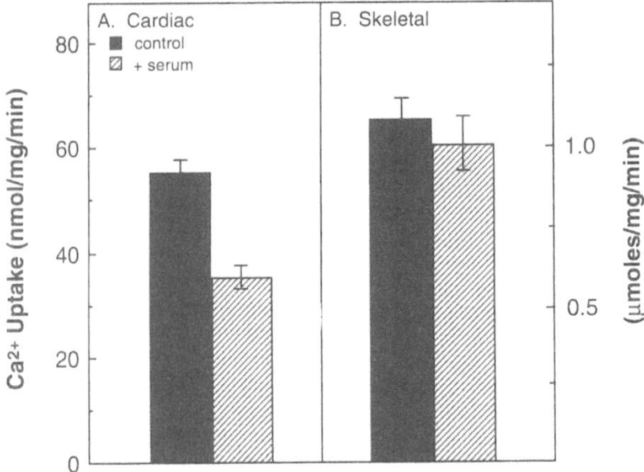

Fig. 3. Comparative effect of serum (50-fold dilution) from cardiomyopathy patients on cardiac (A) or skeletsal (B) sarcoplasmic reticulum. Results represent mean $\pm$ SE for 12 comparisons.

autoantibodies act predominantly at the β-receptor level and that cyclic AMP/PKA regulation of phospholamban may remain intact. Movesian et al. have also reported (18) that the ability of anti-phospholamban antibodies to stimulate Ca^{2+} uptake was unaffected in failing human hearts. It should be borne in mind, however, that the β-receptor/adenylate cyclase system is frequently down-regulated in the failing heart, and this down-regulation may also contribute to defective β-agonist-induced modulation of SR function.

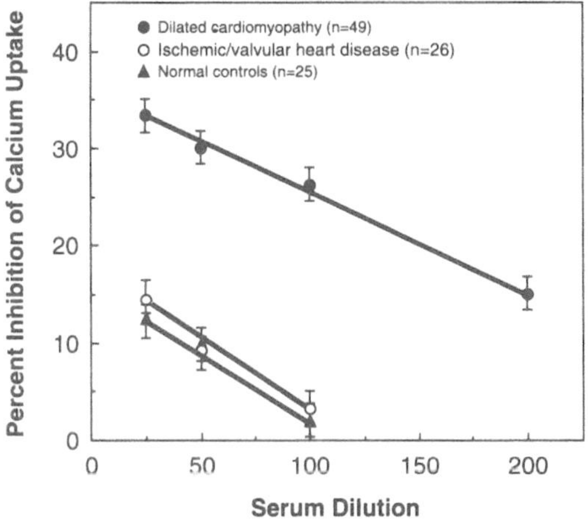

Fig. 4. Comparison of the effects of sera from dilated cardiomyopathy (●) or noncardiomyopathy (○) patients or normal controls (▲) on the Ca^{2+} uptake by cardiac sarcoplasmic reticulum. Results represent mean $\pm$ SE.

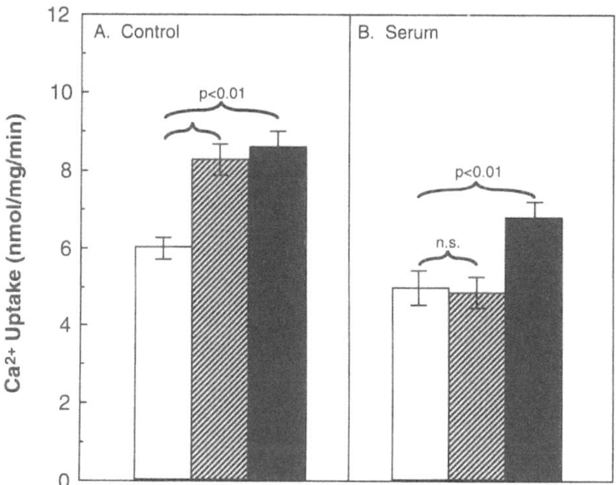

Fig. 5. Isoproterenol- and protein kinase A-mediated stimulation of Ca^{2+} uptake in cardiac myocytes in the absence (A) or presence (B) of serum (1:50) from dilated cardiomyopathy patients. Results represent mean $\pm$ SE for eight comparisons. (□) = control; (▨) = isoproterenol; (■) = PKA

It has been proposed that immunological mechanisms play an important role in the initiation and/or progression of dilated cardiomyopathy. This suggestion is supported by the presence of abnormal lymphocyte function and circulating autoantibodies against a variety of cardiac cell constituents (2, 19, 21). Our results would suggest that the sarcoplasmic reticulum may be a novel target for immune-mediated damage. It is still unknown whether there is, in each patient, a wide spectrum of

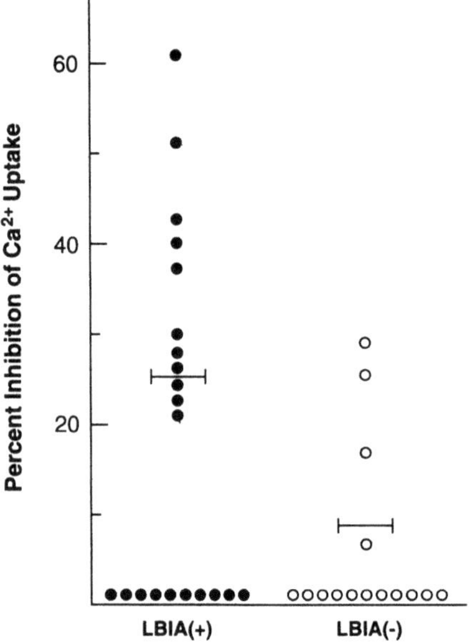

Fig. 6. Distribution of Ca^{2+} inhibitory activity in sera from dilated cardiomyopathy with or without anti-β-receptor antibodies as reflected in the ligand binding inhibition assay (LBIA).

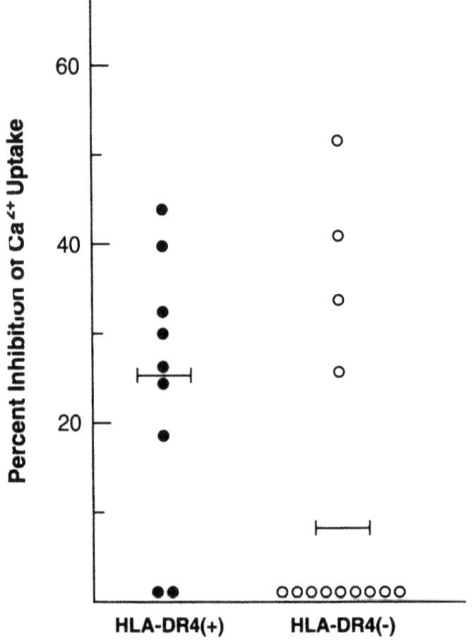

Fig. 7. Comparison of the prevalence of anti-sarcoplasmic reticulum antibodies in HLA-DR4-positive and - negative patients with dilated cardiomyopathy.

immunological disturbances directed against several autoantigens. That this may indeed be the case as suggested by the considerably higher prevalence of anti-β-receptor antibodies (Fig. 6) in patients whose sera inhibited Ca^{2+} transport (50% compared to 20% of those without inhibitory activity, p < 0.01). Additional support comes from a consideration of the HLA-DR antigen distribution. We have previously shown (12) that anti-β-receptor antibodies are restricted to specific HLA-DR phenotypes, notably-DR4/DR1. As Fig. 7 indicates, there is a significant preponderance of Ca^{2+} uptake inhibitory sera in patients with the HLA-DR4 phenotype compared to the HLA-DR4 negative patients. This may be interpreted as evidence that this particular phenotype identifies individuals with dilated cardiomyopathy who are prone to develop autoantibodies.

The antigenic epitopes on the sarcoplasmic reticulum responsible for the observed inhibition of Ca^{2+} transport by cardiomyopathic sera remain to be identified. It is reasonable to assume that the Ca^{2+} transport ATPase itself may be the target of immunological modification and, indeed, preliminary immunoblotting experiments in our patients suggest that this is the case. The absence of reactivity with skeletal muscle sarcoplasmic reticulum is not surprising considering the immunological (7) and structural (1, 8) differeces between cardiac and skeletal Ca^{2+} -pump ATPases. However, experience with monoclonal antibodies raised against ATPase tryptic fragments (16) indicates that most have little functional impact on calcium transport. Those with inhibitory activity appear to do so by preventing ATPase-ATPase interactions that are crucial to calcium translocation. Further studies are in progress to elucidate this issue.

References

1. Brandl CJ, Green NM, Korczak B, MacLennan DH (1986) Two Ca^{2+}-ATPase genes: Homologies and mechanistic implications of deduced amino acid sequences. Cell 44:459–607
2. Caforio ALP, Bonifacio E, Stewart JT, Neglia D, Parodi O, Bottazzo GF, McKenna WJ (1990) Novel organ-specific circulating cardiac autoantibodies in dilated cardiomyopathy. J Am Coll Cardiol; 15:1527–1534
3. Grossman W, McLaurin LP, Rolett EL (1979) Alterations in left ventricular relaxation and diastolic compliance in congestive cardiomyopathy. Cardiovasc Res 13:514–522
4. Gwathmey JK, Copelas L, MacKinnon R, Schoen FJ, Feldman MD, Grossman W, Morgan JP (1987) Abnormal intracellular calcium handling in myocardium from patients with end-stage heart failure. Circ Res 61:70–76
5. Harigaya S, Schwartz A (1969) Rate of calcium binding and uptake in normal animal and failing human cardiac muscle; membrane vesicles (relaxing system) and mitochondria. Circ Res 25:781–794
6. Holmberg SRM, Williams AJ (1989) Single channel recordings from human cardiac sarcoplasmic reticulum. Circ Res 65:1445–1449
7. Jorgensen AO, Arnold W, Pepper DR, Kahl SD, Mandel F, Campbell KP (1988) A monoclonal antibody to the Ca^{2+}-ATPase of cardiac sarcoplasmic reticulum cross-reacts with slow type I but not with fast type II canine skeletal muscle fibers: an immunocyto-chemical and immunochemical study. Cell Motil Cytoskel 9:164–174
8. Komuro I, Kurabayashi M, Shibazaki Y, Takaku F, Yazaki Y (1989) Molecular cloning and characterization of a $Ca^{2+} + Mg^{2+}$-dependent adenosine triphosphatase from rat cardiac sarcoplasmic reticulum. Regulation of its expression by pressure overload and developmental stage. J Clin Invest 83:1102–1108

9. Limas CJ (1978) Calcium transport ATPase of cardiac sarcoplasmic reticulum in experimental hyperthyroidism. Am J Physiol 235:H745–H751
10. Limas CJ, Olivari MT, Goldenberg IF, Levine TB, Benditt DG, Simon A (1987) Calcium uptake by cardiac sarcoplasmic reticulum in human dilated cardiomyopathy. Cardiovasc Res 21:601–605
11. Limas CJ, Goldenberg IF, Limas C (1989) Autoantibodies against β-adrenoceptors in human idiopathic dilated cardiomyopathy. Circ Res 64:97–103
12. Limas CJ, Limas C, Kubo SH, Olivari MT (1990) Anti-beta-receptor antibodies in human dilated cardiomyopathy and correlation with the HLA-DR antigens. Am J Cardiol 65:483–487.
13. Limas CJ, Goldenberg IF, Limas C (1990) Influence of anti-beta-receptor antibodies on cardiac adenylate cyclase in patients with idiopathic dilated cardiomyopathy. Am Heart J 119:1322–1328
14. Mercadier JJ, Lompre AM, Duc P, Boheler KR, Fraysse JB, Wisnewsky C, Allen PD, Komajda M, Schwartz K (1990) Altered Sarcoplasmic reticulum Ca^{2+} ATPase gene expression in the human ventricle during end-stage heart failure. J Clin Invest 85:305–309
15. Miyakoda G, Yoshida A, Takisawa H, Nakamure T (1987) β-Adrenergic regulation of contractility and protein phosphorylation in spontaneously beating isolated rat myocaridal cells. J Biochem (Tokyo) 102:211–224
16. Molnar E, Seidler NW, John I, Martonosi AN (1990) The binding of monoclonal and polyclonal antibodies to the Ca^{2+}-ATPase if sarcoplasmic reticulum: effects on interactions between ATPase molecules. Biochem Biophys Act 1023:147–167
17. Movsesian MA, Bristow MR, Krall J (1989) Ca^{2+} uptake by cardiac sarcoplasmic reticulum from patients with idiopathic dilated cardiomyopathy. Circ Res 65:1141–1144.
18. Movsesian MA, Colyer JJ, Wang JH, Krall J (1990) Phospholamban-mediated stimulation of Ca^{2+} uptake in sarcoplasmic reticulum from normal and failing hearts. J Clin Invest 85:1698–1702
19. Neumann DA, Burek CL, Baughman KL, Rose NR, Herskowitz A (1990) Circulating heart-reactive antibodies in patients with myocarditis or cardimyopathy. J Am Coll Cardiol 16:839–846
20. Ogawa Y (1968) The apparent binding constant of glycoletherdiaminetetraacetic acid for calcium at neutral pH. J Biochem (Tokyo) 64:255–257
21. Schulze K, Becker BF, Schauer R, Schultheiss HP (1990) Antibodies to ADP-ATP carrier-an autoantigen in myocarditis and dilated cardiomyopathy-impair cardiac function. Circulation 81:959–969

Author's address:
C. J. Limas, M. D.
Cardiovascular Division
Department of Medicine
University of Minnesota Medical School
Box 19, UMHC
Minneapolis, Minnesota 55455, USA

Calcium uptake by sarcoplasmic reticulum and its modulation by cAMP-dependent phosphorylation in normal and failing human myocardium.

M. A. Movsesian

Cardiology Division, University of Utah Medical Center, Salt Lake City, Utah, USA

Summary: ATP-dependent, oxalate-supported Ca^{2+} uptake by cardiac sarcoplasmic reticulum was examined in microsomes prepared from left-ventricular free wall myocardium obtained from the explanted failing hearts of transplant recipients with idiopathic dilated cardiomyopathy and the non-failing hearts of kidney donors for whose hearts no suitable recipients were available. There were no significant differences between the two groups with respect to values for V_{max}, $K_{0.5}$ (for Ca^{2+}) or n_{Hill} of basal Ca^{2+} uptake. The stimulation of Ca^{2+} uptake associated with cAMP-dependent phosphorylation of phospholamban could be reproduced by incubation of microsomes with a monoclonal antibody to phospholamban. Stimulation resulted from a decrease in $K_{0.5}$, with no changes in V_{max} or n_{Hill}. The magnitude of stimulation of Ca^{2+} uptake following incubation with anti-phospholamban monoclonal antibody was identical in preparations from normal and failing hearts. Finally, sarcoplasmic reticulum-associated cGMP-inhibited cAMP phosphodiesterase activity in these preparations was characterised. Measurement of steady-state kinetics and pharmacologic sensitivity indicated that this activity was functionally homogeneous. Preparations from failing and non-failing hearts did not differ with respect to either values for V_{max} and K_m or susceptibility to inhibition by the cilostamide derivative OPC 3911. These observations indicate that abnormalities in the regulation of intracellular $[Ca^{2+}]$ in failing human myocardium cannot be ascribed to changes in the level or function of the Ca^{2+}-transporting ATPase, phospholamban or cGMP-inhibited cAMP phosphodiesterase in the sarcoplasmic reticulum.

Key words: Heart failure; sarcoplasmic reticulum; Ca^{2+} uptake; Ca^{2+} ATPase; phospholamban; phosphodiesterase; phosphodiesterase inhibitors; cAMP; cGMP; cilostamide; rolipram

Abnormalities in the regulation of intracellular $[Ca^{2+}]$ have been reported in muscle strips isolated from failing human myocardium (6). It was suggested that these alterations might have resulted from impaired uptake of Ca^{2+} by the sarcoplasmic reticulum. Studies demonstrating diminished rates of ATP-dependent, oxalate-supported Ca^{2+} uptake in homogenates of right-ventricular endomyocardial biopsy specimens raised the possibility that changes in the level or function of the Ca^{2+}-transporting ATPase of the sarcoplasmic reticulum may have been involved (7).

To pursue this issue, ATP-dependent, oxalate-dependent Ca^{2+} uptake was examined in microsomes prepared by homogenisation and differential sedimentation of left-ventricular free wall myocardium obtained from the explanted failing hearts of transplant recipients with idiopathic dilated cardiomyopathy (age 49.1 $\pm$ 5.4 years) and from the non-failing hearts of kidney donors of whose hearts no suitable recipients were available (age 40.3 $\pm$ 6.7 years) (8). The recovery of protein in the microsomes was 0.38 $\pm$ 0.05 mg per g wet weight of tissue in preparations from failing hearts and 0.42 ± 0.07 mg per g in preparations from non-failing hearts. The phospholipid:protein ratio was 33.0 ± 3.5 nmol/mg in preparations from failing hearts and 36.9 ± 4.5 nmol/mg in preparations from non-failing hearts. These observations indicated that our preparative methods resulted in the isolation of comparable microsomal fractions from the two groups. In contrast, the β-adrenergic receptor density in preparations from non-failing hearts was 99.8 ± 14.4 fmol/mg, twice the value of 49.8 ± 5.0 fmol/mg obtained in preparations from failing hearts. This confirmed that the specimens chosen for the preparations were in fact representative of normal and failing human myocardium.

ATP-dependent, oxalate-supported Ca^{2+} uptake in these preparations was studied as a function of extravesicular $[Ca^{2+}]$ at saturating concentrations of ATP. A double-reciprocal plot of the data from a representative preparation is shown in Figure 1. The non-linearity of the relation between $1/[Ca^{2+}]$ and $1/Ca^{2+}$ uptake is consistent with the presence of two cooperatively interacting Ca^{2+} binding sites. The data were fitted by least-squares regression to second order polynomials, and the reciprocals of values for $1/v$ at $1/[Ca^{2+}] = 0$ were taken as values for V_{max}. The data were then re-plotted in Hill form and again fitted by least-squares regression to second order polynomials. Values of $[Ca^{2+}]$ at which $\log v/(V_{max} - v) = 0$ were taken as values for $K_{0.5}$; values for the first derivatives of the polynomials at this point were taken as values for n_{Hill}.

Values for V_{max}, $K_{0.5}$ and n_{Hill} in the preparations from normal and failing hearts are shown in Table 1. There were no significant differences between the two groups with respect to values for these parameters. This lack of difference, together with the comparable recovery of microsomal protein in the two populations, makes it difficult

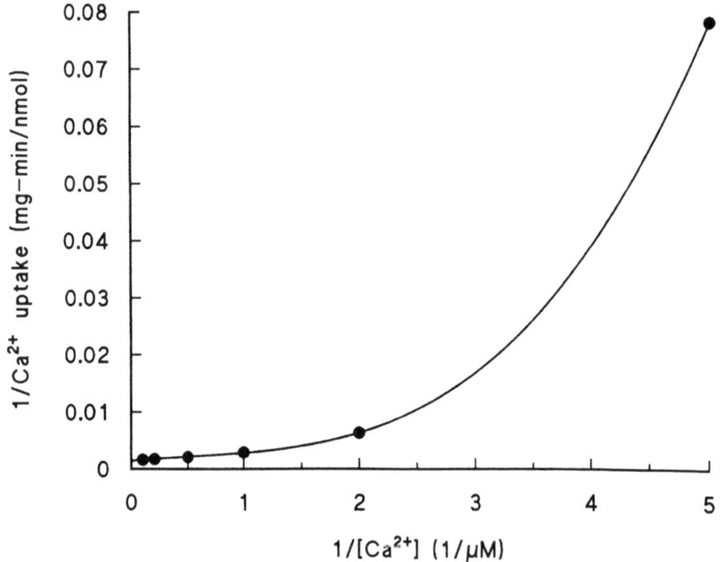

Fig. 1. ATP-dependent, oxalate-supported Ca^{2+} uptake by human cardiac sarcoplasmic reticulum. Measurements were made in a preparation from non-failing left ventricular free wall myocardium. Data (mean values) are plotted in double reciprocal form.

Table 1. Values for steady-state kinetic parameters of ATP-dependent, oxalate-supported Ca^{2+} accumulation by myocardial sarcoplasmic reticulum from normal and failing human hearts.

	V_{max}: (nmol/mg-min)	$K_{0.5}$ (Ca^{2+}): (μM)	n_{Hill}:
Normal: (n = 6)	593 ± 82	0.68 ± 0.07	1.7 ± 0.1
Failing: (n = 9)	593 ± 36	0.63 ± 0.03	1.6 ± 0.1

Values (mean ± standard error) were calculated as described in the text. (Data from (8), reprinted with permission.)

to ascribe the abnormalities in intercellular $[Ca^{2+}]$ transients reported in myocardium from failing human hearts to changes in the level or steady-state kinetic properties of the Ca^{2+}-transporting ATPase of the sarcoplasmic reticulum.

Attention was turned to the regulation of the activity of the Ca^{2+}-transporting ATPase through its interaction with phospholamban, whose phosphorylation by cAMP-dependent protein kinase is accompanied by a stimulation of Ca^{2+} uptake rates in non-human myocardium (for review, see (14)). This stimulation (actually, a "de-inhibition") of Ca^{2+} uptake can be reproduced by incubation of sarcoplasmic reticulum with a monoclonal antibody that binds to an epitope in the cytoplasmic domain of phospholamban containing the phosphorylation site and prevents its interaction with the Ca^{2+}-transporting ATPase (13). We found that this monoclonal antibody cross-reacted with human cardiac phospholamban (10), and that incubation of human cardiac sarcoplasmic reticulum with this antibody stimulated Ca^{2+} uptake (Fig. 2) (9). Stimulation resulted from a reduction in $K_{0.5}$ (in the preparation shown, from 0.53 μM to 0.29 μM), with no changes in V_{max} (665 v. 662 nmol/mg-min) or n_{Hill} (1.7 v. 1.8). It is noteworthy that magnitude of stimulation of Ca^{2+} uptake was small at 1.0 μM Ca^{2+}, a value typical of the peak Ca^{2+} concentrations measured in beating cultured cardiac myocytes, but was quite marked at 0.2 μM Ca^{2+}, a value more typical of trough concentrations (12). This suggests that in human myocardium phosphorylation of phospholamban should have a greater effect on the final free cytoplasmic Ca^{2+} concentrations in diastole than on its initial rapid decline from its peak level in systole.

The ability of anti-phospholamban monoclonal antibody to stimulate ATP-dependent, oxalate-supported Ca^{2+} uptake gave us a simple assay sensitive to changes in both the level of phospholamban relative to the Ca^{2+}-transporting ATPase and the functional coupling of the two proteins. Ca^{2+} uptake was therefore measured in microsomes from failing and non-failing hearts at 0.2 μM Ca^{2+} following incubation in the absence and presence of anti-phospholamban monoclonal antibody (9). As shown in Table 2, no significant difference between the two groups was observed. This suggests that heart failure in humans does not entail alterations in the level of phospholamban in cardiac sarcoplasmic reticulum or in its interaction with the Ca^{2+} transporting ATPase that could account for the reported changes in intracellular $[Ca^{2+}]$ transients.

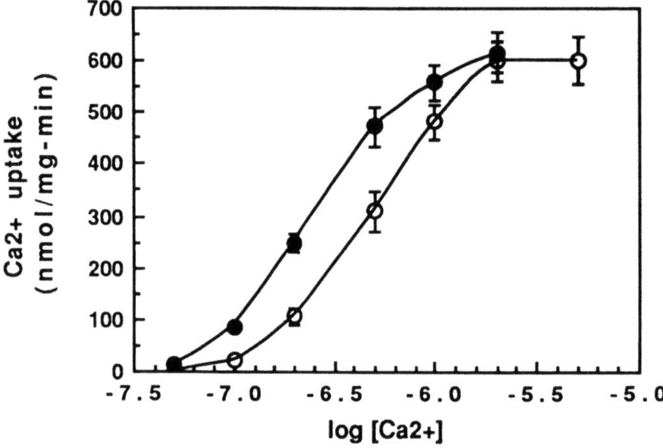

Fig. 2. Phospholamban-mediated stimulation of ATP-dependent, oxalate-supported Ca^{2+} uptake by human myocardial sarcoplasmic reticulum. Ca^{2+} uptake was assayed following incubation of microsomes in the absence (open circles) and presence (filled circles) of anti-phospholamban antibody (mean $\pm$ standard error). Measurements were made in a single preparation from tissue combined from three explanted failing hearts. (Reprinted with permission from (9).)

Table 2. Effect of anti-phospholamban monoclonal antibody on ATP-dependent, oxalate-supported Ca^{2+} accumulation by myocardial sarcoplasmic reticulum from normal and failing human hearts.

	Ca^{2+} accumulation (nmol/mg-min):		Stimulation:
	$-$ Antibody	$+$ Antibody	
Normal: (n = 7)	107.1 $\pm$ 15.5	206.2 $\pm$ 22.1	1.93-fold
Failing: (n = 14)	126.7 $\pm$ 14.0	246.2 $\pm$ 32.3	1.94-fold

Values (mean $\pm$ standard error) were calculated as described in the text. (data from (9), reprinted with permission.)

It remained possible that abnormal intracellular $[Ca^{2+}]$ transients in failing human myocardium were the result of diminished Ca^{2+} uptake by the sarcoplasmic reticulum caused by a reduction in phosphorylation of phospholamban by cAMP-dependent protein kinase. Such changes would be consistent with alterations in β-adrenergic receptor-mediated signal transduction and the consequent impairment of cAMP production (and, hence, cAMP-dependent phosphorylation) in heart failure (2-4). Phosphorylation of phospholamban by cAMP-dependent protein kinase could also be affected, however, by alterations in the activity of cyclic nucleotide phosphodiesterases that hydrolyse cAMP. In animals, a member of the high-affinity cGMP-inhibited cAMP phosphodiesterase family, cGI PDE (often referred to as PDE III) has been found to be associated with the sarcoplasmic reticulum, with the extent of association varying from species to species. The inotropic potency of cGI PDE inhibitors correlates linearly with the potency of their inhibition of sarcoplasmic reticulum-associated cGI PDE activity (16); their inotropic efficacy in different species appears to be a function of the amount of cGI PDE associated with the sarcoplasmic reticulum in each species (15). The correlation of inotropy with inhibition of sarcoplasmic reticulum-associated cGI PDE activity suggests that the enzyme's principal function might be the regulation of phospholamban phosphorylation by cAMP-dependent protein kinase. Moreover, we and others had observed a reduced inotropic response to specific inhibitors of cGI PDE in failing human myocardium (1, 4, 5). This reduced response might have resulted from a diminution in the level or activity of the enzyme in the sarcoplasmic reticulum of myocardium from failing human hearts or in its sensitivity to pharmacologic inhibition.

To test this, cAMP phosphodiesterase activity was studied in the microsomes from human left-ventricular myocardium (11). The steady-state kinetics of cAMP hydrolytic activity showed high affinity for cAMP (approximately 0.15 μM in the example shown), with no evidence for functional heterogeneity (Fig. 3). Identification of this activity as cGI PDE was confirmed by the potent inhibition by cGMP and the water-soluble cilostamide derivative OPC 3911, contrasting with the inhibition by rolipram (a selective inhibitor of cGMP-insensitive high affinity cAMP phosphodiesterases) only at high concentrations of the latter (Fig. 4). Inhibition by cGMP (Fig. 5) and OPC 3911 (data not shown) were competitive with respect to cAMP.

Values for V_{max} and K_m were determined in preparations from normal and failing human left ventricles. As shown in Table 3, these values did not differ in the two groups, even after normalizing values for V_{max} to values for Ca^{2+} uptake rates in each

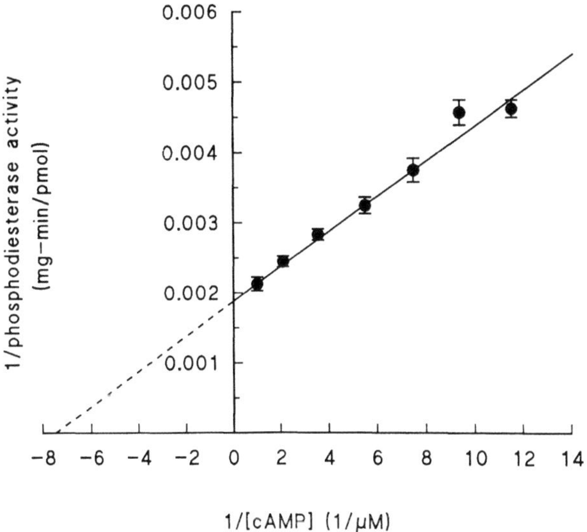

Fig. 3. cAMP phosphodiesterase activity in human cardiac sarcoplasmic reticulum. The microsomal preparation was the same as that used in Fig. 2. Data (mean ± standard error) are plotted in double reciprocal form. (Reprinted with permission from (11).)

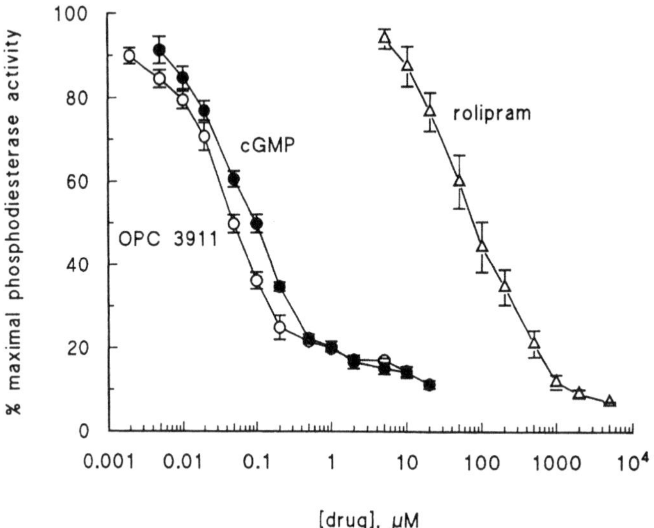

Fig. 4. Inhibition of cAMP phosphodiesterase activity by OPC 3911 (open circles), cGMP (filled circles) and rolipram (open triangles). Measurements were made at 0.2 µM cAMP. The microsomal preparation was the same as that used in Fig. 2. Each point represents the mean $\pm$ standard error. (Reprinted with permission from (11).)

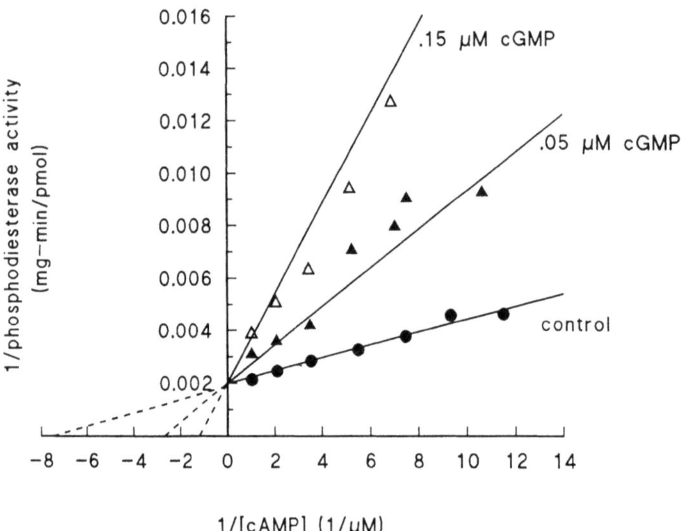

Fig. 5. Inhibition of cAMP phosphodiesterase activity by cGMP. cAMP phosphodiesterase activity was measured in the absence (filled circles) and presence of 0.05 µM (filled triangles) and 0.15 µM (open triangles) cGMP (mean values). The microsomal preparation was the same as that used in Fig. 2. (Reprinted with permission from (11).)

Table 3. Steady-state kinetic parameters of sarcoplasmic reticulum-associated cAMP phosphodiesterase activity in microsomes prepared from normal and failing hearts.

	cAMP phosphodiesterase activity		Ratio of phosphodiesterase activity/	% Inhibition by 0.05 μM OPC 3911
	V_{max} (pmol/mg-min):	K_m (μM):	Ca^{2+} accumulation ($\times 10^3$):	at 0.2 μM cAMP:
Normal (n = 6)	781.7 ± 109.2	0.188 ± 0.031	1.75 ± 0.30	56.6 ± 3.0
Failing (n = 8)	793.9 ± 68.9	0.150 ± 0.027	1.55 ± 0.07	57.4 ± 1.4

Values (mean ± standard error) in the third column were obtained by dividing V_{max} for cAMP phosphodiesterase activity by the rate of ATP-dependent, oxalate-supported Ca^{2+} accumulation (measured at 5.0 μM Ca^{2+}) in each preparation. (Data from (11), reprinted with permission.)

preparation. In addition, no difference between the two groups was observed with respect to inhibition of sarcoplasmic reticulum-associated cAMP phosphodiesterase activity by OPC 3911. On the basis of these observations, the diminished inotropic efficacy of cGI PDE inhibitors in failing myocardium cannot be ascribed to changes in the level, kinetic behavior or pharmacologic sensitivity of sarcoplasmic reticulum-associated cGI PDE.

In summary, our studies of the function of sarcoplasmic reticulum proteins involved in Ca^{2+} uptake and its regulation by cAMP-dependent phosphorylation in preparations from normal and failing human left ventricles have revealed no alterations related to the Ca^{2+}-transporting ATPase, phospholamban or cGMP-inhibited cAMP phosphodiesterase activity that can explain the reported abnormalities in intracellular $[Ca^{2+}]$ transients or the diminished response to phosphodiesterase inhibitors in dilated cardiomyopathy.

References

1. Böhm M, Diet F, Feiler G, Kemkes B, Kreuzer E, Weinhold C, Erdmann E (1988) Subsensitivity of the failing human heart to isoprenaline and milrinone is related to β-adrenoceptor downregulation. J Cardiovasc Pharm 12:726–732
2. Bristow MR, Hershberger RE, Port JD, Minobe W, Rasmussen R (1989) $β_1$- and $β_2$-adrenergic receptor-mediated adenylate cyclase stimulation in nonfailing and failing human ventricular myocardium. Mol Pharm 35:295–303
3. Feldman AM, Cates AE, Veazey WB, Hershberger RE, Bristow MR, Baughman KL, Baumgartner WA, Van Dop C (1988) Increase of the 40,000-mol wt pertussis toxin substrate in the failing human heart. J Clin Invest 82:189–197
4. Feldman MD, Copelas L, Gwathmey JK, Phillips P, Warren SE, Schoen FJ, Grossman W, Morgan JP (1987) Deficient production of cyclic AMP: pharmacologic evidence of an important cause of contractile dysfunction in patients with end-stage heart failure. Circulation 75:331–339

5. Gilbert EM, Hershberger RE, Mealey PC, Volkman K, Weichman RJ, Menlove RL, Movsesian MA, Bristow MR. Pharmacologic and hemodynamic effects of combined β-agonist stimulation and phospodiesterase inhibition in the failing human heart (submitted for publication)
6. Gwathmey JK, Copelas L, MacKinnon R, Schoen FJ, Feldman MD, Grossman W, Morgan JP (1987) Abnormal intracellular calcium handling in myocardium from patients with end-stage heart failure. Circ Res 61:70–76
7. Limas CJ, Olivari MT, Goldenberg IF, Levine TB, Benditt DG, Simon A (1987) Calcium uptake by cardiac sarcoplasmic reticulum in human dilated cardiomyopathy. Cardiovsac Res 21:601–605
8. Movsesian MA, Bristow MR, Krall J (1989) Calcium uptake by cardiac sarcoplasmic reticulum from patients with idiopathic dilated cardiomyopathy. Circ Res 65:1141–1144
9. Movsesian MA, Colyer J, Wang JH, Krall J (1990) Phospholamban-mediated stimulation of Ca^{2+} uptake in cardiac sarcoplasmic reticulum from normal and failing hearts. J Clin Invest 85:1698–1792
10. Movsesian MA, Leveille C, Krall J, Colyer J, Wang JH, Campbell KP (1990) Identification and characterization of proteins in human cardiac sarcoplasmic reticulum. J Mol Cell Cardiol 22:1477–1485
11. Movsesian MA, Smith CJ, Krall J, Bristow M, Manganiello VC (1991) Sarcoplasmic reticulum-associated cAMP phosphodiesterase activity in normal and failing human hearts. J Clin Invest 88:15–19
12. Peeters GA, Hlady V, Bridge JHB, Barry WH (1987) Simultaneous measurement of calcium transients and motion in cultured ventricular cells. Am J Physiol 253:H1400–H1408
13. Suzuki T, Wang JH (1986) Stimulation of bovine cardiac sarcoplasmic reticulum Ca^{2+} pump and blocking of phospholamban phosphorylation and dephosphorylation by a phospholamban monoclonal antibody. J Biol Chem 261:7018–7023
14. Tada M, Kadoma M, Inui M, Fujii J-I (1988) Regulation of Ca^{2+}-pump from cardiac sarcoplasmic reticulum. Methods Enzymol 157:107–154
15. Weishaar RE, Kobylarz-Singer DC, Quade MM, Steffen RP, Kaplan HR (1987) Multiple molecular forms of phosphodiesterase and the regulation of cardiac muscle contractility. J Cyclic Nucleotide Protein Phosphorylation Res 11:513–527
16. Weishaar RE, Kobylarz-Singer DC, Steffen RP, Kaplan HR (1987) Subclasses of cyclic AMP-specific phosphodiesterase in left ventricular muscle and their involvement in regulating myocardial contractility. Circ Res 61:539–547

Author's address:
M. A. Movsesian
Cardiology Division
University of Utah Medical Center
50 North Medical Drive
Salt Lake City, UT 84132
USA

Contractile proteins and sarcoplasmic reticulum calcium-ATPase gene expression in the hypertrophied and failing heart

K. Schwartz, L. Carrier, A.-M. Lompré[1], J.-J. Mercadier, K.R. Boheler

I.N.S.E.R.M. Unité 127, Hôpital Lariboisière, Paris, France
[1]I.N.S.E.R.M. Unité 275, Ecole Polytechnique, Palaiseau, France

Summary: The physiology of myocardial contractility has been studied for over a century, but only recently has molecular biology provided new insights into the mechanisms responsible for the alterations of contraction and relaxation observed during cardiac hypertrophy and heart failure. Pressure and volume overload produce in the myocyte both qualitative changes characterized by protein isoform switches and quantitative changes characterized by modulation of single genes through a mechanogenic transduction the pathways of which are largely unknown. The qualitative changes involve differential expression of multigene families of contractile proteins, especially myosin heavy chain (MHC) and actin. All situations of pressure overload, or of combined pressure and volume overload activate the β-MHC gene and deactivate the α-MHC one, which leads to a slower, more efficient contraction. In rat, pressure overload transitorily activates the α-skeletal actin gene, and both the timing and the distribution of the newly formed β-MHC and α-skeletal actin mRNAs differ. We recently found that the isoactin pattern is the same in patients with end-stage heart failure as that of control human hearts. Moreover, both in rat and human, expression of isomyosins and isoactins are not coordinated, neither during ontogeny nor senescence. All this suggests the existence of several regulatory mechanisms activated during normal cardiac growth or by a mechanical trigger, and preliminary results indicate that it is possible to perform nuclear run-on assays in order to analyze the transcriptional step of these isogenes. Relaxation of the hypertrophied heart is characterized by a relative decrease in the activity of the gene coding for the sarco(endo)plasmic reticulum ATPase (SERCA), both in rat and in man, which can explain some of the alterations of calcium handling by the hypertrophied myocyte. In conclusion, reprogramming of cardiac gene expression during ontogeny and chronic hemodynamic overloading is a complex phenomenon, and it could be hypothesized that further exploration of these genetic events may enable us to better understand how cardiac function is regulated, both in health and in disease.

Key words: Rat heart; human heart; ontogeny; senescence; hemodynamic overload; isomyosins; isoactins; uncoordinated regulation

Introduction

Contractility and function of the hypertrophied heart have been extensively studied for over a century; however, the idea that molecular changes in the proteins synthesized by the myocardium could play a major role in cardiac adaptation to chronic overload was first recognized by the concerned scientific community during a meeting held in Boston in 1987 (17). It is now acknowledged that pressure and volume overload produce in the myocyte qualitative changes, phenotypic conversions characterized by protein isoform switches, and quantitative changes characterized by modulation of individual gene expression; and to describe these phenomena, we proposed the use of the term "mechanogenic transduction" (22). These changes are not genetic defects analogous to the myosin heavy-chain mutations recently demonstrated in hypertrophic cardiomyopathy (7, 23). In what was recently termed by Katz as "cardiomyopathy of overload" (10); the genes are normal, but the way that they are regulated is abnormal. In this context, and over the past several years, our aim has been to try and understand the mechanisms of these dysregulations, while attempting to delineate what changes within the myocyte contribute to the mechanogenic transduction and why they become ineffective during the transition from hypertrophy to failure.

This manuscript will therefore focus on the molecular mechanisms that can explain, at least in part, the alterations of contraction and relaxation that occur during hypertrophy, while also updating the findings in this field.

Contraction

In the heart as in skeletal muscles, the basic unit of contraction is the sarcomere which is composed of a diverse set of proteins working together to generate force and contraction. Two major components of the sarcomere are the thick and thin filaments. The thick filaments consist primarily of myosin composed of two heavy chains and four light chains, while the thin filaments are composed predominantly of actin, tropomyosin, and the troponin complex. The thick filament can contain two possible myosin heavy chain isoforms, α-MHC and β-MHC, and the thin filament can contain two actin isoforms, α-skeletal and α-cardiac. These isogenes are expressed differently with ontogeny, aging, and hypertrophy, and this plays a role in the regulation of contraction (review in (12)). An effect of a permanent hemodynamic overload is activation of the β-MHC gene and a deinduction of the α-MHC one. This change occurs in all tissues and species tested so-far. However, the potential for an increase in β-MHC depends upon the initial phenotype: it is high in rat ventricles and in human atria that normally contain around 90% α-MHC, and it is small in human ventricles, which contain mainly β-MHC. β-MHC is predominant in rat fetal ventricles, from which developed the concept of reactivation of a fetal program with hemodynamic overloading. The change from α-MHC to β-MHC (or isomyosin V1 to V3) results in a slower rate of ATP cycling by myosin, which fully accounts for the slower velocity of contraction of the hypertrophied fiber. The result is an improved economy of force development that is usually considered as adaptative.

Less is known concerning the isoactins. The two isogenes, α-skeletal and α-cardiac are coexpressed at birth in the rodent ventricle and in adult human, pig, and cow hearts (14, 16, 24). In adult rat, the α-cardiac actin (α-card act) isoform is almost exclusively present and, in collaboration with M. Buckingham, we showed that α-skeletal actin (α-skel act) mRNA accumulates with the onset of pressure-overload-induced hypertrophy (21). This was the second example of the reactivation of a fetal program by hemodynamic overload. The time-course and the cellular distribution of the two newly formed transcripts, β-MHC and α-skel act mRNAs, are different (20).

Given these results, we have tried to address the question of whether or not there is a common regulatory pathway for myosin and actin multigene families. Since myosin heavy-chain isogene expressions have been the focus of numerous studies, our recent experiments were designed to determine the pattern of expression for the two sarcomeric actins, α-skeletal and α-cardiac, in different conditions of cardiac growth, in rat and human hearts.

Rat ventricle

The respective levels of skeletal and cardiac actin mRNA were precisely determined by a primer extension assay. This technique that takes advantage of the 9 bp difference in length between α-skel act and α-card act mRNAs allows an unambiguous and precise quantification of the two mRNA isoforms (2). From results published by Yaffe (14) and Buckingham (16), it was known that α-skel act is present in neonates, and Bishopric et al. (2) estimated that it amounted to 50% of the total in one rat heart. By primer extension, we observed that neonatal rat ventricles contain 30–40% α-skel act mRNA (Carrier et al., submitted). This relatively high level was maintained much longer than previously thought, decreasing slowly only after 3 weeks of age, and only at 2 months of age were the transcripts for cardiac actin almost uniquely present. This figure does not change thereafter, even in 2-year-old rats. We did not find any difference between male or female rats, suggesting that expression of sarcomeric actin in the heart is not regulated by sex-related hormones.

Comparison of the time-course of α-skel act and β-MHC mRNA accumulations shows an apparent dissociation. Both β-MHC and α-skel act mRNA accumulate to high levels at birth, but at 3 weeks, while α-skel act is still high, β-MHC has almost completely disappeared. β-MHC mRNAs slowly reaccumulate in adults and with senescence, and attain levels of as much as 80% of total sarcomeric MHC in 2-year-old animals (19). In contrast, α-skel act is not reinduced during senescence (this study). Thus, in the rat, during both normal cardiac growth and with hypertrophy, expression of the multigene families encoding two of the main contractile proteins is not coordinated.

Human ventricle

What occurs in the human heart? Gunning and Kedes reported 5 years ago (in a patient suffering from hypertrophic obstructive cardiomyopathy) that the amounts of α-skel act and α-card act were nearly identical (8); whereas, Bennetts et al. (1) reported in a control patient that α-skel act mRNA represents approximately 20% of total sarcomeric actin. At the protein level, Vanderckerckove et al. (24) found α-skel act accumulations of 20% in both a control and a failing heart. It was thus unclear whether or not cardiovascular diseases in man are accompanied by isoactin switches, and it was completely unknown whether actin gene switching occurs with development. We have recently completed an analysis of the isoactin pattern of a series of normal and pathologic human hearts by two different types of assays, primer extension and dot blots with cloned human sequences specific for each actin isomRNA (3). The primer extension assay was based on the same principle as the one used for the rat, but with an oligonucleotide common to the two human sequences. In the fetal human heart, α-skel act mRNA is a minor component, it increases significantly after birth and there is a dramatic reversal of the isoactin pattern after the first decade. Skeletal actin becomes the major isoform of adults, representing approximately 60% of the total. Cardiac tissue samples obtained from patients undergoing cardiac transplantation due to advanced stages of heart failure were also analyzed. The percentage of α-skel act mRNA of all pathologic hearts was the same as that found in control left ventricles, and we saw no differences between patients with dilated cardiomyopathy, ischemic heart disease or any other cardiac abnormality. Thus, α-actin mRNA accumulations in man do not change from the normal state with end-stage heart failure, although we are unable to say if human hearts synthesize an even higher percentage of skeletal actin isomRNA in earlier stages of heart disease. The molecular basis of a potential thin filament dysregulation in human end-stage heart failure is, therefore, not actin.

Thus, comparisons of the expression of MHC and actin multigene families in man show, as in rat, a complete dissociation. Skeletal actin is upregulated during development, whereas β-MHC more or less roughly remains the same, around 80–90%. Moreover, the sarcomeric actins are regulated in man in a manner apparently opposed to that seen in the rodent heart.

Regulational level of isoactins and isomyosins

The expression of both MHC and actin multigene families is species-specific; the next question is therefore to determine at which level each of these families is regulated: transcriptionally, post-transcriptionally or translationally. With the team of Mahdavi, we showed several years ago during rat development and after a pressure overload that the regulation of β-MHC expression is pretranslational (9). The next step was to analyze the transcription of each isogene by run-on assays, but due to the inherent difficulties of working with intact cardiac tissue from adult as well as from neonatal rats, such studies had not been conducted. We have recently isolated myocyte nuclei from neonatal to age 24-day rat hearts (5) and performed a series of in vitro run-on assays (4). The principle of this technique is based upon the continued ability of isolated cardiac nuclei to transcribe nascent RNAs, but in the presence of radioactive nucleotides. The labeled nascent transcripts can be isolated and hybridized to specific cDNA probes from which the transcriptional activity of individual genes can be determined. Preliminary results indicate that neonatal nuclei transcribe both α- and β-MHC and α-card and α-skel act. The transcriptional activity of several of these genes is dramatically altered by age 24 days. It is thus now possible to explore transcription of individual isogenes and, of course, our immediate aim is to analyze transcriptional activity with ontogeny and during the early phases of cardiac hypertrophy.

Relaxation

Calcium uptake by the sarcoplasmic reticulum is the main mechanism responsible for cardiac relaxation. Transport of calcium ions takes place against a concentration gradient, owing to hydrolysis of ATP. This hydrolysis is mediated by a sarco(endo)plasmic reticulum ATPase (SERCA), whose activity in the heart is regulated by phospholamban phosphorylation. Like myosin and actin, SERCA is a multigene family (review in (12)). One of the isogenes, SERCA2, is alternatively spliced at its 3′end and one of the mRNAs, SERCA2a, is present in adult cardiac muscles. Most authors agree that hemodynamic overload is accompanied by reduced calcium transport by the sarcoplasmic reticulum, and we and others showed last year that, in parallel, there is a relative decrease in the amount of SERCA2a mRNA and no isoform switching (6, 11, 15, 18). Very recently, we studied expression of this gene during ontogeny and senescence (13). The amount of SERCA2a mRNA increases at the end of fetal life and in the early postnatal period; a stable high level is observed during adulthood, and it decreases substantially during aging. In the fetal-neonatal period, the increase in SERCA2a mRNA is parallel to the increase in myosin heavy chain mRNA, but during senescence the two mRNAs do not evolve in parallel, showing that during this late period of life the decrease is specific to SERCA2a and cannot be attributed to myocyte loss. All these results indicate that cardiac growth and hypertrophy do not require expression of specific isoforms of SERCA, but rather quantitative modulation of the expression of a single isoform.

Concluding remarks

These studies demonstrate that, whatever the type of cardiac growth and whatever the animal species, the molecular mechanisms that result in a given cardiac phenotype are not unique and, most probably, multiple factors act at different times and in different fibers. It is striking that both myosin and actin multigene families are expressed in a species-specific fashion and that, while contraction seems to be regulated mainly by isogene switches, relaxation in contrast depends upon the quantitative modulation of the same gene. It is now possible to explore the entire sequence of activation or deactivation of contractile proteins isogenes, which should permit the development of approaches to discover specific regulatory factors and facilitate our understanding of why they do not operate during heart failure.

*Acknowledgements.*This work was supported by I.N.S.E.R.M. and the Association Française contre les Myopathies. Drs. K. R. Boheler and L. Carrier are recipients of fellowships from l'Association Française contre les Myopathies.

References

1. Bennetts BH, Burnett L, dos Remedios CG (1986) Differential co-expression of α-actins genes within the human heart. J Mol Cell Cardiol 18:993–996
2. Bishopric NH, Simpson PC, Ordahl CP (1987) Induction of the skeletal α actin in α1 adrenoreceptor-mediated hypertrophy of rat cardiac myocytes. J Clin Invest 80:1194–1199
3. Boheler KR, Carrier L, de la Bastie D, Allen PD, Komajda M, Mercadier JJ, Schwartz K (1991) Skeletal actin mRNA increases in the human heart during ontogenic development and is the major isoform of control and failing human hearts. J Clin Invest 88:323–330
4. Boheler KR, Carrier L, Chassagne C, de la Bastie D, Mercadier JJ, Schwartz K (1991) Regulation of myosin heavy chain and actin isogenes expression during cardiac growth. Molec and Cell Biochem 104:101–107
5. Chassagne C, Boheler KR, Schwartz K (1991) Description of an *in vitro* transcription assay in nuclei isolated from control and hemodynamically overloaded rat cardiac myocytes. C R Acad Sci Paris, 312:7–12
6. de la Bastie D, Levitsy D, Rappaport L, Mercadier JJ, Marotte F, Wisnewsky C, Brovkovich V, Schwartz K, Lompre AM (1990) Function of the sarcoplasmic reticulum and expression of its calcium ATPase gene in pressure overload-induced cardiac hypertrophy in the rat. Circ Res 66:554–564
7. Geistefer-Lowrance AAT, Kass S, Tanigawa T, Vosberg HP, McKenna W, Seidman CE, Seidman JG (1990) A molecular basis for familial hypertrophic cardiomyopathy: a β cardiac myosin heavy chain gene missense mutation. Cell 62:999–1006
8. Gunning P, Ponte P, Blau H, Kedes L (1983) α-skeletal and α-cardiac actin genes are co-expressed in adult human skeletal muscle and heart. Mol Cell Biol 3:1985–1995

9. Izumo S, Lompre AM, Matsuoka R, Koren G, Schwartz K, Nadal-Ginard B, Mahdavi V (1987) Myosin heavy chain messenger RNA and protein isoform transitions during cardiac hypertrophy. J Clin Invest 79:970–977

10. Katz A (1990) Cardiomyopathy of overload. A major determinant of prognosis in congestive heart failure. N Engl J Med 322:100–110

11. Komuro I, Kurabayashi M, Shibazaki F, Takaku F, Yazaki Y (1989) Molecular cloning and characterization of a Ca/Mg-dependent adenosine triphosphatase from rat cardiac sarcoplasmic reticulum. J Clin Invest 83:1102–1108

12. Lompre AM, Mercadier JJ, Schwartz K (1990) Changes in gene expression during cardiac growth. Inter Rev Cytol 124:137–186

13. Lompre AM, Lambert F, Lakatta EG, Schwartz K (1991) Expression of sarcoplasmic reticulum Ca-ATPase and calsequestrin genes in rat heart during ontogenic development and aging Circ Res 69:1380–1388

14. Mayer Y, Czosneck H, Zeelon PE, Yaffe D, Nudel U (1984) Expression of the genes coding for the skeletal muscle and cardiac actins in the heart. Nucleic Acids Res 12:1087–1100

15. Mercadier JJ, Lompre AM, Duc P, Boheler KR, Fraysse JB, Wisnewsky C, Allen PD, Komajda M, Schwartz K (1990) Altered sarcoplasmic reticulum Ca-ATPase gene expression in the human ventricle during end-stage heart failure. J Clin Invest 85:305–309

16. Minty AJ, Alonso S, Caravatti M, Buckingham M (1982) A fetal skeletal actin mRNA in the mouse and its identity with cardiac actin mRNA. Cell 30:185–192

17. Nadal-Ginard B, Ingwall JS (Chairmen): Scientific Conference on "The molecular biology of the cardiovascular system". 8–12 September, Boston (1987)

18. Nagai R, Zarain-Herzberg A, Brandl CJ, Fujii M, Tada D, McLennan N, Alpert N, Periasamy M (1989) Regulation of myocardial Ca-ATPase and phospholamban mRNA expression in response to pressure overload and thyroid hormones. Proc Natl Acad Sci USA 86:2966–2970

19. O'Neill L, Holbrook N, Lakatta EG (1991) Progressive changes from young adult age to senescence in mRNA for rat cardiac myosin heavy chain genes. Cardiosciense 2:1–5

20. Schiaffino S, Samuel JL, Lompré AM, Garner I, Marotte F, Buckingham M, Rappaport L, Schwartz K (1989) Non synchronous accumulation of α-skeletal actin and β myosin heavy chains mRNAs during early stages of pressure overload-induced cardiac hypertrophy demonstrated by in situ hybridization. Circ Res 64:937–948

21. Schwartz K, de la Bastie D, Bouveret P, Oliviero P, Alonso S, Buckingham ME (1986) α-skeletal muscle actin mRNAs accumulate in hypertrophied adult rat hearts. Circ Res 59:551–555

22. Schwartz K (1990) Phenoconversion and mechanogenic transduction of the mammalian heart. Medecine/Science 6:664–673

23. Tanigawa G, Jarcho JA, Kass S, Solomon SD, Vosberg H-P, Seidman JG, Seidman CE (1990) A molecular basis for familial hypertrophic cardiomyopathy: an α/β cardiac myosin heavy chain hybrid gene. Cell 62:991–998

24. Vandekerckhove J, Bugaisky G, Buckingham M (1986) Simultaneous expression of skeletal muscle and heart actin proteins in various striated muscle tissues and cells. J Biol Chem 261:1838-1843

Author's address:
Ketty Schwartz
INSERM Unité 127
Hôpital Lariboisière
41 Blvd de la Chapelle
F-75010 Paris
France

Extracellular matrix

Factors associated with reactive and reparative fibrosis of the myocardium

K.T. Weber, C.G. Brilla

Division of Cardiology, University of Missouri-Columbia, USA

Summary: Myocardial fibrosis can be defined as an abnormal increase in collagen concentration of either ventricle. This accumulation of collagen, represented predominantly by fibrillar type I collagen, can occur a) on a reactive basis in the interstitial space and adventitia of intramyocardial coronary arteries and does not require myocyte necrosis, or b) as a replacement for necrotic myocytes, where it is considered a scar. Both forms can be found in the same ventricle.

Various factors have been found to contribute to the reactive and reparative fibrosis that appears in both ventricles in acquired hypertension. In the case of microscopic scarring, myocyte necrosis is related to catecholamine or angiotensin II- mediated toxicity, reduced potassium stores that accompany chronic mineralocorticoid excess, and coronary vascular remodeling. Reactive fibrosis is associated with elevations in plasma aldosterone concentrations that are inappropriate relative to dietary sodium intake.

These findings set the stage for additional in vivo and in vitro studies that may shed more light on our understanding of the factors that regulate the accumulation of fibrous tissue in the myocardium – a major determinant of pathologic structural remodeling which enhances its susceptibility to reentrant arrhythmias and ventricular dysfunction.

Key words: Myocardial fibrosis; collagen; connective tissue; aldosterone

Introduction

Heart failure is a well recognized major health problem of increasing proportions. It is also now recognized that one-third of patients with symptomatic heart failure have primary diastolic dysfunction of the left ventricle as the pathophysiologic basis of their symptomatology (11, 34). Observations from this (6, 7, 13, 20, 27) and other (3, 25, 50) laboratories have shown that a pathologic structural remodeling of the myocardium, mediated by a disproportionate accumulation of connective tissue, or fibrosis can account for diastolic dysfunction. The factors that regulate the fibrous tissue response in the myocardium are therefore of considerable interest, particularly since their identification could lead to the development of preventive and reparative therapeutic strategies (54). Herein we review our current understanding of the pathogenesis of myocardial fibrosis wherein we distinguish between reactive and replacement components of the fibrosis tissue response, particularly as they occur in acquired arterial hypertension.

This work was supported in part by NIH grant # R01-31701

Definition and components of myocardial fibrosis

Fibrosis Defined

A disproportionate accumulation of fibrous tissue in the myocardium, as well as other organs, is represented by an elevation in collagen *concentration*. For any given organ weight it naturally follows that when collagen concentration is increased, so too will be its collagen content (e.g., collagen concentration × ventricular weight). An elevation in collagen *content* alone, without a corresponding increase in collagen concentration, on the other hand, does not define fibrosis. This is exemplified by the hypertrophied ventricle seen with atrial septal defect, anemia, or arteriovenous fistula, where collagen content is increased while collagen concentration remains normal (55).

Collagen concentration can be measured by biochemical and morphometric methods. An assay of hydroxyproline, the amino acid specific for collagen, provides a reliable measure of collagen concentration. Refinements to this approach can be used to identify soluble and insoluble fractions of the total collagen pool (14). This simple assessment of collagen concentration, on the other hand, does not identify the morphologic presentation of fibrous tissue. As a result, it neglects important inferences regarding pathogenetic mechanisms. Morphometric techniques also identify collagen concentration, but their sensitivity will depend on both the histochemical approach that is used (e.g., collagen specific vs nonspecific stains) and the nature of the quantitative analysis. A point counting method, for example, depends on the density of the grid utilized to identify collagen fibers. Alternatively, videodensitometry measures the entire volume fraction of fibrillar collagen. In using videodensitometric analysis to distinguish each morphologic presentation of the fibrous tissue response, one obtains a comprehensive analysis of fibrosis (7, 9).

Components of Myocardial Fibrosis

Fibrillar collagen is formed as scar tissue in response to parenchymal cell loss. It is a well recognized component of wound healing that occurs in any organ to preserve its structural integrity. This *replacement fibrosis* proceeds in an orderly manner, having a clear onset to its formation and a definite termination. Cardiac myocyte loss, for example, is replaced by fibrillar collagen. Its formation begins within days of cell death, becomes an identifiable meshwork within the first week (41) and fully established scar tissue beyond day 14 (13), after which it does not increase further (33, 45, 53). In contrast to the reactive expression of fibrosis that is discussed below, the replacement fibrous tissue response is primarily confined to a single area of the myocardium (i.e., the site of necrosis); it does not appear in interstitial spaces or involve intramyocardial coronary arteries in areas remote from the site of cell loss (33, 45). Scarring can occur on a micro- or macroscopic scale, depending on the extent of cell loss.

Another fibrous tissue response that can occur in any organ, including the heart, is a progressive accumulation of fibrillar collagen which first appears within the adventitia of intramural arterioles (i.e., a perivascular fibrosis or perivascular granulomas) and, subsequently, the extracellular space (i.e., an interstitial fibrosis) (5, 44). It does not require parenchymal cell loss. This *reactive fibrosis*, which is also relevant to

Table 1. Stages of normal wound healing that occur after trauma or in association with abnormal blood vessel permeability

---

1. disruption of vascular-interstitial space barrier and escape of plasma proteins (fibrinogen, plasminogen, fibronectin)
2. extravascular coagulation and formation of hydrophobic fibrin-fibronectin gel matrix
3. formation of granulation tissue
   - replication and migration of inflammatory cells and fibroblasts
   - formation of new capillaries
   - degradation of gel matrix
   - synthesis of interstitial collagens
4. formation of fibrous tissue with resorption of capillaries and fibroblasts

---

the formation of solid tumors (21), appears to be a wound-healing response that has gone out of control in that it does not terminate. (See Table 1 for an overview of the stages of normal wound healing.) This suggests that either a perpetual stimulus to collagen formation is present or collagen degradation has been suppressed. This fibrous tissue response occurs on either a microscopic scale, where it may be limited in its distribution within systemic organs (e.g., morphea), or it can be expressed on a macroscopic scale with multiorgan involvement (e.g., systemic sclerosis). In the myocardium such a progressive perivascular fibrosis of intramyocardial coronary arteries, arterioles and interstitial fibrosis has been observed in acquired hypertension associated with unilateral renal ischemia (9, 20, 27, 38, 50) and hyperaldosteronism (9).

In addressing the regulation of myocardial fibrosis it is therefore necessary to examine whether the abnormal rise in collagen concentration is secondary to myocyte necrosis or a reactive fibrosis, or both.

Factors associated with the appearance of replacement fibrosis

Myocyte Necrosis and Ischemia

It is well recognized that a critical reduction in either epicardial (1) or intramyocardial (13) coronary artery luminal area, together with a corresponding reduction in oxygen delivery, will lead to cardiac myocyte necrosis and subsequent scarring of the myocardium. This fibrous tissue response is represented by the in-series addition of fibrillar collagen between remaining viable myocytes. In response to this structural remodeling and deposition of type I collagen, myocardial stiffness rises, particularly at larger strains or filling volumes (13). This wound healing process also includes the extension of collagen fibers from the body of the scar into neighboring viable myocardium, where they envelope viable myocytes (28). This serves to anchor and stabilize the scar. In so doing, however, myocytes so encased become electrically isolated from one another and the resultant heterogeneity in tissue composition, seen in the transition region between the scar and viable myocardium, fosters the development of reentrant ventricular arrhythmias (37, 46, 47). In addition, myocytes surrounded by fibrillar collagen become progressively atrophic (28).

The factors that govern fibrous tissue formation in the setting of myocyte necrosis secondary to ischemia remain to be elucidated. It has been reported that the expression of transforming growth factor-beta (TGF-beta) is increased in the rat heart following coronary ligation (51). TGF-beta is a known mitogenic stimulus to skin fibroblasts and promoter of collagen synthesis in these mesenchymal cells (43). Whether this growth factor is the sole stimulus, however, is unclear since fibroblast proliferation has also been seen in noninfarcted regions of the rat left ventricle, interventricular septum, and right ventricle following coronary ligation (53). Circulating hormones of the renin-angiotensin-aldosterone system (RAAS), particularly aldosterone (ALDO), could be involved given that the fibroplasia seen in the noninfarcted myocardium could be prevented by pretreatment with captopril, but not by an angiotensin II receptor antagonist (53). Circulating ALDO would likely be increased in this experimental model given that a thoracotomy was needed to produce coronary ligation (24) and an impairment in cardiac output following anterior infarction was likely present (19). These issues, however, need to be addressed.

Myocyte Necrosis and Circulating Hormones

Marked elevations in plasma *catecholamines*, such as occurs following head trauma or subarachnoid hemorrhage, are known to be associated with the appearance of contraction band necrosis of cardiac myocytes (52). An abnormal increase in sarcolemmal permeability (2) is an early event that prestages myocyte necrosis following isoproterenol administration in rats. Myocyte necrosis is followed by interstitial edema, enhanced formation of glycosaminoglycans (31), fibroblast proliferation, and enhanced collagen synthesis (32). A fibrillar collagen meshwork is evident by day 7 (41).

The factors involved in mediating the fibrous tissue response that occurs in response to catecholamine-induced myocyte necrosis remain to be elucidated. We find it of interest that the fibrosis associated with isoproterenol administration is largely confined to the endomyocardium of both ventricles, where it appears to be disproportionate to the degree of myocyte necrosis. This synthetic catecholamine is known to also disrupt the structural integrity of the endothelial surface of the endocardium. This leads to the adhesion of platelets to fibrillar collagen in the subendothelial space (4) and raises the intriguing possibility that this abnormal endocardial permeability permits known mitogens, such as platelet derived growth factor, to promote fibroblast proliferation and/or enhanced collagen synthesis with corresponding endomyocardial fibrosis. This hypothesis remains to be tested.

Elevations in plasma *angiotensin* II (AII) in man are thought to be associated with myocyte necrosis (23). Experimental studies have indicated that on day-1 administration of a nonhypertensive intraperitoneal dose of AII there was an abnormal permeability of myocyte sarcolemma to monoclonal antimyosin antibody (49). This early marker of myocyte necrosis was not influenced by pretreatment with phenoxybenzamine or propranolol, while a similar labeling of myocytes was seen on day 1 of unilateral renal ischemia, not thereafter, and could be prevented by captopril pretreatment (49). The response to an antagonist type I AII receptor has not yet been examined. This early myocyte necrosis in renal ischemia was followed by fibroblast proliferation on day 2 (49) and enhanced expression of type-I and type-III collagen

REACTIVE & REPARATIVE MYOCARDIAL FIBROSIS

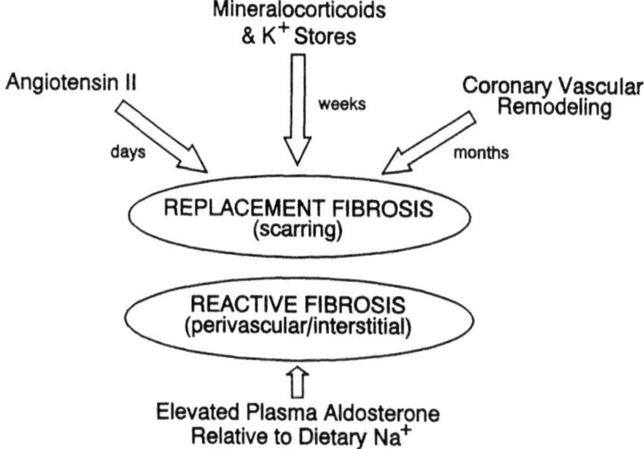

Fig. 1. An overview of the pathophysiologic mechanisms involved in scarring and perivascular/interstitial fibrosis of the myocardium in states of primary or secondary hyperaldosteronism.

mRNAs on day 3 (15). Resultant scars, evident on day 14, involved less than 1% of the left ventricle and had no discernible distribution within the myocardium (49). An infusion of ALDO, on the other hand, was not associated with in vivo immunolabeling with this antimyosin antibody (unpublished observations).

Regulatory mechanisms involved in mediating scar tissue formation following AII-induced necrosis are unclear. A direct influence of ALDO, which would be elevated in association with increased plasma concentration of AII, cannot be discounted at the present time.

Chronic elevations in circulating *mineralocorticoids*, for 4 weeks or more, are associated with myocyte necrosis and replacement fibrosis in the rat and feline myocardium (17). This scarring was related to a reduction in myocardial potassium stores and could be prevented by supplementing the diet of these animals with potassium chloride before they received deoxycorticosterone (17). The ALDO receptor antagonist spironolactone was also effective in preventing scarring in our rat model of hyperaldosteronism (10). In this connection, myocardial scarring was observed in the postmortem heart of a young woman with Bartter's syndrome having hyperplasia of the juxtaglomerular complex, hyperaldosteronism, chronic hypokalemic metabolic alkalosis, and anatomically normal epicardial coronary arteries (42).

Whether elevations in plasma ALDO would influence the fibrous tissue response following myocyte loss and enhanced potassium excretion cannot be addressed at the present time.

Thus, in disease states where the circulating RAAS in activated myocyte necrosis and scarring (see Fig. 1) can occur over time leading to a progressive structural remodeling and compromise in the mechanical and electrical behavior of the myocardium. Medial thickening of coronary resistance vessels may also occur over months to attenuate myocyte blood flow and coronary vasomotor reactivity.

Factors associated with the appearance of reactive fibrosis

Mineralocorticoid Excess

A reactive perivascular and interstitial fibrosis of the hypertrophied left ventricle has been observed in the rat with unilateral renal ischemia (9, 20, 26, 27, 38, 50), where it was also noted to become progressively more severe over time (45, 50). It was also possible to prevent this fibrosis by pretreating such animals with captopril. However, for the dose of this ACE inhibitor used, hypertension and left-ventricular hypertrophy (LVH) were also prevented (29). To determine whether this reactive fibrous tissue response was related to hemodynamic or effector hormones of the RAAS, or to the hypertrophic process itself, a series of in vivo studies were undertaken. Various animal models having acquired arterial hypertension and a diverse profile of the various components of the RAAS were examined (9, 10). Moreover, the collagen volume fractions of the right and left ventricles were measured as they related to interstitial and perivascular fibrosis while scars were excluded. The right ventricle served as a negative internal control for ventricular systolic pressure since it was neither pressure overloaded nor hypertrophied in these models of arterial hypertension, while it was a positive control for circulating hormones involved in the fibrous tissue response.

We found that, in contrast to the reactive fibrosis seen in both ventricles in unilaterial renal ischemia, such a fibrous tissue response did not occur in either ventricle associated with banding the abdominal aorta below the renal arteries, despite comparable hypertension and left-ventricular hypertrophy (LVH). Because the RAAS was not activated with the infrarenal band model, it appeared that either elevated plasma AII or ALDO, or both, were involved. To distinguish among these possibilities a model of hyperaldosteronism was created in uninephrectomized animals maintained on enhanced sodium diet by administering ALDO subcutaneously by osmotic minipumps. In this model of mineralocorticoid excess, associated with comparable hypertension and LVH to that seen with infrarenal band or renal ischemia, plasma AII was suppressed while plasma ALDO was increased. In hyperaldosteronism, we found the reactive fibrosis in the right and left ventricles, indicating that elevations in plasma ALDO, that were disproportionate for the level of sodium intake were likely involved in mediating the fibrous tissue response.

In subsequent studies the importance of circulating ALDO was further examined by pretreating animals with the ALDO receptor antagonist spironolactone (see Fig. 2) prior to the induction of hyperaldosteronism. A small dose of spironolactone that did not prevent arterial hypertension or LVH was used in one group of rats while another group received a larger dose that had these effects. We found that with either dose of spironolactone the reactive perivascular and interstitial fibrosis were not seen and this was true in each ventricle. This was also the case when rats with unilateral renal ischemia were pretreated with the smaller dose of spironolactone. It was not true, however, when rats with hyperaldosteronism were pretreated with a dose of captopril that prevented hypertension and reduced circulating AII; here the fibrous tissue response was again found.

Collectively, these findings emphasize the following concerning the reactive fibrosis of the myocardium seen in acquired hypertension: a) elevations in left-ventricular systolic pressure are not responsible for reactive fibrosis, given that the normotensive right ventricle was also involved; b) arterial hypertension alone,

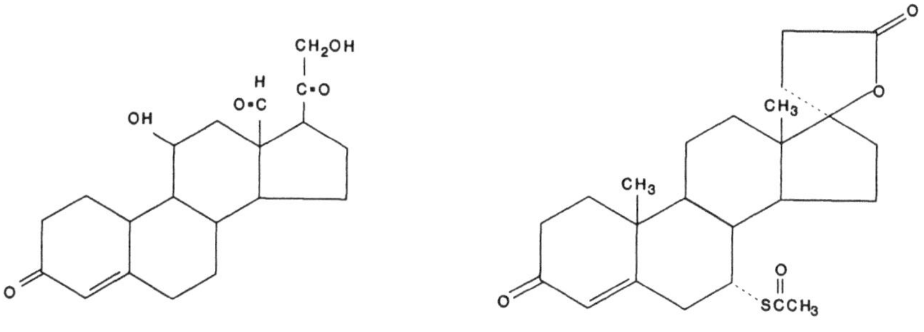

Aldosterone   Spironolactone

Fig. 2. Chemical structure of aldosterone and its pharmacologic antagonist spironolactone.

representing an elevation in coronary perfusion pressure for both ventricles, was not involved, as demonstrated by the lack of fibrosis in the infrarenal band model and the presence of fibrosis in normotensive, captopril-treated animals with hyperaldosteronism; c) the hypertrophic growth of myocytes was not needed to induce fibrosis, as evidenced by the remodeling of the nonhypertrophied right ventricle, indicating that myocyte and fibroblast involvement are based on separate controls (i.e., hemodynamic and hormonal, respectively); and d) elevated plasma ALDO, relative to sodium intake, was in some manner associated with perivascular and interstitial fibrosis. In examining this latter point further uninephrectomized animals receiving exogenous ALDO were maintained on a low sodium diet. This circumstance led to elevations in plasma ALDO that were even greater than those seen with hyperaldosteronism; however, the reactive fibrosis was not seen (8).

These findings also support the benchmark studies of Selye (44), who identified that mineralocorticoid excess (secondary to chronic administration of deoxycorticosterone in uninephrectomized animals maintained on elevated sodium intake) produced perivascular and interstitial fibrosis in the myocardium and multiple systemic organs, such as the pancreas, kidney, and adrenal glands. He termed this structural remodeling that accompanied chronic elevations of this steroid hormone as a disease of adaptation. Preliminary (unpublished) findings from this institution indicate that this reactive fibrosis is seen in the postmortem human myocardium and systemic organs obtained from patients with adrenal adenoma.

Nonapparent mineralocorticoid excess

A similar perivascular and interstitial fibrosis of the myocardium to that found with mineralocorticoid excess has been found in the spontaneously hypertensive rat (7), where it also has been reported to be progressive over time (40, 50). The circulating RAAS is said not to be activated in this model of genetic hypertension. However, the fact that plasma renin activity is "normal" (vis-à-vis suppressed) suggests that this view may need to be reexamined and the contribution of tissue hormones considered. Several lines of evidence can be cited in drawing an association between tissue RAAS and fibrosis. These are presented below.

Messenger RNA for angiotensinogen is found in the adventitia of the aorta, where it is localized within the loose connective tissue cells (i.e., brown adipocytes and fibroblasts) (12). In the heart, tissue angiotensin converting enzyme (ACE) activity is found to be greatest in its valves, which consist largely of fibrillar collagen and fibroblast-like cells; it is also found in the adventitia of intramyocardial coronary arteries and the atria (56). Four weeks following myocardial infarction in the rat, tissue ACE activity is prominent in the resultant scar (30). Moreover, fibroblast proliferation that occurs during the first 2 weeks following infarction in areas remote from necrosis is prevented by ACE inhibition (53). Fibroblast-like cells (i.e., myo-fibroblasts) that are an integral component of granulation tissue, such as that which forms in the region of myocyte loss, contract in response to AII (22), suggesting they have receptors to this peptide hormone. The connexon which joins adjacent myo-fibroblasts to one another, and which promote cell-cell signalling, is regulated by mineralocorticoid-like substances (18). Finally, we find it of interest that endothelin produced by endothelial cells stimulates ALDO production in isolated zona glomer-ulosa cells (16) and promotes growth of cultured fibroblast-like cells (48).

Collectively, these various lines of evidence implicate the tissue RAAS in wound healing secondary to myocardial infarction and suggests it may influence fibroblast behavior. In this regard, we found that in treating SHR in utero and for a subsequent 14 weeks with the ACE inhibitor lisinopril it was possible to not only prevent hypertension and LVH, but the reactive myocardial fibrosis (unpublished findings). This was not the case for hydralazine, which was equally effective in preventing hypertension and LVH in young SHR (39). Thus, the tissue RAAS may be contri-butory to the pathologic structural remodeling of the myocardium in hypertensive heart disease.

Future directions

In cultured fibroblasts obtained from skin or vein, Maquart et al. (35, 36) found that mineralocorticoid-like substances enhance collagen synthesis without altering their growth potential. Future studies regarding the regulation of myocardial fibrosis will need to follow a similar line of inquiry. Moreover, it will be necessary to undertake in vitro studies of cultured cardiac fibroblasts obtained from the same animal species used in the in vivo studies. The growth of such cells and their response in collagen metabolism (synthesis and degradation) to ALDO will need to be addressed. Furthermore, it will be necessary to determine if such a nonclassic response (e.g., collagen metabolism) to ALDO is observed in these nonclassical mineralocorticoid target cells and whether it would be mediated by corticoid receptors as trans-criptional or nontranscriptional events.

References

1. Anversa P, Alden VL, Levicky V, Guideri G (1985) Left ventricular failure induced by myocardial infarction. 1. Myocyte hypertrophy. Am J Physiol 248:H876–H882
2. Benjamin IJ, Jalil EJ, Tan LB, Cho K, Weber KT, Clark WA (1989) Isoproterenol-induced myocardial fibrosis in relation to myocyte necrosis. Circ Res 65:657– 670
3. Bing OHL, Fanburg BL, Brooks WW, Matsushita S (1978) The effect of the lathyrogen

β-amino proprionitrile (BAPN) on the mechanical properties of experimentally hypertrophied rat cardiac muscle. Circ Res 43:632–637

4. Boutet M, Turcotte H, Bazin M, Bourque L (1983) An ultrastructural study of endocardial endothelium alterations in catecholamine-induced infarct-like necrosis. Rev Can Biol Exp 42:87–99

5. Braley-Mullen H, Sharp GC, Bickel JT, Kyriakos M (1991) Induction of severe granulomatous experimental autoimmune thyroiditis in mice by effector cells activated in the presence of anti-interleukin 2 receptor antibody. J Exp Med 173:899–912

6. Brilla CG, Janicki JS, Weber KT (1991) Cardioreparative effects of lisinopril in rats with genetic hypertension and left ventricular hypertrophy. Circulation 83:1771–1779

7. Brilla CG, Janicki JS, Weber KT (1991) Impaired diastolic function and coronary reserve in genetic hypertension: role of interstitial fibrosis and medial thickening of intramyocardial coronary arteries. Circ Res 69:107–115

8. Brilla CG, Pick R, Janicki JS, Weber KT (1990) Myocardial fibrosis in either renovascular or mineralocorticoid hypertension [Abstract]. Clin Res 38:588A

9. Brilla CG, Pick R, Tan LB, Janicki JS, Weber KT (1990) Remodeling of the rat right and left ventricle in experimental hypertension. Circ Res 67:1355–1364

10. Brilla CG, Weber KT (1991) Prevention of myocardial fibrosis in hypertension: role of fibroblast corticoid receptors and spironolactone [Abstract]. FASEB J 5:A1256

11. Brutsaert DL, Sys SU (1989) Relaxation and diastole of the Heart. Physiol Rev 69:1228–1315

12. Campbell DJ, Habener JF (1987) Cellular localization of angiotensinogen gene expression in brown adipose tissue and mesentery: quantification of messenger fibonucleic acid abundance using hybridization in situ. Endocrinology 121:1616–1626

13. Carroll EP, Janicki JS, Pick R, Weber KT (1989) Myocardial stiffness and reparative fibrosis following coronary embolization in the rat. Cardiovasc Res 23:655–661

14. Caulfield JB, Wolkowicz PE (1990) A mechanism for cardiac dilatation. Heart Failure 6:138–150

15. Chapman D, Weber KT, Eghbali M (1990) Regulation of fibrillar collagen types I and III and basement type IV collagen gene expression in pressure overloaded rat myocardium. Circ Res 67:787–794

16. Cozza EN, Gomez-Sanchez CE, Foecking MF, Chiou S (1989) Endothelin binding to cultured calf adrenal zona glomerulosa cells and stimulation of aldosterone secretion. J Clin Invest 84:1032–1035

17. Darrow DC, Miller HC (1942) The production of cardiac lesions by repeated injections of desoxycorticosterone acetate. J Clin Invest 21:601–611

18. Davidson JS, Baumgarten IM (1988) Glycyrrhetinic acid derivatives: a novel class of inhibitors of gap-junctional intercellular communication. Structure-activity relationship. J Pharmacol Exp Ther 245:1104–1107

19. Davis JO, Hartroft PM, Titus EO, Carpenter CCJ, Ayers CR, Spiegel HE, Caspar A, Cavanaugh E (1962) The role of the renin-angiotensin system in the control of aldosterone secretion. J Clin Invest 41:378–389

20. Doering CW, Jalil JE, Janicki JS, Pick R, Aghili S, Abrahams C, Weber KT (1988) Collagen network remodeling and diastolic stiffness of the rat left ventricle with pressure overload hypertrophy. Cardiovasc Res 22:686–695

21. Dvorak HF (1986) Tumors: wounds that do not heal. Similarities between tumor stroma generation and wound healing. N Engl J Med 315:1650–1659

22. Gabbiani G (1981) The myofibroblast: a key cell for wound healing and fibrocontractive diseases. In: Deyl Z, Adam M (eds) Connective Tissue Research: Chemistry, Biology, and Physiology. New York: Liss, pp 183–194

23. Gavras H, Brunner HR, Laragh JH, Vaughan ED, Koss M, Cote LJ, Gavras I (1975) Malignant hypertension resulting from deoxycorticosterone acetate and salt excess. Circ Res 36:300–309

24. Hodsman GP, Kohzuki M, Howes LG, Sumithran E, Tsunoda K, Johnston CI (1988) Neurohumoral responses to chronic myocardial infarction in rats. Circulation 78:376–381
25. Holubarsch C, Holubarsch T, Jacob R, Medugorac I, Thiedemann K (1983) Passive elastic properties of myocardium in different models and stages of hypertrophy: a study comparing mechanical, chemical and morphometric parameters. In: Alpert NR (ed) Perspectives in Cardiovascular Research, Vol 7: Myocardial Hypertrophy and Failure. Raven Press, New York, pp 323–336.
26. Jalil JE, Doering CW, Janicki JS, Pick R, Clark WA, Weber KT (1988) Structural vs contractile protein remodeling and myocardial stiffness in hypertrophied rat left ventricle. J Mol Cell Cardiol 20:1179–1187
27. Jalil JE, Doering CW, Janicki JS, Pick R, Shroff SG, Weber KT (1989) Fibrillar collagen and myocardial stiffness in the intact hypertrophied rat left ventricle. Circ Res 64:1041–1050
28. Jalil JE, Janicki JS, Pick R, Abrahams C, Weber KT (1989) Fibrosis-induced reduction of endomyocardium in the rat after isoproterenol treatment. Circ Res 65:258–264
29. Jalil JE, Janicki JS, Pick R, Weber KT (1991) Coronary vascular remodeling and myocardial fibrosis in the rat with renovascular hypertension: response to captopril. Am J Hypertens 4:51–55
30. Johnston CI, Mooser V, Sun Y, Fabris B (1991) Changes in cardiac angiotensin converting enzyme (ACE) after myocardial infarction and hypertrophy in rats. Clin Exp Pharmacol Physiol 18:107–110
31. Judd JT, Wexler BC (1969) Myocardial connective tissue metabolism in response to injury. Histological and chemical studies of mucopolysaccharide and collagen in rat hearts after isoproterenol-induced infarction. Circ Res 25:201–214
32. Judd JT, Wexler BC (1975) Prolyl hydroxylase and collagen metabolism after experimental myocardial infarction. Am J Physiol 228:212–216
33. Jugdutt BI, Amy RWM (1986) Healing after myocardial infarction in the dog: changes in infarct hydroxyproline and topography. J Am Coll Cardiol 7:91–102
34. Kessler KM (1988) Heart failure with normal systolic function: update of prevalence, differential diagnosis, prognosis, and therapy. Arch Intern Med 148:2109–2111
35. Maquart FX, Bellon G, Brieu MA, Borel JP (1986) Titrated extract from *Centella asiatica* stimulates collagen biosynthesis in human vein wall fibroblast cell cultures. In: Negus D, Jantet G (eds) Phlebology '85. London: J Libbey pp 843–845
36. Maquart FX, Bellon G, Gillery P, Wegrowski Y, Borel JP (1990) Stimulation of collagen synthesis in fibroblast cultures by a triterpene extracted from *Centella asiatica*. Connect Tissue Res 24:107–120
37. McLenachan JM, Dargie HJ (1990) Ventricular arrhythmias in hypertensive left ventricular hypertrophy: relationship to coronary artery disease, left ventricular dysfunction, and myocardial fibrosis. Am J Hypertens 3:735–740
38. Michel JB, Salzmann JL, Ossondo Nlom M, Bruneval P, Barres D, Camilleri JP (1986) Morphometric analysis of collagen network and plasma perfused capillary bed in the myocardium of rats during evolution of cardiac hypertrophy. Basic Res Cardiol 81:142–154
39. Narayan S, Janicki JS, Shroff SG, Pick R, Weber KT (1989) Myocardial collagen and mechanics after preventing hypertrophy in hypertensive rats. Am J Hypertens 2:675–682
40. Pfeffer JM, Pfeffer MA, Fishbein MC, Froehlich ED (1979) Cardiac function and morphology with aging in the spontaneously hypertensive rat. Am J Physiol 6:H461–H468
41. Pick R, Jalil JE, Janicki JS, Weber KT (1989) The fibrillar nature and structure of isoproterenol-induced myocardial fibrosis in the rat. Am J Pathol 134:365–371
42. Potts JL, Dalakos TG, Streeten DHP, Jones D (1977) Cardiomyopathy in an adult with Bartter's syndrome. Am J Cardiol 40:995–999
43. Roberts AB, Sporn MB, Assoian RK, Smith JM, Roche NS, Wakefield LM, Heine VI, Liotta LA, Falanga V, Kehrl JH, Fauci AS (1986) Transforming growth factor type β: rapid induction of fibrosis and angiogenesis in vivo and stimulation of collagen formation in vitro. Proc Natl Acad Sci USA 83:4167–4171

44. Selye H (1946) The general adaptation syndrome and the diseases of adaptation. J Clin Endocrinol 6:117–230
45. Silver MA, Pick R, Brilla CG, Jalil JE, Janicki JS, Weber KT (1990) Reactive and reparative fibrosis in the hypertrophied rat left ventricle: two experimental models of myocardial fibrosis. Cardiovasc Res 24:741–747
46. Strain JE, Grose RM, Factor SM, Fisher JD (1983) Results of endomyocardial biopsy in patients with spontaneous ventricular tachycardia but without apparent structural heart disease. Circulation 68:1171–1181
47. Sugrue DD, Holmes DR Jr., Gersh BJ, Edwards WD, McLaran CJ, Wood DL, Osborn MJ, Hammill SC (1984) Cardiac histologic findings in patients with life-threatening ventricular arrhythmias of unknown origin. J Am Coll Cardiol 4:952–957
48. Takuwa N, Takuwa Y, Yanagisawa M, Yamashita K, Masaki T (1989) A novel vasoactive peptide endothelin stimulates mitogenesis through insitol lipid turnover in Swiss 3T3 fibroblasts. J Biol Chem 264:7856–7861
49. Tan LB, Jalil JE, Pick R, Janicki JS, Weber KT (1991) Cardiac myocyte necrosis induced by angiotensin II. Circ Res (in press)
50. Thiedemann KU, Holubarsch C, Medugorac I, Jacob R (1983) Connective tissue content and myocardial stiffness in pressure overload hypertrophy. A combined study of morphologic, morphometric, biochemical and mechanical parameters. Basic Res Cardiol 78:140–155
51. Thompson NL, Bazoberry F, Speir EH, Casscells W, Ferrans VJ, Flanders KC, Kondaiah P, Geiser AG, Sporn MB (1988) Transforming growth factor beta-1 in acute myocardial infarction in rats. Growth Factors 1:91–99
52. Todd GL, Sterns DA, Plambeck RD, Joekel CS, Eliot RS (1986) Protective effects of slow channel calcium antagonists on noradrenaline induced myocardial necrosis. Cardiovasc Res 20:645–651
53. Van Krimpen C (1991) Cardiac remodeling and angiotensin II after an experimental myocardial infarction. Thesis. University of Limburg
54. Weber KT, Brilla CG (1991) Pathological hypertrophy and cardiac interstitium: fibrosis and renin-angiotensin-aldosterone system. Circulation 83:1849–1865
55. Weber KT, Brilla CG, Janicki JS (1990) Structural remodeling of myocardial collagen in systemic hypertension: functional consequences and potential therapy. Heart Failure 6:129–137
56. Yamada H, Fabris B, Allen AM, Jackson B, Johnston CI, Mendelsohn FAO (1991) Localization of angiotensin converting enzyme in the rat heart. Circ Res 68:141–149

Authors' address:
Karl T. Weber, M.D.
Rm MA432 Medical Science Building
University of Missouri-Columbia
Columbia, MO 65212, USA

The extracellular matrix in the failing human heart

Jutta Schaper, B. Speiser

Max-Planck-Institute, Department of Experimental Cardiology, Bad Nauheim, FRG

Summary: The composition of the extracellular matrix was investigated in eight human hearts explanted at the time of transplantation surgery because of endstage cardiomyopathy. All patients showed clinical signs of heart failure. The tissue was investigated by electron microscopy and immunofluorescence microscopy using monoclonal antibodies against collagen I, III, VI, and IV, fibronectin, laminin, and vimentin. All matrix proteins occurred in increased amounts in the extracellular space separating the myocardial cells by septa of enlarged thickness. Laminin and collagen IV surrounded myocardial and endothelial cells as layers of increased thickness. Vimentin localization was normal in individual cells, but occurred more often and corresponded to the numerous fibroblasts as observed by electron microscopy. It is concluded that an excessive deposition of extracellular matrix material in addition to myocyte degeneration (as reported previously (9)) are the structural correlates of cardiac failure.

Key words: Extracellular matrix; endstage cardiomyopathy; myocyte degeneration; cardiac failure

Introduction

Chronic dilative cardiomyopathy leads to heart failure that can only be treated by transplantation of a donor heart. This therapy allows for long-term survival of many patients, but also for the investigation of the explanted hearts obtained in the operating theater by modern morphological methods. Electron microscopy, immunofluorescence microscopy using monoclonal antibodies, as well as in situ hybridization for the detection and localization of mRNA offer new possibilities for the study of the failing heart.

Previously, changes occurring in the contractile apparatus and in the cytoskeleton in endstage cardiomyopathic human hearts were described (9). The lack of myofibrillar mass and the presence of an increased amount of cytoskeletal elements were the most important findings.

In the present work, we describe changes of the extracellular matrix in these same failing hearts obtained at the time of transplantation surgery. It will be shown that most of the elements of the extracellular matrix, such as collagen I and III, collagen VI and fibronectin, as well as laminin and collagen IV were present in increased amounts, and that activated macrophages and proliferating fibroblasts occurred frequently. It is concluded that both the degenerative changes observed in the myocytes, and the deposition of excessive matrix material jointly contribute to the reduction of cardiac function.

Material and Methods

Eight explanted human hearts with endstage cardiomyopathy and heart failure (EF < 20%, NYHA class IV) were obtained in the operating theatre at the time of transplantation surgery; they were quickly dissected and frozen in liquid nitrogen. The tissue then was stored at $-70\,^{\circ}$C.

Additionally, small tissue samples for electron microscopy were prepared and immediately fixed in cold 3% glutaraldehyde buffered with 0.1 M Na cacodylate. These were embedded in Epon using routine procedures.

Transmural needle biopsies removed from normal appearing left-ventricular myocardium from patients undergoing coronary bypass surgery (informed consent was obtained) served for control purposes for both morphological techniques.

Immunofluorescent staining

Commercially available antibodies against the following matrix proteins were used in this study:
— fibronectin (ICN Immuno Biologicals, Lisle, UK)
— laminin (Dianova, Hamburg, FRG)
— vimentin (Dianova, Hamburg, FRG)
— collagen I and III (Bioscience Products, Switzerland)
— collagen VI (Telios, San Diego, USA)
— collagen IV (Dianova, Hamburg, FRG).

Frozen tissue sections were air dried and fixed for 10 min in acetone at $-20\,^{\circ}$C. After rinsing in PBS the tissue was incubated with the first antibody. The sections were then washed three times for 2 min each with PBS, incubated with biotinylated (anti-mouse, anti-rat, or anti-rabbit IgG, Amersham, UK) second antibody for 1 h, rinsed and treated with FITC labeled streptavidin. Nuclei were counterstained with 0·002% propidium iodide (modified from (7)). The sections were examined in an Olympus Vanox T fluorescence microscope and the results were recorded by photography on Kodak 200 ASA color film.

Electron microscopy

Sections of 50 nm thickness were prepared, stained with uranylacetate and lead citrate, then viewed and photographed in a Philips CM 10 electron miscroscope.

Results

Fibronectin was found to be present throughout the entire extracellular matrix in increased amounts as compared to control tissue (Fig. 1) and the myocardial cells were separated, and sometimes "encapsulated" by the excessive matrix material. The fibronectin filled extracellular space also contained collagen I, III, and VI in substantial amounts (Figs 2 and 3). Laminin surrounded the myocardial cells and the capillaries, not as a fine line corresponding to the basement membrane as seen in normal tissue, but instead it was present as rather thick sheaths (Fig. 4). Vimentin positive structures, i.e., fibroblasts and endothelial cells, were more abundant than in

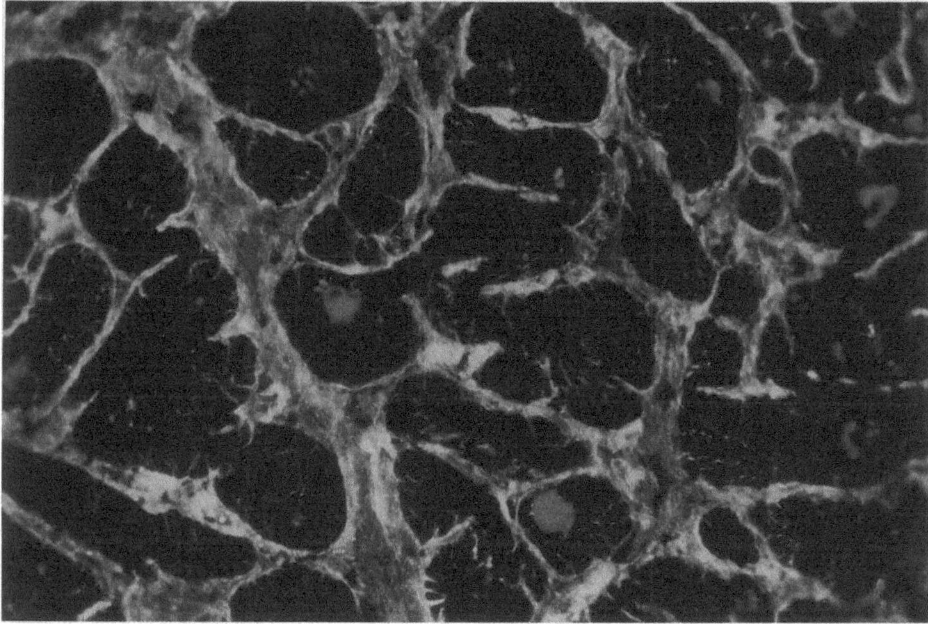

Fig. 1. The green fluorescence for fibronectin in diseased human myocardium is evident as a rather thick layer separating the myocardial cells. Slight intracellular labeling is present in myocardial cells which is due to staining of the T-tubular system. Nuclei are stained red. The yellowish granular material is lipofuscin.

normal tissue (Fig. 5), but vimentin was found in normal amounts in each individual cell.

In the electron microscope the presence of cellular debris and of activated phagocytosing macrophages and numerous proliferating fibroblasts between myocytes was obvious (Fig. 6). The myocytes showed degenerative changes such as lack of contractile material, myelinization of mitochondria, dilatation of the T-tubular system, and sequestration of parts of the cells into the extracellular space.

Discussion

From the immunofluorescent results presented here it becomes obvious that almost all proteins normally present in the extracellular matrix were significantly increased in diseased human myocardium.

Fibronectin is the major protein of the extracellular matrix. It possesses numerous binding sites for collagen, heparin, fibrin, factor XIIIa, proteoglycans, and for cells (for review see (5)). Fibronectin, even though localized in the extracellular space, is closely connected with the cytoskeleton and actin via its binding to integrins (6). Integrins are proteins that cross the cell membrane similar to receptors and that bind fibronectin, laminin, vitronectin, ICAM 1, and other proteins in the extracellular

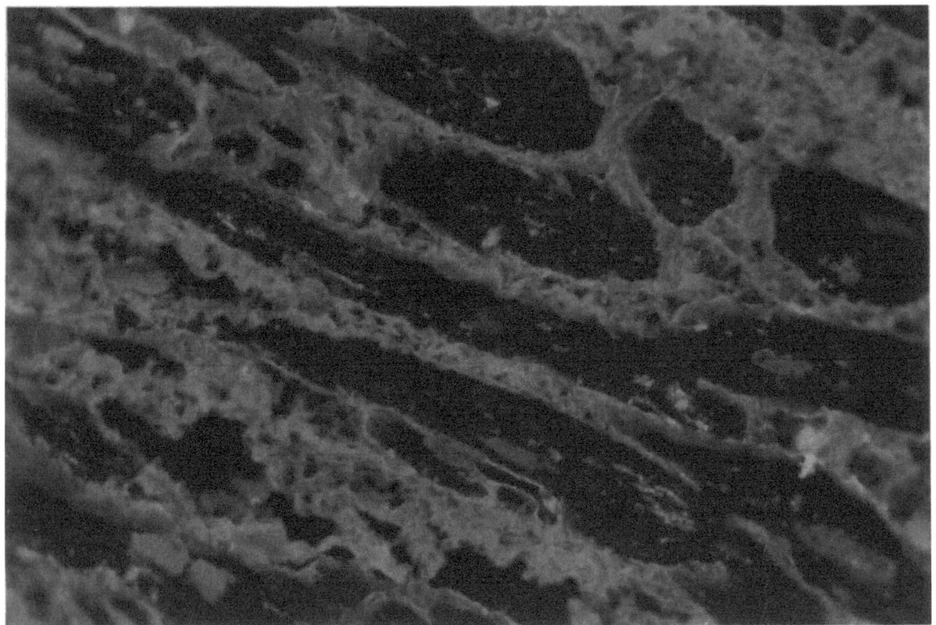

Fig. 2. Collagen I in diseased human myocardium is deposited between myocytes in large amounts. Note the wavy appearance of collagen fibers.

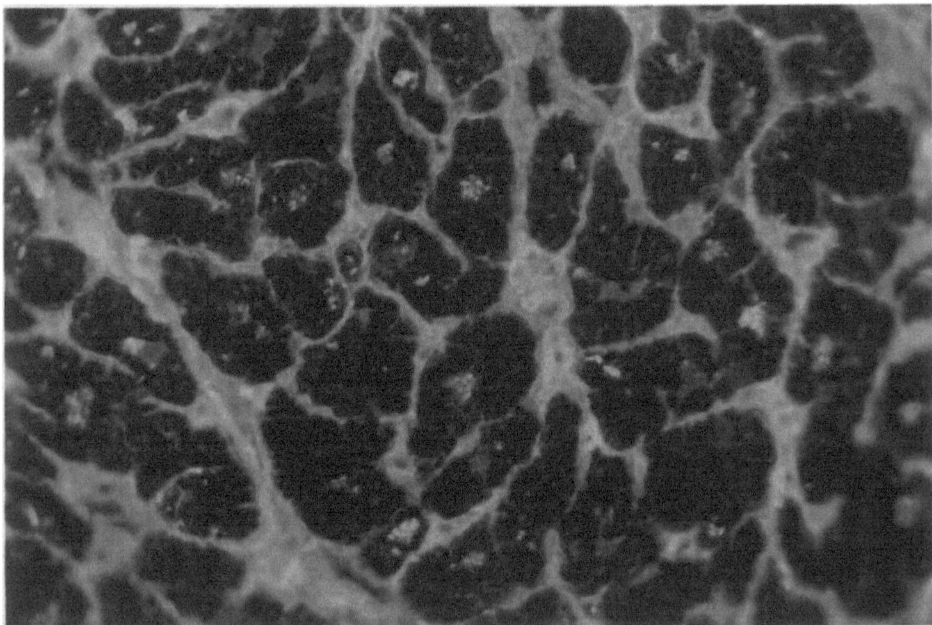

Fig. 3. Collagen III in diseased human myocardium is present in greater amounts than in normal tissue. In contrast to collagen I, collagen III is arranged more like a network in the enlarged extracellular space. Nuclei are red, lipofuscin is yellow and granular.

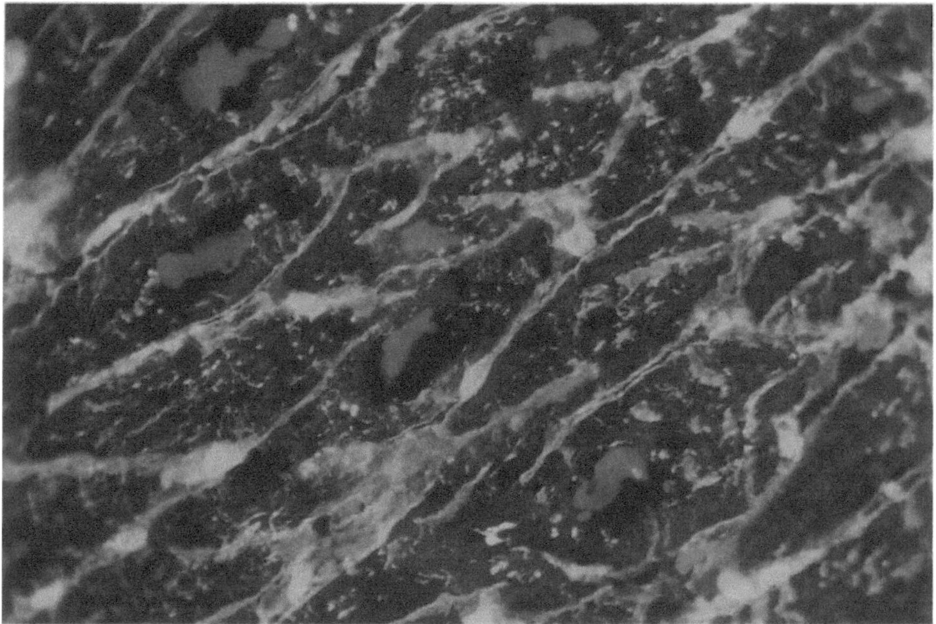

Fig. 4. Laminin is present in large amounts in diseased human myocardium, whereas, normally, only the basal laminae of myocytes and endothelial cells are stained. Intracellular labeling of T-tubules is evident. The enlarged nuclei are stained red.

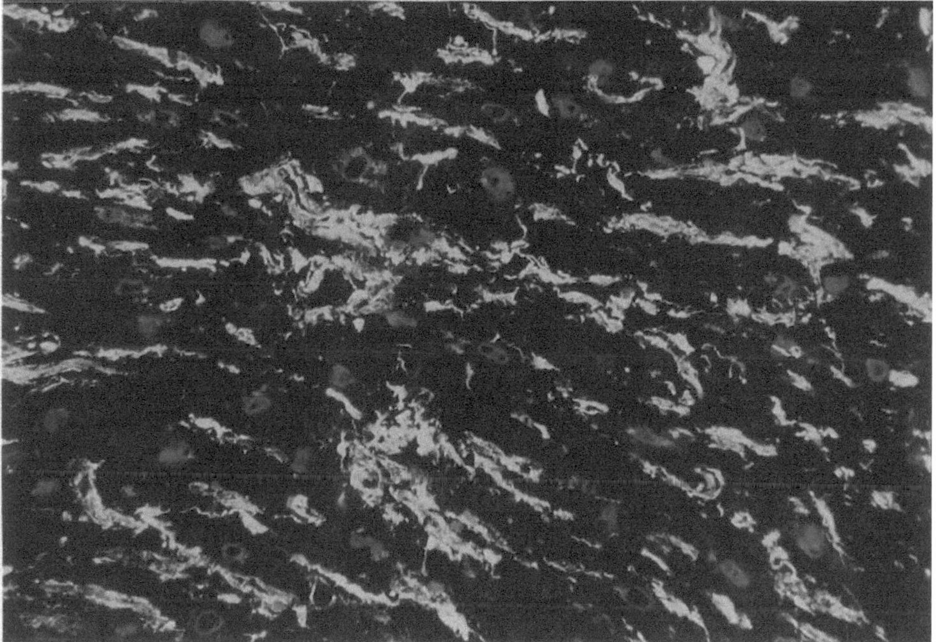

Fig. 5. Vimentin-labeled interstitial cells are abundant in failing myocardium, indicating the increased amount of fibrotic tissue.

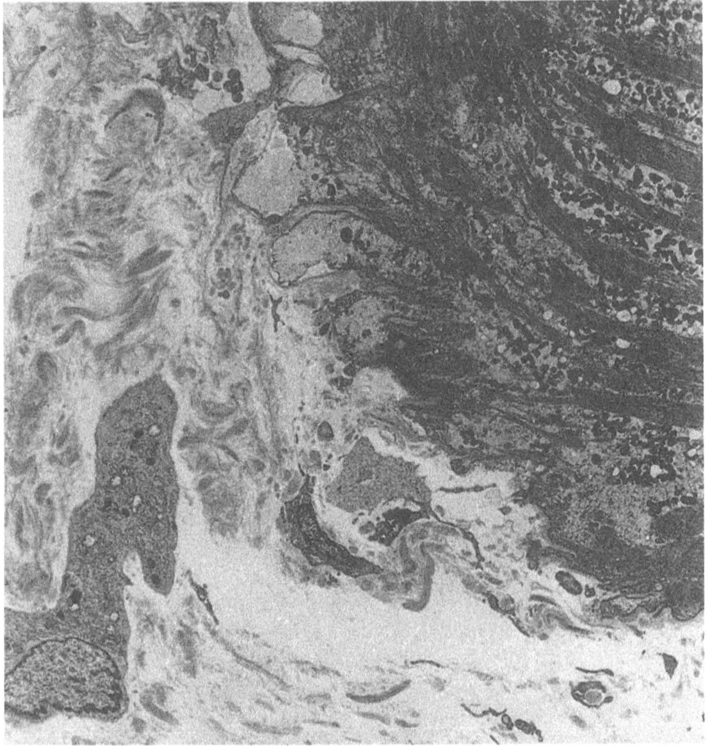

Fig. 6. Electron micrograph of the extracellular matrix and two myocardial cells in a cardiomyopathic human heart. The lack of contractile material is obvious, as are the sequestration of cellular material into the extracellular space. An activated macrophage and abundant collagen fibrils are present. Fibronectin and laminin, however, are not visible in the electron microscope. Magnification 6000

space. Within the cells, integrins bind via talin-vinculin-α-actinin-actin to the contractile apparatus and the cytoskeleton (1). Therefore, any disturbances of the myofilaments and of the cytoskeleton in human cardiomyopathy as described previously (9) may also influence the extracellular matrix and vice versa.

Fibronectin and the collagens I and III have been shown to colocalize in the extracellular matrix (4; 5) and this finding is confirmed here. When fibronectin was observed in increased amounts, the collagens I, III, and VI (3) were also increased in the interstitium. This is easily explained by the numerous binding sites for collagen in fibronectin and by the fact that fibroblasts are known to produce most extracellular matrix proteins (2).

Laminin and collagen IV are the major components of the basement membrane of endothelial cells and of myocytes (8; 10). These proteins are connected closely by binding to the entactin/nidogen system and they have binding sites for heparin, collagen and numerous cell attachment domains, and they share binding to the integrins with fibronectin (1; 5; 8). Comparable to the effects of fibronectin, laminin is involved in the regulation of cell attachment, growth, migration, and proliferation.

In the diseased human myocardium, both laminin and collagen IV were observed to enclose the myocardial and endothelial cells in much greater amounts than in normal tissue.

Vimentin was normal in individual cells, but it was increased in total number corresponding to the increased number of fibroblasts and macrophages in the tissue.

The accumulation of all these proteins and of fibroblasts and macrophages in the extracellular space leads to separation of the myocardial cells, not only morphologically, but certainly functionally as well. The electrical coupling of the myocytes will be inhibited, the diffusion distance for oxygen and substrates will be enhanced, and the myocardial compliance will be reduced due to the increased stiffness of the enlarged interstitium, i.e., to the increased content of fibrosis. Since the contractile apparatus of the myocardial cells is disturbed by disappearance of the myofilaments, and since the elements of the cytoskeleton are increased in amount (9) the changes of the myocytes proper may further contribute to a reduction of contractile function in hearts with endstage cardiomyopathy and, finally, to heart failure.

References

1. Albelda S, Buck A (1990) Integrins and other cell adhesion molecules. FASEB J 4:2868–2880
2. Eghbali M, Blumenfeld O, Seifter S, Buttrick P, Leinwand D, Robinson T, Zern M, Giambrone M (1989) Localization of types I, III and IV collagen mRNAs in rat heart cells by in situ hybridization. J Mol Cell Cardiol 21:103–115
3. Engvall E, Hessle H, Klier G (1986) Molecular assembly, secretion and matrix deposition of type VI collagen. J Cell Biol 102:703–710
4. Furcht L, Smith D, Wendelschafer-Crabb G, Mosher D, Foidart J (1980) Fibronectin presence in native collagen fibrils of human fibroblasts: Immunoperoxidase and immuno-ferritin localization. J Histochem Cytochem 28:1319–1333
5. Hynes R (1990) Fibronectins. Springer, New York, Berlin, Heidelberg
6. Hynes R, Destree A, Wagner D (1982) Relationships between fibronectin, actin, and cell substratum adhesion. Cold Spring Harbor Symp. Quant Biol 46:659–670
7. Jones K, Kniss D (1987) Propidium iodide as a nuclear counterstain for immunofluorescence studies of cells in culture. J Histochem Cytochem 35:123–125
8. Martin G, Timpl R (1987) Laminin and other basement membrane components. Ann Rev Cell Biol 3:57–85
9. Schaper J, Froede R, Hein S, Buck A, Hashizume H, Speiser B, Friedl A, Bleese N (1991) Impairment of the myocardial ultrastructure and changes of the cytoskeleton in dilated cardiomyopathy. Circulation 83:504–514
10. Yurchenco PD, Schittny J (1990) Molecular architecture of basement membranes. FASEB J 4:1577–1590

Author's address:
Dr. Jutta Schaper
Max-Planck-Institute
Department of Experimental Cardiology
Benekestrasse 2
D-6350 Bad Nauheim
Federal Republic of Germany

Mitochondrial function

Dysfunction of the ADP/ATP carrier as a causative factor for the disturbance of the myocardial energy metabolism in dilated cardiomyopathy

H.-P. Schultheiss

Medical Clinic B, Dept. for Cardiology, Pneumology, and Angiology, Heinrich-Heine University, Düsseldorf, FRG

Summary: The adenine nucleotide translocator (ADP/ATP carrier) plays a key role in nucleotide transport across the mitochondrial membrane. The quantity and function of this transport protein were investigated in myocardium from hearts with endstage failing dilated and ischemic cardiomyopathy, and were compared to measurements in nonfailing myocardium. In addition, lactate dehydrogenase (LDH) isoenzymes were determined. The concentration of the ADP/ATP carrier was significantly increased by 48% in myocardium from dilated cardiomyopathic hearts compared to control myocardium. The concentration of the carrier in explanted hearts with ischemic cardiomyopathy did not differ from values in the normal human hearts. Analysis of carrier function revealed similar nucleotide exchange rates in control hearts and hearts with ischemic cardiomyopathy, whereas carrier function was reduced in most hearts with dilated cardiomyopathy. Compared to control hearts, in hearts with dilated cardiomyopathy and decreased nucleotide exchange rate, the carrier content was significantly higher, whereas the carrier content was only slightly increased compared to control in cardiomyopathic hearts with unchanged transport activity. Compared to control hearts, in dilated cardiomyopathy there was a significant increase in LDH_5 and a decrease in LDH_1 isoforms, indicating more anaerobic metabolism in failing dilated cardiomyopathic hearts.

In summary, in hearts with dilated cardiomyopathy disturbed function of the ADP/ATP-carrier may result in altered myocardial energy metabolism and, thus, may be the cause of impaired myocardial function.

Key words: ADP/ATP carrier; mitochondria; lactate dehydrogenase; dilated cardiomyopathy; nucleotide transport

Introduction

Although the pathogenesis of dilated cardiomyopathy is not known, it seems to be evident that chronic heart failure is the result of several factors at the cellular level (1). One main question is the energetic state of the myocardium. There has been considerable debate as to whether there is depression of mitochondrial function in heart failure (1, 7, 27, 28). Changes in the creatine kinase system showing a lower creatine kinase activity, higher MB creatine kinase isoenzyme content and activity, and lower creatine content in pressure-overloaded hypertrophy may represent an adaptation toward anaerobic metabolism (6). The main stimulus for the changes in energy metabolism in the failing heart may be a decrease in coronary perfusion which is profound at the endocardial surface (29).

Experimental studies of chronic heart failure showed that the endocardial to epicardial ratio of coronary perfusion is significantly reduced under baseline conditions. Furthermore, the coronary reserve assessed as the increase in flow after

This work was supported by grants from the Deutsche Forschungsgemeinschaft intravenous dipyridamole was shown to be diminished in patients with idiopathic dilated cardiomyopathy (31). The alterations in coronary perfusion are a direct result of the underlying compensatory mechanisms during the failing process (30). Elevated catecholamine levels that produce tachycardia and increase myocardial contractility will result in limited diastolic coronary blood flow and, consequently, in myocardial ischemia. Impaired diastolic relaxation may prolong compression of the intra-myocardial vessels and thus further reduce the blood flow to the subendocardial layers. The elevated left-ventricular diastolic pressure will also reduce the sub-endocardial blood flow. The reduction in coronary reserve and the decrease in myocardial perfusion could result in a continuous state of subendocardial ischemia that could lead to myocardial necrosis and fiber replacement and, ultimately, to a decompensation of the left ventricle. This will induce a further activation of the neuroendocrine systems with the consequence of an increasing oxygen demand of the myocardium on the one hand, and a decrease in coronary perfusion on the other hand. This vicious circle of a self-perpetuating process might end with a further progression of heart failure due to disturbance of the myocardial energy balance.

Significance of the lactate dehydrogenase and the adenine nucleotide translocator for energy metabolism

Lactate dehydrogenase isoenzymes

Since direct methods of determining the intracellular distribution of oxygen are not available, we looked for indirect means of detecting reduced availability of oxygen. The association of the isoenzymes of lactate dehydrogenase (LDH) with aerobic and anaerobic metabolism and the shifts in isoenzyme distribution, which have been shown to occur with altered environmental oxygen availability, suggested a possible means for evaluating myocardial intracellular oxygen availability (4, 13). The structure of LDH is well known. It exists as two genetically determined subunits which have been designated as heart (LDH-H) and muscle (LDH-M) types corresponding to their relative preponderance in these tissues. Each isoenzyme is formed by four monomeres. LDH_1 consists of H-subunits only, LDH_5 of M-subunits. The other three isoenzymes consist of H- and M-subunits in a random association. The LDH-isoenzyme pattern is closely related to the redox state of the pyridine nucleotides, which is a sensitive marker of the energy state of the cell. In aerobically metabolizing tissues, LDH is active as lactate dehydrogenase. In this condition, the heart is a lactate consumer with the NADH quickly reoxidated by oxidative phosphorylation via mitochondrial shuttle systems. In anaerobically metabolizing tissues, pyruvate formation by glycolysis is enhanced. NADH reoxidation occurs by LDH_5. Now, the LDH serves as pyruvate reductase. In this condition, the heart is a lactate producer. Therefore, it has to be expected that any significant change in the myocardial energy state should alter the LDH-isoenzyme pattern.

Adenine nucleotide translocator

The adenine nucleotide translocator (ANT) – also called ADP/ATP carrier – is an intrinsic hydrophobic protein located within the inner mitochondrial membrane (9,

10). Since this membrane is a priori impermeable to hydrophilic metabolites, the transfer of ATP to the cytosol with its energy-consuming processes and the return of ADP to the inner mitochondrial space for regeneration by oxidative phosphorylation require a specific transport system. The ADP/ATP shuttle being the only active nucleotide transport system in mitochondria, is highly specific, and under normal physiological conditions its activity corresponds exactly to the requirements of ATP production. Of great metabolic significance is the differentiation between ADP and ATP in the transport by the energization of the mitochondrial membrane, which makes the nucleotide strongly exchange asymmetrically, preferring uptake of ADP and release of ATP. As a result, the concentration ratio of ATP/ADP is higher in the extra- than in the intramitochondrial space. Furthermore, the energization of the inner mitochondrial membrane modulates the carrier so as to operate in the manner required for oxidative phosphorylation (9). Although the ADP/ATP carrier translocation is not the only rate-limiting step in oxidative phosphorylation, there is a clear indication for a reversible interaction of the ADP potential difference between the external and mitochondrial compartment with the oxidation-reduction state of the respiratory chain. Thus, the adenine nucleotide translocator seems to play a key role in the regulation of oxidative phosphorylation.

In light of the current discussion of changes in energy metabolism in heart failure, it is of great interest to determine whether the adenine nucleotide translocator, a key protein in cell energy metabolism, is changed in its function and whether there is a down- or upregulation of this transport protein as a consequence of a disturbed myocardial energy metabolism.

Material and methods

Heart tissue was obtained intraoperatively immediately after explantation from patients with proven dilated cardiomyopathy (n = 26) and from patients with end-stage coronary heart disease (n = 8) undergoing heart transplantation. As controls (n = 5), tissue samples were analyzed from two healthy heart donors, whose hearts could not be used for transplantation because of technical problems, from one patient with an open ductus Botalli, one patient with aortic stenosis, and one patient with right-ventricular dysplasia. The hemodynamic characteristics of the DCM patients before transplantation were: left-ventricular end-diastolic pressure 24.3 $\pm$ 7.1 mmHg, cardiac index 1.9 $\pm$ 0.6 ml/min/m^2, ejection fraction 17.6 $\pm$ 5.6%, end-diastolic volume 356 $\pm$ 108 ml.

Tissue preparation and assay procedure

Mitochondria were prepared according to Smith (26). After the isolation of the mitochondria, one part of the suspension was prepared for the quantification of the carrier protein. For that purpose, the ADP/ATP carrier was stabilized with its specific ligand carboxyatractylate (5 nmol/mg protein). The mitochondria (0.5 mg) were suspended in 500 µl 0.25 M sucrose, 10 mM Tris, and 1 mM EDTA, pH 7.4. The quantification of the carrier protein was performed by an immunobinding assay on nitrocellulose membranes using antibodies raised against the isolated ADP/ATP-carrier protein (18). The specific binding of the antibodies to the ADP/ATP carrier bound on the nitrocellulose was detected by ^{125}I-protein-A (150,000 cpm/ml in 1% (v/v) Tween 20/PBS). Afterwards, the radioactivity was assayed in a gamma counter.

Aliquots were removed for measurement of non-collagen protein according to the method of Lowry (11).

The second part of the mitochondrial suspension was taken immediately after the isolation procedure to determine the nucleotide exchange rate of ATP by the inhibitor stop method combined with the back exchange (15, 18).

For LDH-isoenzyme analysis, the tissue samples were immediately placed in ice-cold 0.01 M potassium-phosphate buffer, pH 7.5. After gentle microhomogenization in the same buffer in a ratio of 1:10 (w/v) at 4°C with a pestle tipped by Dynal, and centrifugation at 25 000 × g for 30 min at 4°C, aliquots were taken for enzymatic tests. The pellet containing the endocardium was discarded. For determination of lactate dehydrogenase (LDH) enzymes, the supernatant was diluted in a ratio of 1:100 (w/v) with 30% glycerine to give a final concentration of 25% glycerine. The electrophoretic isolation of the isoenzymes was performed by microisoelectric focusing (IEF) according to (17). The relative amounts of H and M subunits were calculated as described by Markert (12). Values reported are the mean of five IEF-gels for each sample.

Statistical analysis was performed with Student's t-test.

Results

Lactate dehydrogenase isoenzyme pattern

The lactate dehydrogenase isoenzyme pattern was significantly altered in the DCM hearts in comparison to the control hearts. The samples from patients with end-stage coronary heart disease were not analyzed for the LDH-isoenzyme pattern, as the suspected chronic ischemia of the myocardium will consequently lead to an altered isoenzyme pattern. In comparison to the controls, where LDH_1 exhibits the greatest enzyme activity, in the patients with chronic heart failure there was a significant increase in LDH_5 paralleled by a decrease of LDH_1 (Fig. 1). This shift of the isoenzyme pattern towards those isoenzymes containing M-LDH is also expressed by the significant decrease of the percentage of the H-subunits (Fig. 1).

This significant isoenzyme shift toward those isoenzymes containing M-LDH lends support to the hypothesis that there is an adaptive metabolic change in the myocardial cell from patients with dilated cardiomyopathy.

ADP/ATP-carrier concentration

Under optimal solubilization conditions (4% Triton X-100) and an antibody dilution of 1:80, tissue probes from the explanted hearts were estimated for antibody binding in the immunodot assay. Purified ADP/ATP carrier standards were included in each assay. Linearity was observed when tissue probes were assayed at several dilutions. From the slopes of the curves obtained with the heart-tissue probes and the reference curves for the ADP/ATP carrier the relative concentrations of the ADP/ATP carrier per μg of protein were calculated. In comparison to the controls and to the explanted heart with coronary heart disease, the content of the ADP/ATP carrier was significantly increased in the human hearts from patients with dilated cardiomyopathy (Fig. 2). In the normal tissue (controls), the concentration was 50.2 ± 2.1 µg/mg protein, while in the DCM hearts the concentration was 74.3 ± 17 µg/mg protein. However, the carrier concentration in the explanted ischemic hearts did not differ

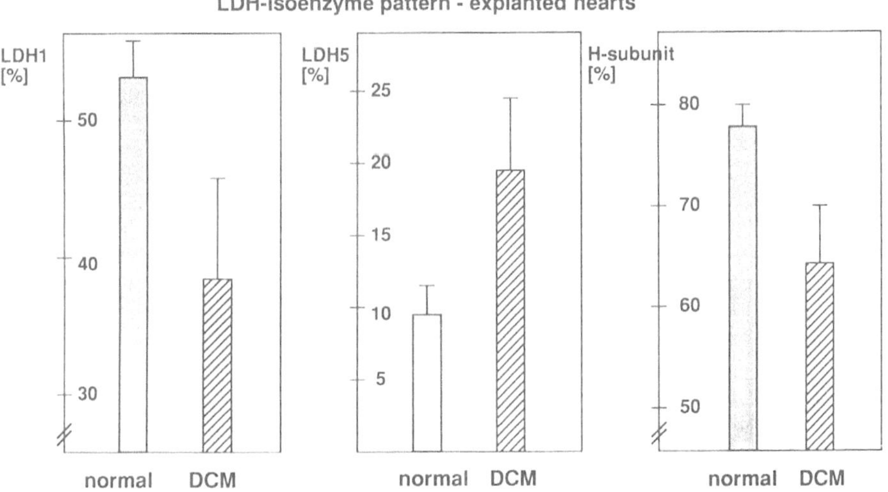

Fig. 1. Disturbance of the myocardial energy metabolism in dilative cardiomyopathy.

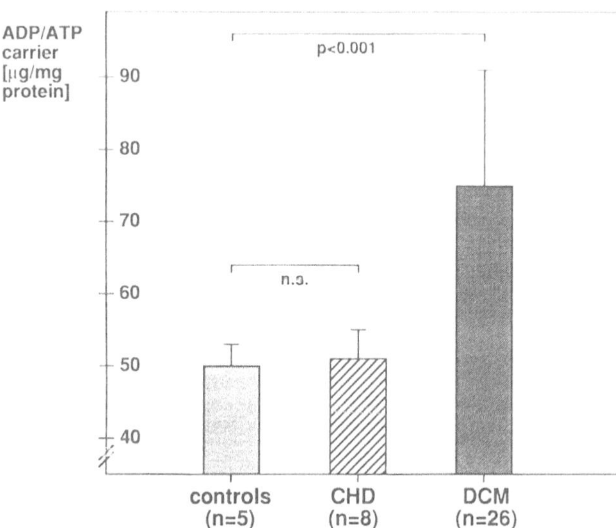

Fig. 2. Quantification of the ADP/ATP carrier in explanted hearts.

from the normal human hearts (51.8 ± 3.8 µg/mg protein), although the severity of the hemodynamic dysfunction of the patients undergoing heart transplantation was comparable in both groups.

Determination of the carrier function

To answer the question whether the function of the ADP/ATP carrier might be disturbed, we measured the nucleotide exchange rate of the isolated mitochondria from the control hearts, the coronary heart disease hearts, and from the DCM hearts. While the exchange rate was the same in mitochondria isolated from the control hearts and from the coronary heart disease hearts, it was significantly lowered in 16 out of 25 of the DCM hearts. These data indicate that the total transport capacity for ATP might be decreased in DCM patients, implicating a diminished energy supply to the cytosol. Interestingly, the carrier content was significantly higher in those hearts in which a decreased nucleotide exchange rate was observed (83.0 ± 12.5), while in those hearts in which the transport activity of ADP/ATP carrier was not changed in comparison to the controls the carrier concentration was only slightly increased (64.8 ± 12.5 µg/mg protein) (Fig. 3).

Discussion

There are several indications for an imbalance between energy production and energy use in the failing human heart (1, 2). The primary findings of this study are that

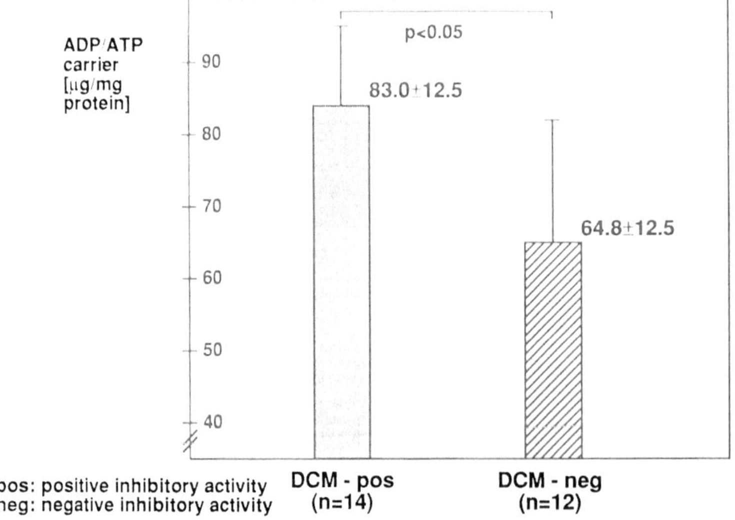

Fig. 3. Quantification of the ADP/ATP carrier in explanted hearts.

the lactate dehydrogenase isoenzyme pattern, the ADP/ATP carrier content, and the exchange rate of the ADP/ATP carrier differ in left-ventricular myocardial tissue from patients with dilated cardiomyopathy and persons with no abnormality affecting the left ventricle. There was a significant increase of LDH_5 concomitant with a decrease of LDH_1 in the myocardium from DCM patients. Consequently, the relative percentage of the H-subunits decreased significantly. These findings support the hypothesis of adaptive metabolic changes in dilated cardiomyopathy. This is in accordance with previous results describing increased myocardial lactate production and increased levels of total LDH activity in severe heart failure (3, 16, 32). Based on the regulatory mechanisms of the LDH isoenzymes, these data indicate that, in chronic heart failure due to dilated cardiomyopathy, the delivery of ATP to the energy-consuming processes in the cytosol by oxidative phosphorylation is augmented through utilization of the more inefficient anaerobic glycolytic pathway (4, 13). Thus, the shift of the LDH-isoenzyme pattern can be taken as an indication for a more anaerobic metabolism in DCM.

Concomitant with the shift of the LDH isoenzyme pattern a significant increase of the ADP/ATP carrier content was observed in the myocardium from patients with DCM in comparison with the controls. This increase of the carrier concentration may indicate an up-regulation of this protein as a consequence of an energy deficit in the cytosol of the myocardial cells. The stimulus for this up-regulation might be a decreased phosphorylation potential of ATP in the cytosol as it is indirectly indicated by the increase of LDH_5.

Several factors can serve as an explanation of a chronic hypoxia of the myocardium – and, here, predominantly at the endomyocardial layer – in chronic heart failure: an increase in wall stress greater at the endomyocardial surface than at the epicardial surface and an elevated LVEDP will decrease the perfusion of the endocardium. Coronary blood flow to the endocardium is also decreased by tachycardia because of shortened diastole and, therefore, shortened duration of perfusion. Impaired diastolic relaxation may prolong compression of the intramyocardial vessels and, thus, further reduce the blood flow. These patients are also more prone to develop myocardial ischemia because of a decreased capillary density. All these factors together may contribute to a state of chronic hypoxia, which then may lead to cellular damage and progressive deterioration of cardiac function.

However, it is doubtful whether this interpretation reflects the "whole truth" concerning the energy metabolism in DCM. Surprisingly, the ADP/ATP carrier content was not increased in the myocardium from patients with severe coronary heart disease. As it seems quite probable that the myocardium of these patients was chronically hypoxic, one would have suspected – under the assumption that tissue hypoxia is a main stimulus for the upregulation of the carrier protein – that the carrier content would significantly be increased in end-stage coronary heart disease. Therefore, on the basis of these results, it seems quite unlikely that the elevated carrier concentration found in DCM represents an adaptation toward anaerobic metabolism. On the contrary, this upregulation of the carrier protein may reflect a disease-specific alteration of the expression of the nuclear DNA encoded carrier genes. This assumption is supported by the fact that the nucleotide transport across the mitochondrial membrane is decreased in a high percentage of the DCM hearts, while an alteration of the nucleotide transport was not seen in coronary heart disease.

Differences in the antigenic and electrophoretic properties of heart, kidney, and liver ANTs imply the presence of ANT isoforms (18, 20). The existence of such isoforms has been confirmed by the recent description of three different ANT cDNAs (2, 5, 14). Furthermore, changes in the amounts of ANT transcripts have been reported in cultured human cells stimulated to replicate or differentiate.

Although dilated cardiomyopathy by definition is a heart muscle disease of unknown cause, several lines of evidence suggest an enteroviral infection as one possible initiating event in a significant portion of genetically predetermined patients. Hereby, the viral infection may act as a trigger that induces an autoimmunological process. With regard to dilated cardiomyopathy, highly specific autoantibodies against the ADP/ATP carrier from heart were found (19, 21, 22). Furthermore, we could show that either immunization with the isolated carrier protein or infection of A/J mice with Coxsackie-B$_3$ virus induced heart reactive antibodies against the ANT and led to a disturbance of the cellular energy metabolism by an alteration of the carrier function (24, 33). In the isolated perfused working heart preparation the hemodynamic records of these hearts showed a more than 50% reduction of external heart work, a parameter reflecting best cardiac performance (25). The amount of the decrease of the external heart work and the magnitude of reduction of the cytosolic-mitochondrial difference of the phosphorylation potential of ATP ($\Delta G_{cyt-mit}$) showed a very close correlation (correlation coefficient of r = 0.87), indicating that the imbalance of the myocardial energy metabolism is responsible for the impairment of cardiac function (23).

In conclusion, these data allow to suggest that the virus infection of the myocardium and/or the virus persistence – at least that shown for dilated cardiomyopathy – may change the expression of the nuclear DNA-encoded ANT genes, with the consequence of a switch of the ADP/ATP carrier isoforms, or may induce an autoimmune reaction against this protein (Fig. 4). As a consequence, the

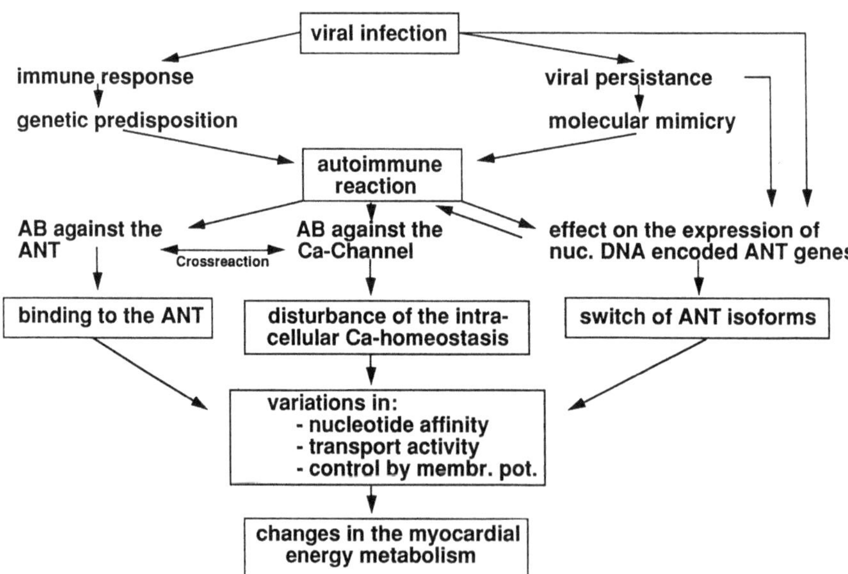

Fig. 4. Possible pathomechanism of dilated cardiomyopathy.

ADP/ATP carrier may exhibit variations in nucleotide affinity, transport activity or the control by the membrane potential. These changes of the functional characteristics of the carrier protein will then lead to the disturbance of the myocardial energy metabolism and, consequently, to an impairment of cardiac function.

References

1. Bashore TM, Magorien DJ, Letterio J, Shaffer P, Unverferth DV (1987) Histologic and biochemical correlates of left ventricular chamber dynamics in man. JACC 9:734–742
2. Battini R, Ferrari S, Kaczmarek L, Calabretta B, Chen S, Baserga R (1987) J Biol Chem 262:4355–4359
3. Bishop SP, Altschuld RA (1971) Evidence for increased glycolytic metabolism in cardiac hypertrophy and congestive heart failure. In: Alpert NA (ed). Cardiac Hypertrophy. New York: Academic Press pp 567–585
4. Everse J, Kaplan NO (1975) Mechanisms of action and biological functions of various dehydrogenase isoenzymes. In: Markert CL (ed). Isoenzymes. II. Physiological functions. New York: Academic pp 29–43
5. Houldsworth J, Attardi G (1988) Two distinct genes for ADP/ATP translocase are expressed at the mRNA level in adult human liver (cDNA library/pEX1 vector/HeLa cells/multiple mRNAs). Proc Natl Acad Sci 85:377–381
6. Ingwall JS, Kramer MF, Fifer MA, Lorell BH, Shemin R, Grossman W, Allen PD (1985) The creatine kinase system in normal and diseased human myocardium. N Engl J Med 313:1050–1054
7. Katz AM (1988) Cellular mechanisms in congestive heart failure. Am J Cardiol 62:3A–8A
8. Katz AM (1990) Cardiomyopathy of overload. A major determinant of prognosis in congestive heart failure. N Engl J Med 322:100–110
9. Klingenberg M, Heldt HW (1982) The ADP/ATP translocation in mitochondria and its role in intracellular compartmentation. In: Sies H (ed). Metabolic Compartmentation. London: Academic Press pp 101–122
10. Klingenberg M (1985) The ADP/ATP carrier in mitochondrial membranes. In: Martonosi AN (ed). The Enzymes of Biological Membranes. Vol. 4. New York: Plenum Publishing pp 511–553
11. Lowry OH, Rosebrough NJ, Farr AL, Randall RJ (1951) Protein measurement with a folin phenol reagent. J Biol Chem 193:265–275
12. Markert CL (1963) Lactate dehydrogenase isoenzymes: dissociation and re-combination of subunits. Science 140:1329–1330
13. Nadal-Ginard B, Markert CL (1975) Use of affinity chromatography for purification of lactate dehydrogenase and for assessing the homology and function of the A and B subunits. In: Markert CL (ed). Isoenzymes. II. Physiological Functions. New York: Academic pp 45–67
14. Neckelmann N, Li K, Wade RP, Shuster R, Wallace DC (1987) cDNA sequence of a human skeletal muscle ADP/ATP translocator: lack of a leader peptide, divergence from a fibroblast translocator cDNA, and coevolution with mitochondrial DNA genes. Proc Natl Acad Sci 84:7580–7584
15. Palmieri F, Klingenberg M (1979) Direct methods for measuring metabolite transport and distribution in mitochondria. In: Sydney F, Lester P (eds). Methods in Enzymology. Vol. 56. New York: Academic Press pp 279–301
16. Peters TJ, Wells G, Oakley CM, Brooksby IAB, Jenkins BS, Webb-People MM, Coltart DJ (1977) Enzymic analysis of endomyocardial biopsy specimens from patients with cardiomyopathies. Br Heart J 39:1333–1338
17. Schultheiss HP, Bispink G, Neuhoff V, Bolte HD (1981) Myocardial lactate dehydrogenase isoenzyme distribution in the normal heart. Basic Res Cardiol 76:681–689

18. Schultheiss HP, Klingenberg M (1984) Immunochemical characterization of the adenine nucleotide translocator: organ- and conformation specificity. Eur J Biochem 143:599–605
19. Schultheiss HP, Bolte HD (1985) Immunological analysis of auto-antibodies against the adenine nucleotide translocator in dilated cardiomyopathy. J Mol Cell Cardiol 17:603–617
20. Schultheiss HP, Klingenberg M (1985) Immunoelectrophoretic characterization of the ADP/ATP carrier from heart, kidney and liver. Arch Biochem Biophys 239:273–279
21. Schultheiss HP (1987) The mitochondrium as antigen in inflammatory heart disease. Eur Heart J 8:203–210
22. Schultheiss HP (1989) The significance of autoantibodies against the ADP/ATP carrier for the pathogenesis of myocarditis and dilated cardiomyopathy – clinical and experimental data. Springer Semin Immunopathol 11:15–30
23. Schultheiss HP (1991) Disturbance of the myocardial energy metabolism in dilated cardiomyoapthy due to autoimmunological mechanisms. Circulation (in press)
24. Schulze K, Becker BF, Schultheiss HP (1989) Antibodies to the ADP/ATP carrier – an autoantigen in myocarditis and dilated cardiomyopathy – penetrate into myocardial cells and disturb energy metabolism in vivo. Circ Res 64:179–192
25. Schulze K, Becker BF, Schauer R, Schultheiss HP (1990) Antibodies to the ADP/ATP carrier – an autoantigen in myocarditis and dilated cardiomyopathy – impair cardiac function. Circulation 81:959–969
26. Smith AD (1967) Preparation, properties and conditions for assay of mitochondria: slaughterhouse material, small scale. In: Sydney F, Lester P (eds). Methods in Enzymology. Vol. 10. New York: Academic Press pp 81–86
27. Unverferth DV, Magorien RD, Lewis RP, Leier CV (1983) The role of subendocardial ischemia in perpetuating myocardial failure in patients with nonischemic congestive cardiomyopathy. Am Heart J 105:176–179
28. Unverferth DV, Lee ShW, Wallick ET (1988) Human myocardial adenosine triphosphatase activities in health and heart failure. Am Heart J 115:139
29. Vatner SF (1988) Reduced subendocardial myocardial perfusion as one mechanism for congestive heart failure. Am J Cardiol 62:94E–98E
30. Weber KT, Janicki JS, Campbell C, Replogle R (1987) Pathophysiology of acute and chronic heart failure. Am J Cardiol 60:3C–9C
31. Weiss MB, Ellis K, Sciacca RR, Johnson LL, Schmidt DH, Cannon PJ (1976) Myocardial blood flow in congestive and hypertrophic cardiomyopathy. Relationship to peak wall stress and mean velocity of circumferential fiber shortening. Circulation 54:884–494
32. Wendt VE, Stock TB, Hayden RO, Bruce TA, Gudbjarnason S, Bing RJ (1962) The hemodynamics and the cardiac metabolism in cardiomyopathy. Med Clin North Am 46:1445–1469
33. Witzenbichler B, Christmann C, Schulze K, Schultheiss HP, Strauer BE (1991) Disturbance of the myocardial energy metabolism in A.SW/SN.J mice after viral infection with Coxsackie B_3 virus. J Am Coll Cardiol 17:264A

Author's address:
Prof. Dr. H.-P. Schultheiss
Medizinische Klinik und Poliklinik B
Abteilung für Kardiologie, Pneumologie u. Angiologie
Heinrich-Heine-Universität Düsseldorf
Moorenstraße 5
D-W 4000 Düsseldorf 1, FRG

Adenine nucleotide metabolism and contractile dysfunction in heart failure – Biochemical aspects, animal experiments, and human studies

V. Regitz, E. Fleck

Department of Internal Medicine and Cardiology, German Heart Institute Berlin, FRG

Summary: In myocardial hypertrophy and heart failure a series of adaptational changes occur – some multiplying contractile units, others slowing shortening velocity and increasing economy of contraction. The demonstration of energy-saving mechanisms in heart failure has prompted further investigations of energy providing and utilizing metabolic pathways. The use of myocardial ATP as a substrate occurs mainly at the myosin-ATPase and at the Ca-ATPase of the sarcoplasmic reticulum. As the Michaelis constant of both enzymes for ATP is in the micromolar (μM) range, whereas cellular ATP content is about 5000 μM, these enzymes are not controlled by the availability of ATP as a substrate. In experimental heart failure in large animals, normal or reduced creatine phosphate levels (in most cases together with normal adenine nucleotides) have been described. Reduced creatine phosphate is found in models with increased oxygen consumption, and creatine phosphate may buffer the ATP pool in these models. In human heart failure due to dilated cardiomyopathy, where resting oxygen consumption per unit mass and lactate extraction are normal in most patients, normal adenine nucleotides, creatine phosphate, and mitochondrial function have been described in the initial studies. These results have been challenged by one study showing decreased ATP levels in dilated cardiomyopathy, correlating with the decrease in ejection fraction. However, only ATP has been measured in this study, whereas total adenine nucleotides may be a more suitable parameter. Recently published results have again demonstrated normal ATP and total adenine nucleotides in human heart failure. In the same patients, significantly decreased myocardial norepinephrine was measured, indicating that metabolic changes had occurred in these hearts, but were independent of adenine nucleotides. In conclusion, the fact that normal myocardial adenine nucleotide levels are found in heart failure, whereas other major biochemical changes are simultaneously present, indicates that reduced myocardial energy production is not the reason for contractile dysfunction.

Key words: Myocardial energy metabolism; adenine nucleotides; heart failure; dilated cardiomyopathy

Metabolic alterations in myocardial hypertrophy and heart failure – myocardial adaptations to chronic mechanical overload

The triggering event leading to contractile dysfunction in dilated cardiomyopathy still remains unknown. However, once mechanical overload occurs in a heart, either due to functional deficits because of unknown metabolic changes due to pressure overload, or due to increased wall stress because of contractile dysfunction in infarcted areas, it is now well recognized that a series of adaptational changes occurs that remodels the heart (for review see (35)). These consequences are qualitative as

Grant support: DFG Re 665/1-2, DFG FL 165/1-2

well as quantitative in nature. In the mechanically overloaded heart there is enhanced, coordinated expression of cardiac genes for new sarcomeres, new mitochondria, calcium channels of the sarcolemma, or atrial natriuretic factor. Diminished gene expression has been observed for the calcium-ATPase (Ca-ATPase) of the sarcoplasmic reticulum, the sodium calcium exchange, and the beta-1-receptors. Shifts in isogene expression have also been described concerning myosin, the Na-ATPase, collagen, and possibly also creatine kinase and lactate dehydrogenase. Slowing of the contraction-relaxation-cycle and an increase in the economy of contraction may result mainly from the alterations in the Ca-ATPase of the sarcoplasmic reticulum, from the decrease in Na-Ca-exchange, from the decrease in beta-receptor density, as well as from the shift in myosin isoforms. The efficiency of these changes is documented by the fact that a hypertrophied heart produces less heat per gram of developed tension, i.e., it uses less energy than does a normal heart for the same amount of work (1). This preponderance of energy-saving mechanisms in cardiac hypertrophy that can also be documented in heart failure has prompted some investigators to search for defects in energy metabolism in both conditions.

Myocardial generation of ATP and its use: substrate function and allosteric effects

Since hydrolysis of ATP provides the energy for contraction, for the maintenance of ion gradients and for macromolecular synthesis these metabolic pathways are primary candidates in the search for defects in energy metabolism.

One of the most important functions of ATP is its use as a substrate in the myosin-ATPase reaction. Contraction is activated by calcium binding to tropo-myosin inducing conformational changes in the actin-myosin-ATPase leading to ATP binding, cross-bridge formation, contraction and splitting of ATP. 30–50% of energy is lost in the form of heat; measurement of this heat can be used for sophisticated analysis of the excitation-contraction process (1).

Maintenance of ion gradients also occurs by ATP-dependent processes. The calcium gradient across the plasma membrane is greater than 1000; a calcium-ATPase and Na-Ca-exchange contribute to its formation (5). Intracellularly, calcium is accumulated in an energy-consuming process in the sarcoplasmic reticulum by a Ca-ATPase and is passively released from the sarcoplasmic reticulum with each excitation contraction cycle, leading to changes in cytoplasmic Ca-concentrations from 10^{-7} to 10^{-4} M with each cycle. Thus, it has been discussed if sarcoplasmic Ca-reuptake in diastole may be affected by a loss of ATP.

Allosteric effects of ATP modulate the activity of the Ca-ATPase of the sarcoplas-mic reticulum and of the Na-Ca-exchanger of the sarcolemma (7, 12). Increases in the activity of these enzymes have been observed with a rise in ATP concentrations from $10^{-7} - 10^{-4}$ M (Ca-ATPase) and from 10^{-4} to 10^{-3} M (Na-Ca-exchanger). Thus, allosteric effects of ATP may influence the diastolic calcium content. Finally, the ATPase activity of actomyosin is suppressed in the presence of 1.5 mM ATP in comparison with 1 mM concentrations (19). In this way, high ATP levels may contribute to maintaining the diastolic dissociation of actin and myosin and, thereby, diastolic relaxation.

In macromolecular synthesis ATP is used as an energy-delivering substrate. Normal turnover rates for typical proteins, myosin or cytochromes are in the range of days. So far, it has not been determined under which conditions synthesis is impaired, and if availability of ATP can play a role.

Association between myocardial function and adenine nucleotide levels

Myocardial energy demands and turnover of high-energy phosphates under normal conditions are high and can only be met by adequate rates of oxidative phosphorylation. Under metabolic conditions like ischemia the supply of high-energy phosphates becomes insufficient and cellular ATP and creatine phosphate levels drop (for review see (17)). Cardiac function in this condition is usually impaired. However, it is still unknown which cellular reactions are altered as a direct consequence of reduced ATP levels.

Half-saturation of the myosin-ATP-ase for free ATP is obtained in the range of μM concentrations, and this means that the enzyme is saturated at much lower ATP concentrations than are usually avilable in the cell (Michaelis constant: $< 1 \mu M$, cellular ATP: approx. 5000 μM). This low Michaelis constant of the myosin-ATP-ase may be the reason for the fact that myocardial contractility is unaffected by changes in ATP concentrations between 5000 μM and 30 μM, as has been shown in isolated myofibrils, as well as in cardiac myocytes with hyperpermeable plasma membranes (8, 22, 38). Although rough associations between myocardial ATP content and ventricular function have been reported in myocardial ischemia (30), it has also been documented in more recent studies that accelerated ATP repletion after ischemia did not improve post-ischemic regional myocardial function (13). In addition, functional recovery in reperfused ischemic myocardium can occur despite the presence of low ATP levels (36). There is now accumulated evidence that ATP concentrations per se do not determine the maximal myosin-ATPase activity and, consequently, ATP levels are not a direct determinant of contractile function.

The calcium transport by the sarcoplasmic reticulum is coupled to ATP-hydrolization, but the ATP-dependence of the Ca-ATPase of the sacroplasmic reticulum is complex. The Michaelis constant of the enzyme for ATP is in the range of 1 μM, far below the cellular ATP concentrations. However, the rate of ATP-hydrolization and of Ca-transport is decreased by increasing the ADP concentration, suggesting that changes in the cytosolic phosphorylation potential may impair calcium accumulation in the sarcoplasmic reticulum. In addition, there seems to be an allosteric regulation site in the enzyme which speeds up calcium accumulation in the presence of higher ATP concentrations (34). Therefore, it has been discussed if the impaired diastolic reuptake of calcium in heart failure (12) may be attributed to reduced ATP concentrations, and if this may impair diastolic function. However, as the highest ATP concentrations used experimentally for stimulation of the enzyme were in the range of about 100 μM (34) (still well below the physiological ATP-concentrations in myocytes) there is no good reason that reduced ATP levels limit the sarcoplasmic Ca-ATPase activity in vivo.

Other allosteric effects of ATP possibly interacting with relaxation include the "plasticizing" effects of ATP on contractile proteins, as well as the stimulatory effects of ATP on cellular calcium export. Impairment of both effects would negatively affect diastolic function. However, there is no experimental proof for a linkage between ATP concentrations, relaxation velocity or diastolic function. In addition, there is no good clinical evidence for allosteric effects of reduced ATP concentrations in early heart failure; however, this may be due to the difficulties to assess small changes in diastolic function in otherwise healthy patients.

In conclusion, changes in ATP-concentrations leading to a decrease of ATP well below its physiological levels do not affect the reactions where ATP acts as a substrate. This concerns contraction as well as the diastolic accumulation of calcium in the sarcoplasmic reticulum and is based on the low Michaelis constants of the respective enzymes for ATP. Allosteric effects of ATP occur at higher concentrations and may influence enzyme activities with yet unknown functional consequences.

Adenine nucleotide levels in heart failure

Changes in energy metabolism in heart failure have already been investigated in animal studies. Focusing on large animal models, reduced creatine phosphate concentrations in combination with unchanged ATP levels are the most frequent finding (11, 20, 27). In dogs with left-ventricular hypertrophy and failure after aortic banding (11) normal adenine nucleotides and creatine phosphate levels were found, together with unchanged oxygen and lactate consumption in the resting state. In other experimental models of heart failure oxygen consumption was increased (10, 21). An increase of oxygen consumption from .5 to 12 ml/100 g × min lead to a change of the creatine/creatine phosphate-relation from 1.0 to 3.9, whereas the ADP/ATP ratio remained unchanged (0.17), demonstrating the efficient buffering of the ADP/ATP-pool by the creatine/creatine phosphate system (21).

The mitochondrial production of ATP and its utilization are coupled through the creatine phosphate-creatine system (31). The ATP generated by oxidative phosphorylation is first used to phosphorylate creatine to creatine phosphate, which functions as a storage and transport form for cellular energy. At the ATP-consuming cellular structures, creatine phosphate rephosphorylates ADP to ATP. Thus, the cellular ATP/ADP contents can be maintained inspite of considerable changes in the creatine phosphate stores.

In large animals, basically, two forms of heart failure have been produced experimentally: one with normal creatine phosphate, adenine nucleotides, lactate consumption, oxygen consumption (11), and the other with decreased creatine phosphate, normal adenine nucleotides, and increased oxygen consumption (21, 27). In the models with increased oxygen consumption the ATP/ADP pool has been buffered by creatine phosphate. Whether the differences between the two models can be explained by the degree of failure or by other experimental conditions is not yet clear. In contrast to animal models, the oxygen consumption per unit myocardial mass in human heart failure at rest is normal or rather low (3, 25, 26). The myocardial perfusion is normal or reduced, the extraction of oxygen is normal (25, 26) or may be slightly increased (3). A production of lactate at rest is not a consistent finding (25, 26). The normal resting oxygen consumption in human heart failure together with the buffering of the ADP/ATP-pool by creatine phosphate shown experimentally make an ATP-deficit due to an imbalance of oxygen need and availability unlikely.

It has been postulated that, in human dilated cardiomyopathy, reduced ATP concentrations in right- and left-ventricular endomyocardial biopsies do correlate with the reduction of the left-ventricular systolic function (2, 37). It is of particular interest that these two studies describe higher ATP levels than those published before in studies on animals or man (2). Only ATP and not total adenine nucleotides have been measured. However, to assess the effect of chronic changes in myocardial nucleotide metabolism the total adenine nucleotides represents a more suitable parameter than ATP alone; the total adenine nucleotides are not sensitive to short periods of acute ischemia, whereas ATP is sensitive (4, 16, 32). Short periods of acute ischemia are, however, unavoidable when working with human endomyocardial biopsies, and may be the reason for artificially reduced ATP levels or a high variability in nucleotide content. The time needed for retracting the bioptome, handling and freezing the biopsies requires about 10–20 s. This period of time may be long enough to break down the available creatine phosphate, to increse ATP, and to start ATP-breakdown after creatine phosphate-depletion (4, 16, 32). In contrast, during the first minutes of ischemia total adenine nucleotide levels are probably not altered (23). If, however, ischemia or other metabolic changes leading to decreased levels of ATP persist, the total adenine nucleotides is depleted. Therefore, to determine long-term changes in energy metabolism in human heart failure, total adenine nucleotides represents a useful parameter and should be determined in addition to ATP.

Based on these considerations we conducted a study in which ATP and total adenine nucleotides in endomyocardial biopsies from patients with dilated cardiomyopathy, heart failure of different origins and controls (left-ventricular ejection fraction > 55%) were determined (28). Normal ATP levels as well as normal total adenine nucleotides were found in all groups. The same holds true for patients with severely impaired left-ventricular ejection fraction (LVEF < 30%). ATP and total adenine nucleotide contents were well in agreement with nucleotide concentrations previously determined in experimental animals (16, 32). Thus, a reduction of myocardial adenine nucleotide content in heart failure seems unlikely.

Data on normal adenine nucleotides in human heart failure are in agreement with older studies that measured ATP, inorganic phosphate, and creatine phosphate in surgically obtained biopsies from patients with (n = 14) and without (n = 8) congestive heart failure (6). Nucleotide levels in both patient groups were comparable and identical to well-known concentrations in animal models. In addition, these authors studied mitochondrial function in preparations from left-ventricular papillary muscles in the earlier mentioned patient groups. Electron transport was tightly coupled to oxidative phosphorylation; respiratory control ratio and mitochondrial ATPase activity in heart failure patients were normal. These results suggest that there are no abnormalities of energy production in heart failure that cause defective myocardial performance.

Possible effects of changes in phosphorylation potential

In ischemia there is evidence that changes in the free energy of ATP hydrolysis or the energy availability are associated with contractile dysfunction (18, 24). In human heart failure, only rough estimations of the free energy of ATP hydrolysis may be made, as creatine phosphate levels for the calculation of free ADP are not available. Creatine phosphate levels in experimental heart failure (27) are significantly higher than in ischemia where creatine phosphate drops to less than 10% of normal levels in the first seconds (16, 32). Thus, free energy of ATP hydrolysis in heart failure is probably less depressed than in ischemia and may remain without consequences.

Adaptational metabolic changes in heart failure and their relation to adenine nucleotide content

Shifts in the myosin isoforms and actin isoforms can be observed in some animal models in heart failure and, partially, also in humans. In animal models, nonsynchronous accumulation of α-skeletal actin and β-myosin heavy-chain mRNA is found in early stages of pressure overload (33). Based on the absence of spatial and temporal coordination in their accumulation, different regulatory mechanisms must be implicated in the induction of the two genes. No loss of ATP has been described in early experimental pressure-induced hypertrophy (11, 21). Thus, the described alterations in the contractile proteins are probably not linked to decreased ATP levels.

A decrease in myocardial carnitine content, probably also fatty acid utilization, and an increase in lactate dehydrogenase activity in heart failure have been observed in patients where normal myocardial adenine nucleotides have been determined in parallel (23). Changes in creatine phosphate kinase are supposed to occur already in patients with moderate hypertrophy without failure (15), a group shown by others to have normal myocardial adenine nucleotides (14).

In one series of patients studied by us, myocardial catecholamines and adenine nucleotides could be determined in the same endomyocardial biopsies from heart failure patients (Table 1). A clear reduction in myocardial norepinephrine, indicating increased norepinephrine turnover was found, whereas total adenine nucleotides and ATP remained unchanged when measured in the same biopsies.

In summary, in heart failure several metabolic changes occur that are suspected to be a consequence of reduced myocardial ATP levels (altered myosin isoforms, altered

Table 1. Myocardial total adenine nucleotides (TAN) in consecutive patients with dilated cardiomyopathy and reduced myocardial norepinephrine (NE) content (NE < 7 pg/µg NCP). Mean values of controls are also listed.

Patient Nr.	NE (pg/µgNCP)	TAN (µmol/mgNCP)	LVEF (%)
1	4.1	32.5	28
2	6.2	43.5	39
3	4.4	30.1	31
4	4.5	30.5	37
5	6.1	25.2	25
6	4.9	32.9	44
7	6.6	17.2	31
8	6.9	53.1	29
9	6.5	39.0	22
10	5.7	33.7	33
11	5.0	21.8	25
12	1.0	32.6	28
13	3.5	33.5	31
14	0.2	35.8	20
15	6.8	51.1	27
16	4.5	28.3	41
17	1.3	25.8	18
Mean ± SD	4.3 ± 2.2	33.3 ± 9.4	29.9 ± 7.6
Controls: n = 16	12 ± 3.4	—	65 ± 5
n = 14		36.9 ± 14.6	65 ± 5

                                                      327

creatine phosphate kinase isoforms) or to cause reduced myocardial ATP levels (decreased fatty acid oxidation, increased catecholamine stimulation). However, data obtained in comparable experiments or even measurements in humans in the same endomyocardial biopsies indicate that these metabolic changes occur independent from nucleotide metabolism.

Conclusions

Normal myocardial adenine nucleotide levels in patients with dilated cardiomyopathy or heart failure of different origin indicate that reduced myocardial energy production is not the reason for impaired contraction. This view is supported by findings in animal experiments as well as by biochemical considerations related to enzyme kinetics. Our current view therefore favors that the basic biochemical defects in dilated cardiomyopathy and the major biochemical changes in heart failure of different origin occur independent from ATP-supply.

Acknowledgements: We thank M. Rappolder for expert biochemical assistance and G. Al-Sharif for skilfull secretarial help.

References

 1. Alpert NR, Mulieri LA (1982) Increased myothermal economy of isometric force generation in compensated cardiac hypertrophy induced by pulmonary artery constriction in the rabbit. Circ Res 50:491–500
 2. Bashore ThM, Magorien DJ, Letterio J, Shaffer Ph, Unverferth DV (1987) Histologic and biochemical correlates of left ventricular chamber dynamics in man. J Am Coll Cardiol 9:734–742
 3. Blain JM, Schafer H, Siegel AL, Bing RJ (1956) Studies on myocardial metabolism. VI. Myocardial metabolism in congestive failure. Am J Med 20:820–833
 4. Braasch W, Gudbjarnason S, Puri S, Ravens KG, Bing R (1968) Early changes in energy metabolism in the myocardium following acute coronary artery occlusion in anesthetized dogs. Circ Res 23:429–438
 5. Chapman RA Control of cardiac contractily at the cellular level. Am J Physiol; 245:H335–52
 6. Chidesy CA, Weinbach, Pool, Morrow AG (1966) Biochemical Studies of Energy Production in the Failing Human Heart. J. Clin Invest 45:40–50
 7. DiPolo R (1976) The influence of nucleotides on calcium fluxes. Fed Proc 35:2579–2582
 8. Donaldson SKB, Bond E, Seeger L, Niles N, Bolles L (1981) Intracellular pH vs MgATP concentration: relative importance as determinants of Ca-activated force generation of disrupted rabbit cardiac cells. Cardiovasc Res 15:268–275
 9. Fleck E (1984) Regionale Myokarddurchblutung bei koronarer Herzerkrankung. Habilitationschrifft. Thieme Verlag, Stuttgart, New York
10. Furchgott RF, Lee KS (1961) High energy phosphates and the force of contraction of cardiac muscle. Circ 24:416–428
11. Gaasch WH, Zile MR, Hoshino PK, Weinberg EO, Rhodes DR, Apstein CS (1990) Tolerance of the hypertrophic heart to ischemia. Studies in compensated and failing dog hearts with pressure overload hypertrophy. Circ 81:1645–1652
12. Gwathmey JK, Copelas L, MacKinnon R, Schoen FJ, Feldman MD, Grossman W, Morgan JP (1987) Abnormal intracellular calcium handling in myocardium from patients with end-stage heart failure. Circ Res 61:70–76

13. Hoffmeister HM, Mauser M, Schaper W (1963) Regional function during accelerated ATP repletion after myocardial ischemia. Circulation 68(Suppl III):III 193

14. Horstig G, Mitrovic V, Podzuweit T, Scheld HH (1991) Einfluß von Ischämie auf den myokardialen Stoffwechsel bei hypertrophiertem und normalem Myokard. Zeitschrift für Kardiol. 80:Suppl 3, 73

15. Ingwall JS, Kramer MF, Fifer MA, Lorell BH, Shemin R, Grossman W, Allen PD (1985) The creatinine kinase system in normal and diseased human myocardium. N Engl J Med 313:1050–1054

16. Jennings RB, Reimer KA, Hill ML, Mayer SE (1981) Total ischemia in dog hearts in vitro. I. Comparison of high energy phosphate production, utilization and depletion, and of adenine nucleotide catabolism in total ischemia in vitro vs. severe ischemia in vivo. Circ Res 49:892–902

17. Jennings RB, Steenbergen C (1985) Nucleotide metabolism and cellular damage in myocardial ischemia. Ann Rev Physiol 47:727–749

18. Kammermeier H, Schmidt P, Jungling E (1982) Free energy changes of ATP hydrolysis: a causal factor of early hypoxic failure of the myocardium? J Mol Cell Cardiol 14:267–277

19. Katz AM (1964) Influence of tropomysin upon the reactions of actomyosin at low ionic strength. J Biol Chem 239:3304–3311

20. Katz AM (1988) Cellular mechanisms in congestive heart failure. Am J Cardiol 62:3A–8A.

21. Kuebler W, Dietz R, Maurer W, Schomig A (1984) Metabolic aspects of compensatory mechanisms in cardiac failure. G-Ital-Cardiol 14(9); P 678–684

22. McClellan GB, Winegrad S (1978) The regulation of the calcium sensitivity of the contractile system in mammalian cardiac muscle. J Gen Physiol 72:737–764

23. Nees S (1989) The adenosine hypothesis of metabolic regulation of coronary flow in the light of newly recognized properties of the coronary endothelium. Z Kardiol 78: Suppl. 6, 42–49.

24. Neubauer S, Ingwall JS (1991) Energieverfügbarkeit von ATP bestimmt die postischämische funktionelle Erholung des isolierten Frettchenherzens. Zeitschrift für Kardiologie 80:Suppl 3, 15

25. Nitenberg A, Foullt JM, Blanchet F, Zouioueche S (1985) Multifactorial determinants of reduced coronary flow reserve after dipyridamole in dilated cardiomyopathy. Am J Cardiol 55:748–754

26. Pasternac A, Noble J, Streulens Y, Elie R, Henschke C, Bourassa M (1982) Pathophysiology of chest pain in patients with cardiomyopathies and normal coronary arteries. Circ 65:778–789

27. Pool PE, Spann JF, Buccino RA, Sonneblink EH, Braunwald E (1967) Myocardial high energy phosphate stores in cardiac hypertrophy and heart failure. Circ Res 21:365–373

28. Regitz V, Rappolder M, Bossaller C, Strasser R, Fleck E (1990) Myocardial ATP, ADP and AMP in heart failure of different origins. Eur HJ 11, Suppl 189

29. Regitz V (1991) Myokardialer Energiestogffwechsel und Katecholamingehalt bei Herzinsuffizienz. Habilitationsschrift an der Freien Universität Berlin

30. Reibel DK, Rovetto MJ (1978) Myocardial ATP synthesis and mechanical function following oxygen deficiency. Am J Physiol 234:H620–624

31. Saks VA, Kupriyanov VV, Elizarova GV, Jacobus WE (1980) Studies of energy transport in heart cells J Biol Chem 255:755–763

32. Schaper J, Mulch J, Winkler B, Schaper W: Ultrastructural, functional, and biochemical criteria for estimation of reversibility of ischemic injury: A study on the effects of global ischemia on the isolated dog heart. J Molec Cell Cardiol 11:521–541

33. Schiafino S, Samuel JL, Sassoon D, Lompré AM, Garner I, Marotte F, Buckingham M, Rapport L, Schwartz K (1989) Nonsynchronous accumulation of α-Skeletal Actin and β-Myosin Heavy chain mRNAs during early stages of pressure-overload-induced cardiac hypertrophy demonstrated by in situ hybridization. Circ Res 64:937–948

34. Shigekawa M, Dougherty JP, Katz AM (1978) Reaction mechanism of Ca^{2+}-dependent ATP hydrolysis by skeletal muscle sarcoplasmic reticulum in the absence of added alkali

metal salts. I. Characterization of steady state ATP hydrolysis and comparison with that in the presence of KCl. J Biol Chem 253:14442–1450
35. Swynghedauw B (1989) Remodelling of the heart in response to chronic mechanical overload. Eur Heart J 10:935–943
36. Taegtmeyer H, Roberts AFC, Raine AEG (1985) Energy metabolism in reperfused heart muscle: metabolic correlates to return of function. JACC 6:864–870
37. Unverferth DV, Magorien RD, Kolibash AJ, Lewis RP, Lykens M, Altschuld R, Baba N, Leier CV (1981) Biochemical and histologic correlates of ventricular end-diastolic pressure. Intern J Cardiol 1:133–141
38. Weber A (1969) Parallel response of myofibrillar contraction and relaxation to four different nucleoside triphosphates. J Gen Physiol 53:781–791

Authors' address:
Dr. Vera Regitz
Department of Internal
Medicine and Cardiology
Augustenburger Platz 1
1000 Berlin 65, FRG

Achievements of the Symposium

Achievements of the symposium

Cellular and molecular alterations in the failing human heart

Ch. Holubarsch, G. Hasenfuss

Dept. of Cardiology, Medizinische Universitätsklinik, University of Freiburg, Freiburg, FRG

Introduction

Regarding research on heart failure, experimental studies on isolated myocardium have been restricted to animal models for many years. The transfer of results obtained from animal studies to heart failure in the human being is questionable, because of the differences between species on the one hand, and experimental methods of heart failure induction on the other. With increasing frequency of cardiac transplantation during the last few years, studies were performed on isolated myocardium from failing and non-failing human hearts, using different methods to preserve and prepare human myocardium for in vitro studies.

At the same time considerable progress was made in molecular biology and methods in experimental genetics. It is evident that cell function may be modified by altered gene expression: 1) The rate of transcription of a given gene may be increased or decreased. 2) Differential splicing of pre-mRNA may occur. 3) Switching between individual genes is possible, if multigene families exist.

The organizers of the meeting on "Cellular and Molecular Alterations in Human Failing Myocardium" realized these developments and saw the need for bringing together the experts in these fields. The symposium was held in Gargellen, Montafon, Austria, in June 1991, as one of a series of meetings on myocardial and vascular function and pathophysiology.

The sarcolemmal membrane

The first paper of the symposium was given by Brodde regarding the membrane receptor system in the non-failing human myocardium. Positive inotropic effects through β_1- and β_2-adrenoceptors are brought about by an increase in intracellular cyclic AMP concentration. Under physiological conditions, contractility and rate of the human heart are under the control of β_1-adrenoceptors, because noradrenaline is the primary transmitter which stimulates β_1-adrenoceptors and does not influence β_2-adrenoceptors. However, β_2-adrenoceptors may come into play under pathological situations, i.e., stress and shock, when high amounts of adrenaline (which stimulate β_1- and β_2-adrenoceptors) are released from the adrenal gland. In addition, the following receptors are coupled to the adenylate cyclase: 5-hydroxytryptamine, histamine-2, vasoactive intestinal peptide, and glucagon receptors. These receptors, however, cause only submaximal activation when compared to the effects of isoproterenol (β_1-adrenoceptor stimulation).

There is a receptor system present in human myocardium which acts independently of cyclic AMP through the phospholipase C/diacylglycerol/1,4-inositol triphosphate pathway. α_1-adrenergic-, angiotensin-II-, and endothelin-receptors are coupled to this system. Although an increase of contractility can be observed via these receptors, their effects are quantitatively less important compared to the β-receptor system.

Three receptors exist that act by inhibition of cyclic AMP formation: these are the so-called Gi-protein-coupled receptors. Activation of muscarinic-2- and adenosine A_1-receptors leads to inhibition of adenylate cyclase and negative inotropic effects.

In heart failure, the sympathetic system is chronically stimulated and "down-regulation" of the β_1-receptors occurs. Bristow and coworkers studied extensively this phenomenon. It is important to note that, in idiopathic dilated cardiomyopathy, only β_1 but not β_2-receptors are down-regulated. This down-regulation of β_1-receptors has functional consequences. Because all existing β_1-receptors in the normal myocardium are coupled to the adenylate cyclase system, i.e., there are no spare receptors, the decrease in the number of β_1- receptors directly indicates a reduced contractile reserve available by β-receptor stimulation. It is of highest clinical interest that the down-regulation of β_1-receptors can be prevented, or even up-regulation can be induced by chronic therapy with beta-blockers.

Beside down-regulation of the β_1-adrenoceptor system, the G_i-proteins play an important role in regulation of myocardial adenylate cyclase and, thereby, contractility. Boehm et al. investigated the $G_{i\alpha}$-proteins in ischemic and dilated cardiomyopathy using a new and highly specific radioimmunological method. This recently developed method seems to be superior to the pertussis-induced ADP-riboxylation reaction. In dilated cardiomyopathy the increase in $G_{i\alpha}$ was 118%, and in ischemic cardiomyopathy it was 40%. This mechanism, in addition to β_1-adrenoceptor down-regulation, accounts for the decrease in cyclic AMP concentrations in failing myocardium, and may explain the reduced positive inotropic effects of phosphodiesterase inhibition in these tissues.

These results are confirmed and specified by Eschenhagen et al., who analyzed mRNA subtypes of $G_{i\alpha}$ in idiopathic dilated cardiomyopathy and in ischemic heart disease. $G_{i\alpha2}$ and $G_{i\alpha3}$-mRNA were the predominant subtypes in human ventricles. In idiopathic dilated cardiomyopathy and in ischemic heart disease, $G_{i\alpha2}$-mRNA levels were increased by 103% and by 77%, respectively, whereas $G_{i\alpha3}$- as well as $G'_{s\alpha}$-mRNA were not different from controls. The up-regulation of $G_{i\alpha2}$ is interpreted as a mechanism of adaptation in a situation where catecholamine levels are elevated due to heart failure.

Intracellular cyclic AMP levels are regulated by both the adenylate cyclase which forms cyclic AMP from ATP, and by the phosphodiesterase enzymes which degrade cyclic AMP. Inhibition of phosphodiesterase, therefore, increases cyclic AMP and, thereby, contractility.

Schmitz and coworkers demonstrated that positive inotropic effects of phosphodiesterase inhibitors are greatly diminished in isolated strip preparations from human failing myocardium when compared to non-failing myocardium. These authors investigated the phosphodiesterases of heart muscle in greater detail. Regarding substrate specificity, K_m, and V_{max} of the enzymes, as well as IC_{50}-values of phosphodiesterase inhibitors, no differences were found between the enzymes of failing and non-failing myocardium. The authors concluded that: 1) Reduced potency of phosphodiesterase inhibitors does not result from alterations in phosphodiesterases in the failing heart, and 2) an increase in signal transducing inhibitory G-proteins may keep the adenylate cyclase activity reduced (see Eschenhagen et al. and Boehm et al.), so that, even in the presence of phosphodiesterase inhibitors, the levels of cyclic AMP are diminished.

Pagani and coworkers investigated the types of phosphodiesterase isoenzymes existing in cardiac and smooth muscle. They identified five different types of isoenzymes. The isoenzyme which is inhibited by currently used PDE inhibitors is the cGi-cyclic AMP-phosphodiesterase. The pharmacological potency of milrinone with respect to inhibition of the cGi-phosphodiesterase activity was similar in normal and failing human myocardium. It is concluded that the observed ineffectiveness of PDE-inhibitors in the failing heart is related to the decreased basal turnover of cyclic AMP rather than a lack of pharmacological potency (see Schmitz et al.).

Another approach for the treatment of heart failure may be the pharmacological inhibition of the cyclic GMP-phosphodiesterase enzyme. This pharmacological mechanism is of great interest, because cyclic GMP stimulation is involved in the action of nitrates, endothelium-derived relaxing factor, and the atrial natriuretic peptide. The authors compared the hemodynamic responses of cGi-cAMP phosphodiesterase inhibitors (enoximone, milrinone) with those of the specific cGMP-phosphodiesterase inhibitor zaprinast. Inhibitors of cGi-cAMP phosphodiesterase increase cardiac contractility and heart rate, decrease vascular resistance, and leave renal function (natriuresis and diuresis) unchanged. In contrast, the cGMP phosphodiesterase inhibitor zaprinast reduces vascular resistance, tends to reduce heart rate, and increases diuresis and natriuresis. Therefore, newer substances that inhibit cGMP-phosphodiesterase may be advantageous in long-term treatment of heart failure.

Another system which may be altered in heart failure and cause changes in excitation-contraction coupling is Na-K-ATPase. This enzyme is also the receptor for digitalis glycosides which increase contractility by inhibiting the Na-K-ATPase and cause intracellular Na^+ to rise. Therefore, studies on the Na-K-ATPase are of interest for two different reasons: 1) Alterations of Na-K-ATPase may occur in congestive heart failure and thereby alter contractility, and 2) up-regulation of this enzyme may develop due to chronic digitalis therapy. Allen et al. analyzed mRNA of all three catalytic Na-K-ATPase α-subunits and found no differences between human failing and non-failing myocardium, neither qualitatively nor quantitatively. Digitalis receptor concentration was also found to be unaltered in heart failure despite chronic glycoside therapy. The authors conclude that no change in the expression of Na-K-ATPase occurs in human heart failure, and that chronic digitalis therapy does not lead to upregulation of the specific receptors.

The contractile proteins

The sarcolemmal membranes and the sarcoplasmic reticulum are responsible for excitation-contraction coupling and thereby regulate the system that has to actually develop force in order to shorten and do the work, i.e., the contractile apparatus. These proteins are composed of myosin and actin, troponin, and tropomyosin. Over many years, we have learned that the contractile proteins are not the "slaves of the membranes", they rather are able to acutely and chronically adapt to certain conditions. In heart failure, especially the chronic alterations of the contractile machinery are of importance. Rupp and Jacob gave a review of the current knowledge on biochemical and functional alterations of the acto-myosin system. There is now convincing evidence that myofibrillar ATPase activity is depressed in human failing myocardium, whereas myosin ATPase is not altered. One might conclude from this data that the defect in the contractile apparatus is related to interaction between myosin and the thin filament, i.e., it is located at the actin site or within the regulatory proteins. This would explain why myosin ATPase activity is observed to be unaltered in human heart failure, whereas myofibrillar ATPase activity is reduced.

Furthermore, there is no doubt about the fact that the α-/β-myosin heavy chain heterogeneity which was first observed in small mammals does not play any role in the ventricular human myocardium. The amount of α-myosin heavy chains in ventricular human myocardium is so small that it cannot account for significant alterations of the myofibrillar ATPase activity. This small proportion of α-myosin heavy chains is replaced by β-myosin heavy chains, already by a hemodynamic overload which does not lead to heart failure. However, Rupp and Jacob did find a new heterogeneity of the heavy chains of myosin by using an improved native gel electrophoresis. A relationship between isoform subtypes and the load of the heart could be demonstrated in mitral valve disease. The first migrating myosin was named V_A and was found to be 25–40% of total myosin. At the moment, it is not established how this $V_{A/B}$ heterogeneity relates to depressed myofibrillar ATPase activity or whether it contributes to heart failure.

A method which allows analysis of excitation-contraction coupling and the function of the contractile apparatus in the intact cardiac muscle is the myothermal approach. Hasenfuss et al. measured the force generated and the heat simultaneously liberated by muscle strip preparations that were obtained from the left ventricle of non-failing hearts and failing hearts, including ischemic as well as dilated car-diomyopathy. A further advantage of this method, which has extensively been used in animal experiments, relies in the capability to separate between those energetic processes that are associated with activation, i.e., mainly calcium release and uptake, and those with contraction, i.e., crossbridge cycling. From the data obtained, the number of calcium ions which are active during one cardiac cycle can be calculated as well as the force-time integral of a single crossbridge. The force-time integral of the crossbridges was increased by 30–40% in dilated and ischemic cardiomyopathy. The authors conclude that – from a functional point of view – the increase in force-time integral is associated with a more economical contraction, but such a change is also combined with prolonged relaxation and decreased maximal power output.

A second finding relates to excitation-contraction coupling: in dilated cardio-myopathy, the activation heat reflecting the number of calcium ions cycling is greatly diminished, thus explaining the fact that peak isometric tension is down to 54% of control. This data is in contrast to that of other scientists, who suggested no difference in peak twitch tension between non-failing and failing human myocardium may exist. However, these discrepancies can be explained by differences in experimental condi-tions on the basis of the inversed force-frequency relationship (see also Mulieri et al.).

Accordingly, differences in peak tension may be present at physiological stimulation rates, but absent at low rates of stimulation. Therefore, in dilated cardiomyopathy, fundamental alterations seem to occur in excitation-contraction coupling, as well as at the level of the contractile proteins.

As discussed above, the biochemical basis of altered crossbridge mechanics may reside in the regulatory proteins. Anderson and coworkers studied the expression of troponin T, a thin-filament regulatory protein, in normal and failing human hearts using western blot analyses. Two isoforms of troponin T were found: TnT_1 and TnT_2. TnT_2 is expressed at higher quantities in the failing hearts. It is of great interest that a negative linear relationship was found between the content of TnT_2 and myofibrillar ATPase. The authors conclude that these alterations are adaptive in character, and are not a cause of heart failure. It can also be concluded that the increased amount in TnT_2 associated with depressed myofibrillar ATPase may be responsible for the increased crossbridge force-time integral found by Hasenfuss et al.

Beside alterations of the regulatory proteins, the phosphorylatable myosin light chains might be altered in heart failure and influence the function of the contractile apparatus. This question was studied by Morano. He found that a myosin light-chain polymorphism exists in human myocardium. The ratio LC-2/LC-2* is found to be 70/30 in most of the patients with left-ventricular dysfunction.

In contrast to the constant isoenzyme pattern of myosin light chains in the failing human myocardium, the phosphorylation level was zero in four patients suffering from idiopathic dilated cardiomyopathy, New York Heart Association classes II–III, whereas values between 15–33% were obtained in other patients. The degree of light-chain phosphorylation may have functional consequences: phosphorylation increased responsiveness and sensitivity of isometric tension generation of skinned fibers to calcium, but left maximum shortening velocity unchanged. This could be explained by an increase in f_{app}, which is the rate-constant for crossbridge transition from the non-force-generating into the force-generating state.

Short- and long-term alterations of the responsiveness of the contractile proteins to calcium have been discussed in the past. Hajjar et al. reported their recent results obtained in skinned myofibrils from non-failing and end-stage failing hearts: no difference was observed with respect to calcium sensitivity. The same question was studied using a tetanization procedure and different extracellular calcium concentrations in the intact cardiac muscles: when peak tetanic force and intracellular calcium concentration, as measured by the aequorin method, were related to each other, no differences could be found between the normal and the myopathic preparation. Interestingly, however, the calcium sensitizer DPI 201-106 was only effective in myopathic preparations by shifting the force -pCa curve to the left by 0.2 pCa units. No effect was seen in the preparations from normal hearts. The authors did not find differences in peak developed force under the mentioned conditions, which is concluded to indicate an unchanged number of crossbridges in myopathic preparations.

Molecular genetics

Altered gene expression may significantly contribute to functional changes of the myocardium under conditions of adaptation and failure. For example, myosin light and heavy chains, as well as actin are encoded by more than one gene, respectively.

Therefore, switching from one gene to another within a multigene family may greatly alter enzymatic as well as mechanical function. In human cardiac tissue, α- and β-myosin heavy chains exist. Whereas α-myosin heavy chains predominate in the atria, β-myosin heavy chains are almost exclusively present in the ventricles of human hearts (Rupp and Jacob). The genes of these two different types of heavy chains are located on chromosome 14. Vosberg and coworkers have sequenced the β-gene in its entire length and have studied its expression in muscle cells. In particular, using DNA-mediated gene transfer in cultures of chicken embryonic myoblasts, they defined the locus of the basic gene promotor, which by itself was found to be almost inactive even in muscle cells. However, by deletion mapping of the region of interest of the β-gene a candidate signal sequence was found which had the capacity to stimulate this promotor and was localized 210 basepairs 5' to the basic promotor. This signal sequence represents an enhancer-like site of the basic promotor. Furthermore, Vosberg et al. identified a protein from nuclear extracts of cardiac and skeletal muscle which binds to the described enhancer region. This study presents the state of our knowledge on gene control regarding skeletal and cardiac muscle growth.

Sideman and Sideman, in the first part of their presentation, review the anatomical and pathophysiological findings of familial hypertrophic cardiomyopathy (FHC) which is – by definition – a genetically inherited, autosomal dominant disorder of cardiac muscle. The occurrence of angina pectoris, heart failure, malignant arrhythmias, sudden cardiac death, cerebral embolism, as well as endocarditis describe the clinical features of this cardiac disorder.

Secondly, the authors report on the complex genetic methods used to clarify the locus of mutation for two families affected with this disease: In one family, a single missense mutation was found on chromosome 14 in exon 13 of the β-cardiac myosin heavy chain. In the myosin heavy chain produced by this altered gene from that family, in one position along the polypeptide pattern an aminoacid is exchanged; instead of arginine, glutamine is present in position 403. It is postulated that this exchange may have structural and functional consequences, because the size and the electrical charge of both amino acids are different.

In the other family, a hybrid alpha plus beta myosin heavy chain gene was found by Southern blot analyses. We know that the relative content of α- and β-myosin heavy chains is constant within each species, and can be altered due to hemodynamic alterations (pressure overload) and neurohumoral changes (thyroxicosis). However, in this family with FHC, the molecule generated by the hybrid gene is an α-myosin heavy-chain-type for the first 27 exons, and β-myosin heavy-chain-type for the rest of the 13 exons. Although the exact consequence of such a structure for function of the complete actomyosin system is at present unknown, such an unusual protein may be speculated to be the cause for irregular assembly of myofibrils, as well as uncoordinated function of the contractile apparatus.

From the presented and some unpublished data, it is concluded that FHC is due to multiple mutations at different times and on different loci among the families affected. Even more important is the second conclusion that knowing more about the structure and function and their relation to each other in the most important regions of the myosin heavy chain may eventually lead us to explain abnormal and failing myocardial function and development of hypertrophy on a molecular level. Furthermore, this may provide the basis for prevention and treatment of FHC in special cases, and of congestive heart failure in general.

Sarcoplasmic reticulum

The sarcoplasmic reticulum is an organelle which plays – together with the sarcolemmal membrane – the most important function in electro-mechanical coupling, which mainly reflects intracellular calcium homeostasis on a beat-to-beat basis. Chevalier et al. compared the molecular alterations of some enzymes involved in calcium regulation between the senescent and the overloaded hypertrophied heart. Cardiac hypertrophy develops in the ageing heart, as can be demonstrated by the increase in left-ventricular weight, as well as by morphometric studies on cardiac myocytes. Furthermore, responsiveness to β-adrenergic modulation is diminished similarly in the ageing and in the overload hypertrophied heart. The authors were able to show that hypertrophied myocytes exhibit unchanged density of sarcoplasmic calcium channels, whereas the density of the sarcoplasmic reticulum calcium ATPase and the sarcolemmal Na-Ca-exchanger is decreased. The author came to the final conclusion that the imbalance between the calcium fluxes controlling enzymes in cardiac hypertrophy leads to fragile calcium homeostasis from which arrhythmogeneity may result.

Mulieri et al., in the first part of their presentation, reported on their methods of preserving and preparing human myocardium. The use of 30 mM 2,3-butanedione monoxime (BDM) in normal Krebs-Ringer solution is shown to preserve the myocardium from cutting injury. This allows the investigators to make very thin preparations from the human left ventricle, so that experiments can be conducted under physiological conditions, i.e., 37°C experimental temperature and stimulation rates between 30 and 200/min. The conclusions which can be drawn from those experiments are the following: 1) The force-frequency relationship is positive in the normal and negative in the failing human myocardium. This can be interpreted as the main contribution to the development and progression of heart failure, because any chronotropic effects in the failing myocardium will lead to a negative inotropic event. Furthermore, this observation may lead to therapeutic consequences after testing bradicardiac substances and in developing them for long-term treatment of heart failure. 2) In idiopathic dilated cardiomyopathy, the drop in systolic function with increasing heart rate is not accompanied by any increase in diastolic tension. Therefore, systolic dysfunction per se is not necessarily associated with diastolic compliance disturbances. 3) Mulieri did not observe increases in resting or diastolic tension in the failing human myocardium.

Pieske et al. have also investigated the force-frequency relationship in isolated human left-ventricular myocardium from normal and failing hearts under physiological conditions. They demonstrated that, in contrast to idiopathic dilated cardiomyopathy where force-frequency relation is inverse, in subacute myocardial failure due to myocarditis, the force-frequency relation is not different from non-failing control. This data indicates that alteration of the force-frequency relationship takes place only in the chronic syndromes of heart failure and/or may be restricted to a subgroup of patients with dilated cardiomyopathy. Abnormalities in the intracellular calcium handling may explain the reversed force-frequency relationship.

Using the bioluminescent calcium indicator aequorin, Meuse and coworkers compared the cyclical changes of intracellular calcium levels between myocardium from normal hearts and that of end-stage failing hearts. The light signal recording from a non-failing heart muscle consists of a single component: The peak of the light signal is reached much faster than peak systolic force development and declines to baseline before peak systolic force is reached. Such types of monophasic aequorin signals are typical for normal heart muscles. In myopathic muscles, calcium transients are different in two ways: Firstly, the light signals are prolonged, and secondly, there is a second light component – called L_2. Therefore, it seems that the prolongation of the action potential and of the time-course of mechanical contraction are associated with abnormalities of calcium handling. The authors also described a reversed force-frequency relation. However, in contrast to the observations of Mulieri and coworkers, Meuse et al. showed that, by increasing the heart rate from 0.5 Hz to 1 Hz, total force development, (i.e., diastolic and systolic force) is increased in myopathic cells. This increase results from a dramatic increase in diastolic tension, whereas systolic force amplitude is somewhat decreased. This data was obtained at 30°C and extracellular calcium concentration was 16 mM. In contrast, the data of Mulieri (37°C, calcium concentration 2.5 mM) show a decrease in systolic force development when increasing heart rate without any change in diastolic force. Abnormalities in calcium handling are likely to explain the reversed force-frequency relation in the failing human myocardium and may contribute to impaired relaxation and decreased diastolic compliance under certain conditions.

Regarding the question of abnormal calcium handling in myopathic myocardium, Beuckelmann and Erdmann presented their work on calcium currents and intracellular calcium concentrations in isolated single cardiac cells from ventricles of normal and failing human hearts. Calcium currents across the membrane were measured using the original voltage-clamp technique, and the intracellular calcium levels were defined using the fluorescent dye fura 2. The authors found unchanged peak calcium-current densities, which indicate normal trigger calcium influx in myopathic preparations in end-stage heart failure. Intracellular calcium concentrations were significantly elevated at rest, and peak calcium transients were significantly smaller in myopathic preparations. The authors conclude an unchanged influx of trigger calcium on the one hand, but impaired calcium release from the sarcoplasmic reticulum.

Calcium homeostasis is a prerequisite for adequate regulation of cardiac contractility and may be altered in heart failure. Peak free calcium concentration is a result of the total amount of calcium released, of the intracellular calcium buffers exchanging calcium, and of the rate of removal from the cytosol. Blanchard and Alpert used the myothermal analysis to evaluate the calcium homeostasis in human left-ventricular myocardium. Tension-independent heat was analyzed from heat and force-time integral measurements in normal solution and one containing 4 mM BDM. Excitation-contraction coupling was altered by using different extracellular calcium concentrations, as well as different stimulation rates. Tension-independent heat was shown to closely correlate with the developed tension-time integral. Under the assumption that two-thirds of calcium transport is achieved by the sarcoplasmic reticulum with a coupling ratio of two calcium ions per ATP molecule, and that one-third of calcium transport is due to sodium-calcium exchanger supported by the sodium-potassium ATPase (one calcium ion per ATP molecule), the measured tension-independent heat values can be converted to calcium cycled: to achieve 50% activation of the contractile proteins, about 28 nmol/g wet weight are required. This value is 35% smaller than the 43 nmol/g value obtained by Fabiato's biochemical analysis. This discrepancy may be explained by absence of buffers with higher calcium affinity than troponin C in the human preparations. Furthermore, the authors provide direct evidence that postextrasystolic twitch potentiation is either due to increased calcium loading of the sarcoplasmic reticulum or due to altered behavior of the sarcoplasmic release site. The myothermal analysis, therefore, seems to be very well suited for further studies investigating abnormalities in calcium handling in preparations obtained from failing hearts.

Because an insufficient activation of the contractile proteins is unlikely to be due to altered L-type calcium currents characteristics (see Beuckelmann and Erdmann), the hypothesis has to be studied of whether the calcium-release channel of the sarcoplasmic reticulum may be altered in structure and function. Holmberg and Williams have been able to isolate sarcoplasmic membrane vesicles and to incorporate them into artificial planar phospholipid bilayers, so that the activity of single calcium-release channels could be investigated under voltage-clamp conditions. It was demonstrated that the characteristics of these channels in failing hearts are not different from those in normal animal cardiac tissue. However, if these preparations were exposed to tryptic or oxidant stress – like in reperfusion after ischemia – an initial increase in channel opening was found to be followed by irreversible loss of channel function. Therefore, disturbance of the calcium-release channel may play an important role in ischemia and reperfusion, but not in chronic congestive heart failure.

The above-mentioned abnormalities of calcium handling in myopathic preparations may be the result of an altered calcium uptake of the sarcoplasmic reticulum. In an early paper, Limas and coworkers (1) have demonstrated that the calcium transport ATPase of the sarcoplasmic reticulum is depressed in human dilated cardiomyopathy. The authors tested the hypothesis that such a depressed activity of calcium uptake by the sarcoplasmic reticulum may be related to immunological mechanisms in dilated cardiomyopathy: the authors are able to show that in 14 out of 49 patients suffering from dilated cardiomyopathy, the serum or immunoglobulin G induced a decline of the calcium transport activity. In 50% of these patients, anti β-receptor antibodies were also found, whereas these antibodies were present in only 20% of patients, whose sera did not show inhibition of the calcium uptake. The authors conclude that immunological mechanisms play an important role in the pathogenesis of dilated cardiomyopathy, in particular, by modifying sarcoplasmic reticulum function.

This view is not supported by the results of Movsesian. This paper studied the ATP-dependent, oxalate-supported calcium uptake in microsomes of the sarcoplasmic reticulum in failing and non-failing left-ventricular myocardium. In contrast to Limas and Limas, he did not find any significant differences between the steady-state kinetic parameters of calcium accumulation by sarcoplasmic reticulum from normal and failing human hearts. Furthermore, Movsesian studied cyclic AMP-dependent phosphorylation of phospholamban with a monoclonal antibody against phospholamban. However, the magnitude of stimulation of calcium uptake following incubation with anti-phospholamban monoclonal antibody was identical in microsomes from normal and failing hearts. Activities of sarcoplasmic reticulum-associated cyclic GMP-inhibited cyclic AMP phosphodiesterase were also unchanged. The author concluded that abnormalities in calcium handling cannot be referred to functional alterations of the calcium transporting ATPase, phospholamban or cyclic GMP-inhibited cyclic AMP phosphodiesterase of the sarcoplasmic reticulum. Differences between his results and those of Limas are difficult to explain. The most likely explanation is that different preparation procedures may yield biochemically and functionally different preparations.

Schwartz and coworkers reported on their studies regarding gene expression of myosin heavy chains, isoactins, and sarcoplasmic reticulum calcium-ATPase. In the rat, pressure overload activated the β-myosin heavy chain and α-skeletal actin. The authors also found that content of the β-myosin heavy chains and the α-skeletal actin is different during life: whereas β-myosin heavy chain mRNA increases during life, α-skeletal actin mRNA is not reinduced.

In contrast to the rat myocardium, myocardium from the human left ventricle of failing and non-failing hearts showed neither differences in the β-myosin heavy chain (pure V3, see Rupp and Jacob) nor alterations of the isoactin pattern. However, alterations were found for the sarcoplasmic reticulum calcium ATPase mRNA: whereas there was no isoform switching of this enzyme present in failing human myocardium, the gene activation was depressed. Therefore, quantitative modulation of single isoform expression may explain some of the observed abnormalities of intracellular calcium handling in diseased human myocardium (see Meuse et al., Mulieri et al., Hasenfuss et al.).

The interstitium

The extracellular space, along with its components, may play an important structural and functional role in normal and failing myocardium. Schaper and Speiser studied morphological changes of the myocardium from patients with idiopathic dilated cardiomyopathy versus normal myocardium using immunofluorescent techniques, as well as electron microscopy. Both myocytes and extracellular matrix were investigated. Fibronectin was found to be increased in large amounts in the extracellular matrix of the diseased hearts. Myocytes sometimes were separated by each other or even "encapsulated" by excessive deposits of fibronectin. Collagen types I, II, and IV were also present as fine lines close to the myocytes and capillaries in normal hearts, and appeared in the form of thick sheets in the diseased hearts. In the extracellular matrix phagocytosing macrophages and proliferating fibroblasts were seen between the myocytes, which showed evidence of degeneration: decreased amounts of contractile proteins, myelinization of mitochondria, and dilatation of the T-tubular system. The authors conclude that the observed excessive deposits of extracellular matrix components may lead to 1) disturbances of the electrical coupling between myocytes, 2) increased diffusion distances, 3) increased diastolic stiffness (diastolic heart failure), and 4) a secondary reduction of myofilaments. Therefore, contribution of the extracellular alterations may greatly contribute to development and progression of heart failure. Weber and Brilla reported on possibilities to prevent extracellular fibrosis. First, these authors review the different pathophysiological mechanisms of fibrosis formation and neurohumoral contributions to two different forms of fibrosis of the heart: The so-called replacement fibrosis, i.e., fibrosis following primary myocardial cell death, and reactive fibrosis, i.e., fibrosis before any myocardial cell damage or death occurs. With respect to reactive fibrosis, these authors clearly demonstrate that 1) elevations of ventricular or aortic pressures alone are not responsible for fibrosis, 2) the growth of myocytes is not necessarily associated with fibrosis, and 3) elevated plasma renin-angiotensin levels and aldosterone levels in particular play a major role in perivascular and interstitial fibrosis. Therefore, interstitial fibrosis may be prevented by 1) angiotensin-converting enzyme inhibition and/or 2) spironolactone, an aldosterone antagonist.

The mitochondrial function

Neither the sarcoplasmic reticulum nor contractile proteins can work properly if the ATP supply is limited. Schultheiss presented some indirect evidence for an imbalance between energy production and energy demand in the failing human heart. There exist five isoenzymes of lactat-dehydrogenase (LDH). Whereas LDH_1 is present in tissues in which aerobic metabolism takes place, LDH_5 is predominant in other tissues in which anaerobic metabolism is needed. Therefore, in the myocardium of normal hearts, LDH_1 exhibits the highest enzyme activity. In the failing hearts, however, there is a shift of the isoenzyme pattern from LDH_1 to LDH_5. Schultheiss interpreted this shift as an adaptive metabolic change of myocardium which may be the consequence of insufficient energy supply. The author investigated another system which is involved in energy supply of the myocytes: the ADP/ATP carrier. This carrier is located at the inner mitochondrial membrane and is responsible for fast exchange of ATP from inside to outside and of ADP from outside to inside. Interestingly, in the failing hearts, the content of ADP/ATP carrier was significantly increased by about 50% when compared to normal hearts. When the function of this carrier system was studied, the nucleotide exchange rate of isolated mitochondria was found to be lowered in failing hearts, despite increased content of the carrier. At first glance, one might interpret this phenomenon as an adaptive reaction to chronic hypoxia or ischemia. However, the author rejected this hypothesis, because the ADP/ATP carrier content was found to be unchanged in ischemic heart disease. Therefore, other pathophysiological mechanisms may come into play, e.g., auto-immunological processes. When the author studied this question, he found highly specific autoantibodies against the ADP/ATP carrier in failing hearts. He concluded that immunological processes may be primary and lead to disturbances of energy metabolism.

If in myocardial hypertrophy and heart failure the energy supply is assumed to be a major problem, the high-energy phosphate concentrations should be found to be reduced in myopathic tissues. This question was carefully studied by Regitz and Fleck in biopsy specimens from patients with and without dilated cardiomyopathy. Because the procedure of retracting the bioptome and freezing the biopsies takes about 20 s, and as ATP is sensitive to ischemia, ATP levels may be artificially low; therefore, the authors also measured total adenine nucleotide levels, which are not sensitive to ischemia which occurs due to biopsy procedure. The authors found no differences in the levels of ATP and total adenine nucleotides between patients with normal ventricles and those with severely impaired ejection fraction. The authors conclude reduced myocardial energy production may not be the cause for impaired contraction.

Conclusion

The results presented by the authors demonstrated fundamental changes in different systems of cardiac muscles in heart failure. However, in many cases it is unclear whether these changes are primary and causal, or secondary and resulting from myocardial failure. Abnormal intracellular calcium handling seems to play a major role in myocardial failure and may be the consequence of incoordinate gene expression for different enzymes controlling calcium homeostasis.

References

Limas CJ, Olivari MT, Goldenberg IF, Levine TB, Benditt DG, Simon A (1987) Calcium uptake by sarcoplasmic reticulum in human dilated cardiomyopathy. Cardiovasc Res 21:601–605

Authors' address:
Dr. Ch. Holubarsch
Dept. of Cardiology
Medizinische Universitätsklinik
University of Freiburg
Hugstetter Strasse 55
7800 Freiburg, FRG

Subject index

MIX
Papier aus verantwortungsvollen Quellen
Paper from responsible sources
FSC® C105338

If you have any concerns about our products,
you can contact us on
ProductSafety@springernature.com

In case Publisher is established outside the EU,
the EU authorized representative is:
Springer Nature Customer Service Center GmbH
Europaplatz 3, 69115 Heidelberg, Germany

Printed by Libri Plureos GmbH
in Hamburg, Germany